OSCAR FABER'S

REINFORCED CONCRETE

Rewritten and extended by

John Faber

B.Sc., M.I.C.E., M.I.Struct.E., M.Am.Soc.C.E.

and

Frank Mead

B.Sc., A.M.I.C.E., A.M.I.Struct.E.

Taylor & Francis
Taylor & Francis Group

LONDON AND NEW YORK

First published 1952
Second edition 1961
Reprinted with amendments 1965
Reprinted 1967, 1970, 1973, 1974
Reissued as a paperback 1977 by
Taylor & Francis
2 Park Square, Milton Park, Abingdon, Oxon, OX14 4RN
Reprinted 1979

Transferred to Digital Printing 2005

© 1965 John Faber

ISBN 0 419 11450 5

Distributed in the U.S.A. by Halsted Press,
a Division of John Wiley & Sons, Inc., New York

PREFACE

SINCE Dr. Oscar Faber's original edition of this book was published in 1952, the art of constructing in concrete, be it reinforced or prestressed, has leapt forward at such a rate that there is now a pressing need for a revised edition incorporating radical changes in many parts. The free exchange of technical knowledge at International Conferences, the operation of professional and commercial bodies in many countries, and the world-wide exchange of technical papers and bulletins has led to a general pooling and sifting of information. Thus there has been a concentration or drawing-together of ideas, particularly from Great Britain, America and the continent of Europe, with a general increase of appreciation of latest research and practical achievement. This has led recently to the publication of many new Codes and Standards.

For example, in 1956 a Joint Committee of the Institution of Civil Engineers and the Institution of Structural Engineers prepared a booklet entitled *The Vibration of Concrete*. And British Standard Codes of Practice have been published as follows: C.P. 114 (1957) – *The Structural Use of Reinforced Concrete in Buildings*; C.P. 2007 (1960) – *Design and Construction of Reinforced and Prestressed Concrete Structures for the Storage of Water and Other Aqueous Liquids*; and C.P. 115 (1959) – *The Structural Use of Prestressed Concrete in Buildings*. Dr. Faber served on the Drafting Committees for all these documents, and in the case of the first three he acted as chairman.

Soil mechanics too, at one time regarded by many as an infant prodigy, has now settled down to a state of reasonable maturity enabling design problems of spread foundations, retaining walls and piling to be understood and treated more rationally than previously: and this has been followed by the issue of Civil Engineering Codes of Practice, No. 2, *Earth-Retaining Structures (1951)*, and No. 4, *Foundations (1954)*.

Accordingly it was felt right that this new edition should be considerably expanded and largely rewritten so that it should cover a far wider field than previously, and at the same time be brought entirely up to date. Whereas the first edition was intended rather as a general introduction to the subject, the present edition aims at dealing more completely with basic theory, and describes also a broad range of essentially practical applications of reinforced and prestressed concrete. American methods and research findings are referred to where it is felt these throw

iii

additional light or knowledge on the subjects under discussion.
New chapters have been added on Slabs, Retaining Walls and
Concrete Roads; and in particular the chapters on Elements of
Reinforced Concrete Design, Foundations, Bunkers and Silos,
Shell-Concrete Roofs, Water-Towers and Reservoirs, Chimneys,
and Prestressed Concrete have been considerably extended, with
numerous examples, design-curves, and diagrams added to make
the work more specific in its approach and of greater use to engineers directly engaged on a wide range of practical engineering
problems. A great number of the design-curves are completely
new, based on original approaches to many of the problems; in
many cases they should make for a considerable saving of labour
in design offices, offering direct methods where before the passage
had in general been tortuous. The practical examples are based
on the authors' own experience, and where more than one method
of design is advocated by various other sources, the methods given
have been chosen deliberately as being regarded the most direct
and appropriate. And no inessential dressings have been added
to the calculations; they are intended to be directly to the point,
and free from irrelevancies.

Many of the chapters conclude with references to other works
where further information can usefully be obtained. The authors
have deliberately avoided giving *long* lists of references, as they
felt it their duty to be selective in the matter. Where appropriate
the references are indicated by superscript numerals within the
text.

It is of course well known that Dr. Faber played an exceptionally
vigorous part in the development of reinforced concrete right up
to the time of his death in 1956: outstanding examples are his
contributions to the understanding of creep, shrinkage and shear.
The present authors had the pleasure of working closely with
Dr. Faber for many years, and have tried all along to select and
present the new material in this edition in a manner which they
feel would have been in accordance with Dr. Faber's own views.

In general the symbols, signs and abbreviations used throughout the book are as recommended in B.S. 1991, particularly
Part 4 (1960). These are not yet everywhere popular or accepted,
but the authors feel sure (having tried them) that they make for
convenience and clarity in the long run. A Glossary of Symbols
is given on page xiii: this refers to the symbols used most
frequently, particularly those relating directly to reinforced
concrete itself. Other symbols, as for example those relating to
soil properties, piling work, pressures in silos, chimney design,
and so on, are defined as and where they occur in the text.

The authors acknowledge with gratitude the valuable assistance they have received from published papers and other literature, and from many friends and friendly organisations. In particular thanks are due to the British Standards Institution,* the Institution of Civil Engineers, the Institution of Municipal Engineers, the Institution of Water Engineers, the Institution of Structural Engineers, the Cement and Concrete Association, the Road Research Laboratory of the D.S.I.R., and Her Majesty's Stationery Office for permission to reproduce various tables. Thanks are also due to Messrs. John Wiley & Sons Inc., New York, for permission to reproduce two figures from Terzaghi and Peck's *Soil Mechanics in Engineering Practice*; and to Messrs. E. & F. N. Spon for permission to reproduce one table from Capper and Cassie's *The Mechanics of Engineering Soils*.

It has required more than two persons to prepare this new edition. The diagrams have been prepared by many of the authors' assistants in the office, though the majority are by Mr. R. Hawken; and in the task of checking the calculations, and preparing the index, a great deal of help has come from Mr. W. Davidson, B.Sc. The draft and final fair-copy typescripts have been prepared by Miss E. Harris with great skill and enormous enthusiasm. Our warm appreciation to all these individuals is here recorded.

JOHN FABER
FRANK MEAD

Clappersgate,
Birchway,
Harpenden.
January, 1960.

---

* Tables are included from the following British Standards. These Standards may be obtained from British Standards House, 2 Park Street, London, W.1, at their published prices.

B.S. 882 (1954) Concrete Aggregates from Natural Sources.

B.S. 785 (1938) Rolled Steel Bars and Hard Drawn Steel Wire for Concrete Reinforcement.

C.P. 114 (1957) Amended 1965. The Structural Use of Reinforced Concrete in Buildings.

C.P. 115 (1959) The Structural Use of Prestressed Concrete in Buildings.

C.P. 2007 (1960) The Design and Construction of Reinforced and Prestressed Concrete Structures for the Storage of Water and Other Aqueous Liquids.

CONTENTS

Contents

GLOSSARY OF MAIN SYMBOLS
(relating to reinforced-concrete members and structures)

Note: Other symbols are defined in the text where they occur, principally as follows:

Soil Mechanics	in Chapters 7 and 8
Piling Work	in Chapter 9
Silo and Bunker Pressures	in Chapter 10
Shell Concrete	in Chapter 11
Chimney Work	in Chapter 13
Prestressed Concrete	in Chapter 15

A	area loaded
A_c	area of concrete
A_E	equivalent area of concrete in reinforced member
A_s	total area of steel
A_{sc}	area of steel in compression
A_{st}	area of steel in tension
A_w	total cross sectional area of stirrup
a_1	ratio $\dfrac{l_a}{d_1}$
b	breadth of rectangular member, or breadth of flange of T-beam or L-beam
b_r	breadth of rib of T-beam or L-beam
D	diameter
d	overall depth of member
d_1	depth to tension steel ("effective" depth)
d_2	depth to compression steel
d_n	depth to neutral axis (depth of concrete in compression)
d_s	depth of slab
E	elastic modulus
E_c	elastic modulus of concrete
E_s	elastic modulus of steel
e	eccentricity
e_b	eccentricity of column load P_b
F	force
F_C	compression force
F_{cc}	compression force in concrete
F_{sc}	compression force in steel
F_T	tension force
f	actual stress
f_{cc}	stress in concrete in direct compression

f_{ct} stress in concrete in direct tension

f_{cb} compressive stress in concrete due to bending (outer fibre)

f'_{cb} compressive stress in concrete at underside of flange in T-beam or L-beam

$f_{c(av)}$ average compressive stress in concrete

f_{st} stress in steel in tension

f_{sc} stress in steel in compression

f_{sy} yield stress of reinforcement

I second moment of area of section

I_C second moment of area of section (ignoring steel)

I_E second moment of reinforced member of equivalent area A_E

K_b stiffness of beam

K_1 stiffness of lower column

K_u stiffness of upper column

L, l length

l_a lever arm of moment of resistance

M bending moment at any section

M_r moment of resistance of a section to bending (safe working)

$M_{r(ult)}$ ultimate moment of resistance of a section to bending

M_b product $P_b e_b$

m modular ratio $= \dfrac{E_s}{E_c}$

n_1 ratio $\dfrac{d_n}{d_1}$

o sum of the perimeters of the bars in the tensile reinforcements

P applied load

P_b load applied eccentrically to column in load factor balanced design

P_o axial load permissible on short column

p total percentage of steel in member subject to torsion

p_{cc} permissible stress in concrete in direct compression

p_{ct} permissible stress in concrete in direct tension

p_{cb} permissible compressive stress in concrete due to bending

p_{sc} permissible stress in steel in compression

p_{st} permissible stress in steel in tension

Q total shear across a section

q shear stress in concrete

R modulus of rupture

r ratio $\dfrac{A_s}{bd} = $ ratio of $\dfrac{\text{total longitudinal steel}}{\text{overall area of member}}$

r_p ratio $\dfrac{A_s}{bd} \times 100 = $ percentage of $\dfrac{\text{total longitudinal steel}}{\text{overall area of member}}$

r_{p1} ratio $\dfrac{A_{st}}{bd_1} \times 100 =$ percentage of $\dfrac{\text{longitudinal tension steel}}{\text{effective area of member}}$

r_s ratio $\dfrac{A_{sc}}{A_{st}}$

s spacing or pitch of stirrups

T torque, or torsional resistance

t_1 ratio $\dfrac{f_{st}}{f_{cb}}$

u cube crushing strength of concrete

u_w u for works test

W applied load

w applied load per unit area

y distance to extreme fibre

z section modulus

α, β empirical bending moment coefficients for slabs spanning in two directions

δ linear deformation

ε strain

ε_c strain in concrete

ε_s strain in steel

CHAPTER 1

COMPOSITION OF REINFORCED CONCRETE

ART. 1.1. INTRODUCTION TO REINFORCED CONCRETE

CONCRETE

CONCRETE is a general term for conglomerates made artificially from cement, with sand and broken stone or similar materials described generally as fine and coarse aggregates. These ingredients, mixed together with water, form a plastic mass which sets on standing into the hard solid material known as concrete. Concrete resembles stone in weight, hardness, brittleness and strength. Depending on the quality and the proportions of ingredients used in the mix, the properties of concrete can vary almost as widely as the different kinds of stone that occur in nature.

Concrete when not reinforced has considerable strength in compression but very little strength in tension; the ratios varying between about 10:1 and 15:1. Having a low tensile strength, it follows that concrete is weak in bending, shear and torsion. The tensile strength besides being low is also unreliable: it may be entirely destroyed by shock or sudden jar, or as a result of shrinkage arising from setting or drying, or due to thermal contraction. Hence the use of unreinforced concrete is normally limited to applications where great compressive strength and weight are the principal requirements, and where tension and bending stresses are either totally absent or, if they occur at all, are extremely small. Examples of the use of unreinforced concrete include thick spread foundations, dock walls, dams, gravity retaining walls, and certain arch types.

REINFORCED CONCRETE

In most structural applications, tension stresses of considerable magnitude have to be accommodated. For this purpose steel *reinforcements* (rods, bars or wires) are embedded in the concrete at the time of casting, so forming the composite material known as *reinforced concrete*. The reinforcements being steel have a high tensile strength and by judicious design they can be so disposed

1

in the concrete as to be available to take all tensile stresses where-ever these occur, whether as a result of direct tension forces or bending, shear or torsion. In this way full advantage is taken of the strength of the concrete in the compression zones of the structure, and the reinforcements provide the tensile strength which unreinforced concrete lacks. Reinforcements suitably disposed can also serve to increase the strength of concrete members in compression, as well as control the effects of shrinkage and temperature changes.

Thus concrete, intelligently reinforced, is transformed from being brittle and unreliable into a composite material having compressive, tensile, bending, shear and torsional strengths which the designer can adjust with economy to suit his require-ments by varying the amount and disposition of the reinforce-ments. In this way works can be constructed in reinforced concrete using members of considerably reduced dimensions with conse-quent savings in weight, space and cost. Indeed the field of application for reinforced concrete extends far beyond the limits where the bulky nature of unreinforced concrete would render its use impracticable.

EARLY HISTORY

Reinforced concrete is not the product of any sudden discovery: rather it is the outcome of gradual and continuous development. The history of this development is traced in National Building Study Special Report No. 24, H.M. Stationery Office (1956)[1] from which much of the following has been taken.

In early days engineers strengthened mass concrete structures by embedding steel rails in the concrete, though at the time there was no clear understanding of the function of the rails. Similarly in 1848 Lambot built a concrete rowing-boat strengthened by the inclusion of a rectangular mesh of iron rods. And in 1855 Coignet took out an English patent for "certain improvements in the use and preparation of plastic materials or compositions to be used as artificial stone, or as concrete or cement for building and other purposes": floors could be made "by burying beams, iron planks, or a square mesh of rods in concrete on falsework".

A better scheme was devised by Wilkinson – a Newcastle plasterer – who in 1854 took out an English patent for embedding in concrete floors or beams a network of flat iron bars or second-hand wire rope raised over the supports and sagging near the bottom of the slab or beam at midspan. In 1872 Monier built a reinforced concrete reservoir of 130 cubic-metre capacity, and in 1892 Coignet's proposal to use reinforced concrete for the new main drainage system of Paris was accepted. The greatest width of tunnel was 17 ft., and Coignet's wells were only $3\frac{1}{8}$ in. thick reinforced with a mesh of round bars $\frac{5}{16}$ in. and $\frac{5}{8}$ in. in diameter. Also in 1892, Hennebique took out a patent for reinforced concrete beams with round main bars fish-tailed at the ends, and with flat hoop-iron stirrups: in 1897 he introduced cranked-up rods, and in 1898 he adopted the tee-beam. Percy constructed the California Academy of Science in 1889, and in 1892 the Museum Building of Leland Stanford Junior University which has spans of 45 ft. and proved resistant to the earthquake shocks of 1906.

Theoretical appreciation appears to have been developing by 1887 when a small book *Das System Monier* was prepared by Wayss and Koenen to back up Monier's practical progress. The three basic principles were that the steel alone took the tensile stress, that the transfer of force to the steel took place through the adhesion between the concrete and the steel, and that volume changes in steel and concrete brought about by changes of temperature were near enough equal. The importance of the relationship between the moduli of elasticity of steel and concrete, and the effect of this on the position of the neutral axis was first recognised by Neumann in 1890. Later, in 1894, Coignet carried out many experiments, and with Tedesco prepared a paper on the calculation of structural members in reinforced concrete. In 1895 Considère began an extensive series of tests on the resistance of beams and columns, which amongst other things led him to condemn the simple rodded column with occasional hoops, and to recommend close helical binding. It was also Considère who recommended forming the ends of the main bars of beams into a hook with a diameter five times that of the bar.

Thus at the start of the present century there were emerging the bones of an understanding of reinforced concrete practice and theory. From that time to the present, development has continued steadily. Materials have become better understood and improved upon, mixing and construction techniques have advanced, the distribution of stress and mode of failure of different members and structural types have received considerable study with the result that calculations can now be made with an increased understanding of the strength and behaviour of the material, with appropriate reduction in margin for errors and consequent saving in weight of material and cost. By considerable ingenuity the scope of structural forms has been broadened with great success to include prestressed-concrete, shell-concrete and other refreshing styles. Up-to-date methods for calculating the various structural forms are given in later chapters in this book.

USES AND APPLICATIONS

Reinforced concrete being a resilient composite material is capable of resisting shocks, blows, compressions, tensions, bending, shear and torsional stresses. In addition to these properties it has the further valuable qualities of permanence (neither rotting like timber, nor rusting like steel), resistance to the teredo navalis and similar pests, fire resistance, "water-retaining" properties, and economy in upkeep (not requiring periodic painting for protection); and being cast in moulds, it can be formed to any desired shape or pattern, offering considerable scope in aesthetic treatment. It is not surprising therefore that reinforced concrete is now being used for many of the purposes for which steel, timber and brickwork were formerly employed. The following list includes some of the many applications:

Foundation piles, foundation slabs, bases, rafts.
Columns, beams and floor slabs in buildings.
Bridges, viaducts, retaining walls.
Piers, jetties, wharves, docks, slipways, and other riverside and marine structures.
Coal bunkers, cement silos, grain silos, flour mills, warehouses, storage structures and other commercial and industrial structures.
Water pipes, tanks, water towers and reservoirs.
Ships, barges, lighters.
Shell roofs.
Tall chimneys.
Roads and pavings.
Railway sleepers.

ART. 1.2. PORTLAND CEMENT

Of the various types of cement available and suitable for making concrete, *Portland cement* is by far the most commonly used. The raw materials used in the manufacture vary and depend on what is available in the locality to be served by any particular cement works. Most frequently these are chalk and clay; but limestone and marl, or limestone and shale are also commonly used. An intimate mixture of the two prime ingredients is burned to a white heat (the point of incipient fusion), and the resultant clinker cooled and ground with gypsum to the extremely fine powder known as Portland cement.

The mixing of the ingredients before burning can either be by a *wet process* with water used as the vehicle, or by a *dry process*. In Great Britain at the present time the wet process is more commonly used, but experience is showing that considerable economies can be achieved with the dry process where the condition of the raw materials is favourable. The reason is as follows. In the wet process, the mixed ingredients enter the kiln with a moisture content of up to about 40 per cent, and the whole of this moisture has to be driven off as steam involving considerable fuel consumption; whereas in the dry process the moisture content to be dissipated in the burning is only about 14 per cent. In the dry process it is also necessary initially to drive off the natural moisture contained in the raw materials as quarried, so that these are in a condition to be ground and intimately mixed and blended in a dry powdery condition. About 14 per cent water has then to be added by spray to the blended raw meal before it is passed to the kiln for burning. Clearly then it is a matter of economics as to whether the raw materials are better handled by the wet or dry process, depending primarily on their natural moisture content.

WET-PROCESS OF MANUFACTURE

A simplified flow-chart for the production of cement by the wet-process is given in Fig. 1.1. The raw materials (chalk and clay) are mixed with water to a creamy consistency in *wash-mills*. These are circular tanks in which a central vertical spindle carries a rotating steel framework from which heavy harrows are suspended. Gratings are provided above the perimeter walls of the wash-mill tanks, and when the harrows have broken the raw materials down to the size of the grating openings, the slurry is splashed through the gratings to an external annular trough. The wash-mills are generally of three grades known as *primary*, *secondary* and *screening* with the roughest harrows and coarsest screens at the primary mills. The different grades of mills are arranged in series so that the slurry which has passed the screens of the primary mills is fed by gravity to the secondary mills and so on, until on leaving the screening mills only 5 per cent of the slurry is retained on a B.S. 170 mesh test sieve.

The slurry is then pumped to tall *mixing tanks* where it is kept agitated by compressed air introduced at the bottom of the tanks. Here the proportions of raw materials are checked and adjusted as required. The slurry is then passed to large *slurry storage tanks* where it is kept stirred by vertical paddles suspended from rotating arms. Here it waits its turn until the kilns are ready to receive it.

The *kilns* are frequently about 300 ft. long, 12 ft. diameter, made up of steel plate and lined with firebrick. They are set at an angle of about $2\frac{1}{2}$ degrees to the horizontal and mounted on rollers at intervals of 60 ft. or so and geared so as to rotate slowly about their axis. The slurry is fed into the upper end of the kilns and, owing to the combined effect of the slope and rotation, gradually works its way down to the lower end where the kiln is fired. Thus the slurry meets the hot gases and flames which pass up

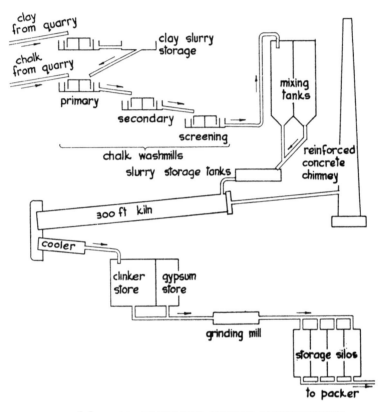

FIG. 1.1. FLOW-CHART FOR CEMENT MANUFACTURE
(WET-PROCESS)

through the kiln and thence to dust collectors and the chimney. The slurry is first dried out at the upper (cooler) part of the kiln, and as it descends and reaches the zone of maximum tempera- ture (about 2800°F) near the lower end, the constituents combine chemically. At this stage the slurry has been burned into extremely hard nodules about the size of walnuts known as *clinker*, which falls out of the lower end of the kiln into a *cooler* where a current of air cools it down from its white heat. The cooling air is thus heated to about 800°F and becomes the secondary combustion air at the lower end of the kiln, so economising in fuel. In Great Britain the fuel is invariably finely powdered coal. In other countries use is made of fuel oil or natural gas where these are in ready supply.

The clinker is then ground in steel tube *grinding mills* with about 3 per cent of gypsum to retard the setting process of the finished cement. The mills are horizontal rotating cylinders often about 30 ft. long and 6½ ft. diameter, divided into three or four compartments: each compartment contains hard-steel and cast-iron balls, which are lifted by the rotation of the mill so as to fall and break the clinker by crushing and attrition. Where there are four compartments in the mill, the first generally contains balls about 4 in. diameter, the next has balls about 3 in., the next 2 in., and the last 1 in. The cement passes from one compartment to another, and grinding continues until for Ordinary Portland cement only about 5 per cent is retained on a B.S. 170 mesh test sieve; and for Rapid-Hardening Portland cement only about 1·0 per cent residue is retained. The cement is then conveyed to large silos where it is stored until required for packing or bulk loading and despatch.

DRY-PROCESS OF MANUFACTURE

A flow-chart for the dry-process of manufacture is given in Fig. 1.2. The raw materials (limestone and shale, etc.) are crushed at the quarry to about 1 in. size and then conveyed to the works where they are air-dried and ground by *raw-meal grinding mills* to an extremely fine powder. The mills are normally of ball or edge-runner type. The raw meal is then transferred to the *raw meal blending silos* where the proportions of raw materials are checked and adjusted as required, the meal being mixed either mechanically or by the vigorous entry in quantity of compressed air through the floors of the blending silos.

The raw meal is then passed to the *granulator* which comprises an inclined rotating dish or drum into which the raw meal is admitted and on to which about 14 per cent of water is sprayed. The addition of the water causes the meal to form into nodules which are passed into a *preheater*, one type of which is a moving grate heated by the hot gases emitted from the kiln. In the preheater the nodules are baked dry before entering the upper end of the kiln.

The kiln itself is similar in construction and in manner of support to that used in the wet-process except that it is shorter, there being far less heat exchange to perform. Clinker is discharged from the lower end of the kiln as in the wet-process, and from this stage onwards the procedure for grinding, storage, and packing is as before.

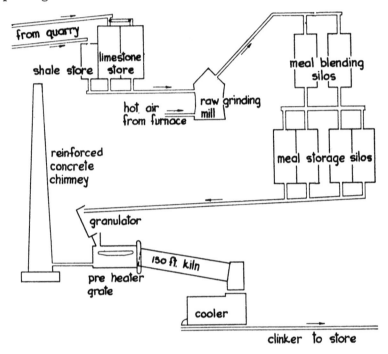

FIG. 1.2. FLOW-CHART FOR CEMENT MANUFACTURE
(DRY-PROCESS)

QUALITY OF ORDINARY PORTLAND CEMENT

Ordinary Portland cement has to comply with the requirements of B.S. 12 (1958) to which reference should be made. The following notes indicate briefly the nature of those requirements.

(i) *Strength*

The physical strength of cement is now determined from compression tests carried out on concrete cubes. The method of tensile tests on mortar briquettes, as given in the old B.S. 12 (1947) has been dispensed with, as giving results which are less consistent and not as closely related to the strength of concrete made with the same cement. The compression tests may be made either on *mortar cubes* of 2·78 in. side or on *concrete cubes* of 4 in.

side, the latter test approaching most nearly the use of the cement in concrete work. The cube strength requirements of the 1958 Standard are an advance on the 1947 values; this is welcome progress seeing that the actual strengths of British cements normally come well above these figures.

For the *mortar compressive strength* the cement is gauged with a standard sand in the proportions by weight of one part cement to three parts sand, and the water/cement ratio is 0·40. This mixture is placed in moulds of 2·78 in. side with machined faces, and compacted by vibration on a machine. The cubes are then kept in air at a temperature of 66°F and relative humidity not less than 90 per cent for 24 hours, after which they are stored in water at 66°F until required for testing. The cubes are tested on their sides in a compression machine with the load applied at the rate of 5000 lb/sq. in. per minute. No packing is used.
Minimum requirements are:

2200 lb/sq. in. at 3 days (72 hours);
3400 lb/sq. in. at 7 days (168 hours) and must show an increase on the 3-day test result.

For the *concrete compressive strength* the cement is mixed in given proportions with any aggregates meeting certain specified requirements; and the water/cement ratio is 0·60. The mixture is placed in moulds of 4 in. side and compacted by standard hand-tamping. The cubes are then stored and tested as previously described for the mortar cubes, except that the rate of application of load is 2000 lb/sq. in. per minute.
Minimum requirements are:

1200 lb/sq. in. at 3 days (72 hours);
2000 lb/sq. in. at 7 days (168 hours) and must show an increase on the 3-day test result.

(ii) *Setting Time*

For determining the *setting time*, a neat cement paste is made up of standard consistence, placed in a circular mould E 80 mm diameter and 40 mm deep, and struck off level and kept at a temperature of 66°F in an atmosphere not less than 90 per cent relative humidity. The needle type C, 1·0 mm square, weighted to 300 gm as shown in Fig. 1.3 (*Vicat Apparatus*), is then applied repeatedly to the surface of the block; and when the needle fails to pierce the block completely, the time interval since the water was first added to the cement is known as the *initial setting time*. Needle F, with an annular attachment, is then substituted for the plain needle, the annular attachment being so arranged that the needle projects 0·5 mm below it. When this needle makes an indentation in the block, but the annular attachment does not, the time interval since the water was first added to the cement is known as the *final setting time*. These times are arbitrary but nevertheless form a useful yardstick. The initial set should be long enough to enable mixing, transporting, placing and tamping

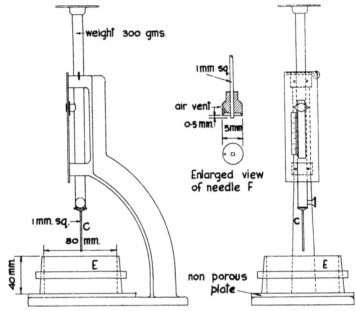

FIG. 1.3. VICAT APPARATUS FOR DETERMINING THE
SETTING TIMES OF CEMENT

of the concrete to be completed before setting starts, and the 1958 B.S. has increased this period 50 per cent over the old 1947 Standard.

Minimum requirements are now as follows:
    Initial setting time not less than 45 minutes;
    Final setting time not more than 10 hours.

(iii) *Soundness*

For soundness (i.e. assurance that the cement will not expand after setting due to the presence of free lime) a neat cement paste is made up of standard consistence, and placed in a brass cylindrical mould, 30 mm diameter and 30 mm deep, cut through on one side and provided with long pointers on either side of the cut as indicated in Fig. 1.4. The mould with glass covers is then submerged in water at a temperature of 66°F for 24 hours, and the distance between the indicator points is measured. The water is then brought to boiling point in half an hour and kept boiling for one hour, after which the mould is removed from the water and allowed to cool. The distance between the points is then

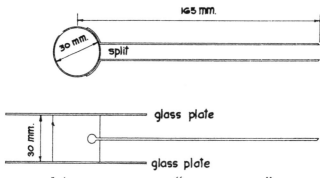

FIG. 1.4. APPARATUS FOR "LE CHATELIER" TEST
FOR SOUNDNESS

measured again, and the difference between the two measurements is required by B.S. 12 not to exceed 10 mm. This is known as the *Le Chatelier* test.

(iv) *Fineness*

The sieve test of the old British Standard has now been deleted, and the specific surface test substituted as a better method of testing the fineness of the cement. The sieve test only measured the proportion of coarse material in the cement, and the advantage of the specific surface test is that sometimes it is important to know the fineness of the other sizes. The fineness of cement as indicated by specific surface is now expressed as total surface area in sq. cm/gm.

The minimum requirement of B.S. 12 for specific surface is 2250 sq. cm/gm.

(v) *Chemical Composition*

For chemical composition B.S. 12 defines certain requirements in regard to the ratios of lime, silica, alumina, iron oxide, and also the amounts of magnesia and sulphur present. The maximum permissible total loss on ignition is also defined. Some people hold that if the cement passes the physical tests it is unlikely to fail in complying with the requirements of the chemical tests. The chemical tests are more difficult to carry out.

QUALITY OF RAPID-HARDENING PORTLAND CEMENT

Rapid-Hardening Portland cement is also covered by B.S. 12 (1958) though certain of the requirements are enhanced as follows:

(i) *Strength*

Minimum requirements for *mortar compressive strength* are:

3000 lb/sq. in. at 3 days (72 hours);
4000 lb/sq. in. at 7 days (168 hours) and must show an increase on the 3-day test result.

Minimum requirements for *concrete compressive strength* are:

1700 lb/sq. in. at 3 days (72 hours);
2500 lb/sq. in. at 7 days (168 hours) and must show an increase on the 3-day test result.

(ii) *Fineness*

The minimum requirement of specific surface is 3250 sq. cm. per gm.

ART. 1.3. SULPHATE-RESISTING PORTLAND CEMENT

Sulphate-Resisting cement is a Portland cement and mainly differs from Ordinary Portland cement in that its chemical components in the raw mix are generally adjusted so that, after burning, the compound tricalcium aluminate is reduced to something less than 5 per cent. It is the tricalcium aluminate that is attacked by sulphates in ground waters, and consequently Sulphate-Resisting Portland cement is able to resist attacks of sulphate concentrations up to certain limited amounts.

Such sulphate concentrations are frequently found in the ground in many parts of Great Britain and elsewhere, and where these are greater than 0·5 per cent in the soil, or 100 parts in 100,000 parts of ground water, are liable to cause disintegration of concrete made with Ordinary Portland cement, and the use of Sulphate-Resisting cement becomes necessary. Where sulphate concentrations exceed 2 per cent in the soil or 500 parts in 100,000 parts of ground water, the protection given by Sulphate-Resisting Portland cement normally ceases to be effective, and consideration has to be given to the use of High Alumina cement. The sulphate concentration of sea-water is generally only about 220 parts per 100,000.

Whenever the use of special cements is considered, it is best to seek advice from the cement manufacturers; and satisfactory results can then only be achieved using compact dense concrete free from voids or excess mixing-water. The use of special cements will not compensate for weak mixes or poor workmanship.

ART. 1.4. PORTLAND BLASTFURNACE CEMENT

Portland Blastfurnace cement is similar in many respects to Ordinary Portland cement except that a proportion of granulated blastfurnace slag (not to exceed 65 per cent of the total) is ground down with Ordinary Portland cement clinker to produce the cement. Portland Blastfurnace cement has to comply with the requirements of B.S. 146 (1958). In general these do not differ substantially from the requirements of B.S. 12 (1958) for Ordinary Portland cement except that the strength requirements are somewhat reduced as follows:

Minimum requirements for *mortar compressive strength* are:
1600 lb/sq. in. at 3 days (72 hours);
3000 lb/sq. in. at 7 days (168 hours) and must show an increase on the 3-day test result.

Minimum requirements for *concrete compressive strength* are:

800 lb/sq. in. at 3 days (72 hours).
1600 lb/sq. in. at 7 days (168 hours) and must show an increase on the 3-day test result.

ART. 1.5. HIGH ALUMINA CEMENT

High Alumina cement is also known as *aluminous cement* or *ciment fondu*. Its manufacture is entirely different from that of Portland cements. The raw materials are chalk and bauxite, the latter being a special clay of extremely high alumina content. The raw materials are mixed together dry, fed into special furnaces, and melted at an extremely high temperature. The molten material is cast into pigs, broken up, and then ground into cement without the addition of any other material whatsoever. This is a much more expensive process than the manufacture of a Portland cement, but there are certain applications where the properties of High Alumina cement have advantages over other types. It produces concretes of far greater strength and in considerably less time even than Rapid-Hardening Portland cement, allowing earlier removal of the formwork. It is also more resistant to certain forms of chemical attack, such as that of sulphates, oils, and sugar juices. These properties make it attractive in spite of its greater cost, when high strengths are required at a very early age or when corrosive conditions are sufficiently severe.

However, great care has to be taken in the use of High Alumina cement. It must not be mixed with any other types of cement, and has to be stored separately and handled by separate mixers, transporters and other equipment. The heat given off on setting is greater than with other cements, and the concrete is best not cast in thicknesses greater than 12 in. to 15 in. unless special precautions are taken. Nor should it be cast when its temperature on setting can rise above 80°F. Thus it is inapplicable to work in tropical countries where the excessive temperature may have a deleterious influence on its strength. In view of the heat given off on setting, it is necessary in all applications to strip formwork at a very early stage and keep the concrete thoroughly saturated with water by continuous hosing for a period of at least 24 hours after placing. High Alumina cement should only be used strictly in accordance with the manufacturer's recommendations.

QUALITY OF HIGH ALUMINA CEMENT

The tests and test requirements for High Alumina cement are covered by B.S. 915 (1947) which are briefly as follows:

(i) *Strength*

Minimum requirements for *mortar compressive strength* are:

6000 lb/sq. in. at 1 day (24 hours);
7000 lb/sq. in. at 3 days (72 hours) and must show an increase on the 1-day test result.

(ii) *Setting Time*

Initial setting time is to be not less than 2 hours and not more than 6 hours.

Final setting time is to be not more than 2 hours after the initial set.

(iii) *Fineness*

Minimum requirement of specific surface is 2250 sq. cm/gm.

(iv) *Soundness*

The opening of the points in the Le Chatelier test is not to exceed 1·0 mm.

(v) *Chemical Composition*

The total alumina content is not to be less than 32 per cent by weight of the whole. The ratio of the percentage by weight of alumina to the percentage by weight of lime is to be not less than 0·85 nor more than 1·3.

ART. 1.6. AGGREGATES

Aggregates for concrete work are divided into two groups, *coarse* and *fine*, and are covered by B.S. 882 (1954). Coarse aggregates are mainly retained on a $\frac{3}{16}$ in. B.S. test sieve, and comprise uncrushed gravel, crushed gravel and crushed stone. Fine aggregates mainly pass a $\frac{3}{16}$ in. B.S. test sieve, and include natural sands and crushed gravel or stone.

Aggregates should be selected with regard to the following:

Strength
Size
Particle shape
Surface texture
Grading
Impermeability
Cleanliness
Chemical inertness
Cost.

STRENGTH

A strong aggregate makes for a strong concrete. Thus granite aggregate will produce a strong concrete whereas pumice or burnt clay aggregates, which are sometimes used where saving of weight is desired, produce concretes of lower strength.

SIZE

Besides the minimum limit of $\frac{3}{16}$ in. for coarse aggregates, there arises the question of maximum size. Clearly the aggregate must be small enough to enable it to be worked in between and around all reinforcements and into all corners of the work. In C.P. 114 (1957)[2] it is recommended that the maximum size should not exceed one-quarter the minimum thickness of any member: also that, for heavily reinforced members, e.g. the ribs of main beams, the maximum size of aggregate should be $\frac{1}{4}$ in. less than the minimum lateral distance between the main bars, or $\frac{1}{4}$ in. less than the minimum cover whichever is the smaller. However, where the reinforcement is widely spaced, as in solid slabs, the maximum size of aggregate may be as great or even greater than the minimum cover.

A maximum size of $\frac{3}{4}$ in. is usually satisfactory. It enables better grading to be achieved than with reduced sizes, and presents less area of aggregate surface needing to be cemented – a given volume of large particles clearly possessing less surface than the same gross volume of smaller particles.

PARTICLE SHAPE

Suitable particle shapes vary between *rounded, irregular* and *angular* as indicated in Fig. 1.5. For a given workability of any

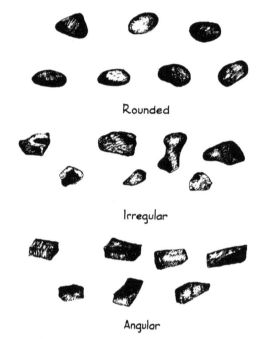

Rounded

Irregular

Angular

FIG. 1.5. PARTICLE SHAPE OF AGGREGATES

concrete mix, angular aggregates require the highest water/cement ratio and rounded aggregates the least. Since the strength of concrete depends so greatly on the water/cement ratio, the selection of particle shape is all-important.

SURFACE TEXTURE

Other things being equal, a rough surface gives a stronger

concrete than a smooth one. It is clear that when concrete is subject to crushing there is a tendency for smooth-surfaced particles to slide past one another and to act as wedges disrupting the concrete. Aggregates with rough surfaces do not lend themselves so readily to these actions. Where the aggregates are smooth and very strong, failure of the concrete usually follows the surface of the stones and is limited by the adhesion of the mortar to the surface, the intrinsic strength of the aggregate not being developed. On the other hand when the aggregates are rough and weaker than the mortar, failure may take place by shearing through more of the stone and less of the mortar. For this reason it sometimes happens that a brick concrete may develop nearly as high a strength as one composed of broken flints, though the intrinsic strength of the flints may greatly exceed that of the brick. Thus in addition to inherent strength, surface texture is an important property of aggregates in determining the crushing strength of a concrete. Granite usually combines both virtues and hence can give a very strong concrete.

GRADING

For a concrete to be dense and impervious, it is necessary for there to be sufficient mortar to fill the voids in the coarse aggregate, and sufficient cement to fill the voids in the fine aggregate. If all the particles were of the same size the percentage of voids

TABLE 1.1. *Gradings for nominal-size graded coarse-aggregates*

B.S. sieve	Percentage by weight passing B.S. sieves		
	$1\frac{1}{2}$ in. to $\frac{3}{16}$ in.	$\frac{3}{4}$ in. to $\frac{3}{16}$ in.	$\frac{1}{2}$ in. to $\frac{3}{16}$ in.
3 in.	100	—	—
$1\frac{1}{2}$ in.	95–100	100	—
$\frac{3}{4}$ in.	30–70	95–100	100
$\frac{1}{2}$ in.	—	—	90–100
$\frac{3}{8}$ in.	10–35	25–55	40–85
$\frac{3}{16}$ in.	0–5	0–10	0–10

would be high, up to about 50 per cent, whereas when the particle sizes are graded in such a manner that the smaller particles can occupy the voids between the larger particles, the percentage of voids is considerably reduced. Such grading results in a reduction in the amount of mortar required to fill the voids, and consequently a reduction in the amount of cement.

3

Suitable gradings for coarse and fine aggregates are given in
B.S. 882 (1954) from which the values in Tables 1.1 and 1.2 have
been taken. For the fine aggregates the gradings are split into
four zones covering the range of coarse sands (Zone 1) to finer
sands (Zone 4). The use of Zone 4 sands in concrete is normally
limited to special mixes. Some aggregates are found in nature

TABLE 1.2. *Gradings for fine-aggregates*

B.S. sieve	Percentage by weight passing B.S. sieves			
	Grading Zone 1	Grading Zone 2	Grading Zone 3	Grading Zone 4
¾ in.	100	100	100	100
3/16 in.	90–100	90–100	90–100	95–100
No. 7	60–95	75–100	85–100	95–100
No. 14	30–70	55–90	75–100	90–100
No. 25	15–34	35–59	60–79	80–100
No. 52	5–20	8–30	12–40	15–50
No. 100	0–10	0–10	0–10	0–15

already well graded. Alternatively they can be graded by screening into various sizes and then remixed on site in the desired
proportions. Where special concrete mixes are used, aggregates of
¾ in. maximum size should be delivered to site in ¾ in. and ⅜ in.
sizes to requirements of B.S. 882 (1954) for remixing. With *nominal concrete mixes* this is often not necessary.

IMPERMEABILITY

It is clearly essential that aggregates used in concrete for
reservoirs, water towers, and other water-retaining structures
should be impermeable. This is equally important in all reinforced
concrete work where a high standard of permanence is required;
otherwise air and moisture will penetrate with consequential
spalling of the outer concrete due either to rusting of the reinforcements or to the wedging action of frost on water contained
within the aggregate.

CLEANLINESS

Aggregates must be free from clay, silt, fine dust or other
adherent coatings likely to interfere with the proper mixing of the
materials or likely to weaken the adhesion between the individual
particles in the hardened concrete. B.S. 882 (1954) prescribes limits
for such impurities. Most gravels and sands from inland pits
require washing before they are suitable.

CHEMICAL INERTNESS

Some materials contain traces of sulphur, or unburnt coal which may be in such small quantities as to appear initially to give good concrete. In time, however, such impurities may cause swelling or disruption of the concrete as a result of internal chemical action, or may attack the reinforcements.

COST

In practice an engineer has to balance the value of the aggregates based on the foregoing considerations against their cost. This generally hinges on their availability and location in relation to the site of the work. Thus a granite which has to be hauled a hundred miles or so may not be economic if a local stone will give 80 per cent as strong a concrete at a greatly reduced cost, even though reduced stresses and larger members have to be adopted in the design for the work.

ART. 1.7. WATER

Water for mixing concrete must be clean and free from harmful materials such as clay, loam or acids. Any town supply is suitable; so are nearly all river waters unless polluted by trade effluents. Although the use of sea-water is not to be recommended, it has been shown generally to have but little deleterious effect, except only that the resulting concrete is liable to be slightly hydroscopic and never becomes bone-dry in a moist atmosphere.

B.S. 3148 (1959) describes tests to determine the suitability of doubtful waters. These tests compare the initial setting times and compressive strengths of blocks and cubes, using the results obtained with distilled water as a control standard.

ART. 1.8. DESIGN OF CONCRETE MIX

It has already been explained that for a concrete to be dense and free from internal voids there has to be sufficient cement to fill the voids between the fine aggregate particles to make a dense mortar, and sufficient mortar to fill the voids between the coarse aggregate particles. Suppose both the fine and coarse aggregates contain 50 per cent voids, then it would seem that one part of cement would just fill the voids in two parts of fine aggregate, making two parts of mortar, and that these two parts of mortar would just fill the voids in four parts of coarse aggregate.

Experiments have shown that with normal techniques one part of cement will not completely fill the voids of two parts of fine aggregate when the latter has 50 per cent voids. This is because some of the cement gets between the aggregate particles holding them further apart than they would otherwise be, so increasing the voids. At the same time the volume of mortar is increased from two parts to about 2·2 parts. In practice, however, aggregates can be selected with voids less than 50 per cent, and it is found generally that in a 2:1 sand/cement mix the voids are completely filled. Similarly in concrete a 2:1 coarse/fine aggregate ratio generally satisfies the requirement of filling the voids; and frequently this ratio is adhered to without variation.

NOMINAL CONCRETE MIXES

Thus in C.P. 114 (1957) all the *nominal concrete mixes* whether using Portland cement or High Alumina cement give the volume of coarse aggregate as twice the volume of fine aggregate. Though where a denser or more workable concrete can be produced by a modification of that ratio it may be varied between the limits 1½ to 1 and 3 to 1, provided the sum of the volumes of coarse and fine aggregates, each measured separately, is not varied from the sum of coarse and fine aggregates given for the nominal mix. The optimum ratio of coarse/fine aggregates will depend principally on the particle shape, grading, and surface texture of the coarse aggregate, as well as on the grading zone of the fine aggregate. As the fine aggregate grading becomes progressively finer, so the ratio of coarse/fine aggregate should be progressively increased. The finest zone of all, Zone 4, is not suitable for use in the nominal concrete mixes.

TABLE 1.3. *Permissible stresses for Portland Cement and Portland Blastfurnace Cement concrete where aggregates comply with B.S. 882*

Nominal mix	Permissible concrete stresses				
	Compression		Shear	Bond	
	Direct	Due to bending		Average	Local
	lb/sq. in.	lb/sq. in.	lb/sq. in.	lb/sq. in.	lb/sq. in.
1:1:2	1140	1500	130	150	220
1:1½:3	950	1250	115	135	200
1:2:4	760	1000	100	120	180

C.P. 114 (1957) sets out *permissible working stresses* for each of the nominal concrete mixes it defines: values for intermediate proportions can be obtained by interpolation. Before applying these stresses in any work, unless satisfactory evidence is produced beforehand from reliable sources, *preliminary cube strength tests* have first to be carried out on 6 in. cubes made with the actual materials and proportions to be adopted, and these have to reach minimum specified values. As the work proceeds *works cube strength tests* have similarly to reach specified values, though somewhat reduced. Where the cube strengths specified are not reached, or where on the other hand the cube strengths are consistently exceeded, the permissible working stresses are to be varied within certain specified limits in certain cases. The C.P. 114 (1957) permissible working stresses for Portland cement concrete and Portland Blastfurnace cement using aggregates complying with B.S. 882 are given in Table 1.3, and the minimum cube strength requirements are given in Table 1.4.

TABLE 1.4. *Proportions and strength requirements for Portland Cement and Portland Blastfurnace Cement concrete where aggregates comply with B.S.* 882

(1)	(2)		(3)		(4)	
Nominal mix	Cubic feet of aggregate per 112 lb. of cement		Cube strength within 28 days after mixing		Alternative cube strength with 7 days after mixing	
	Fine	Coarse	Preliminary test	Works test	Preliminary test	Works test
			lb/sq. in.	lb/sq. in.	lb/sq. in.	lb/sq. in.
1:1:2	1¼	2½	6000	4500	4000	3000
1:1½:3	1⅞	3¾	5000	3750	3350	2500
1:2:4	2¼	5	4000	3000	2700	2000

Thus for example, the permissible working stresses using the nominal mix 1:2:4 with Portland cement and B.S. 882 aggregates are given as follows:

Direct Compression	760 lb/sq. in.
Compression due to Bending	1000 lb/sq. in.
Shear	100 lb/sq. in.
Average Bond Stress	120 lb/sq. in.
Local Bond Stress	180 lb/sq. in.

For the same mix, the preliminary cubes have to reach:

2700 lb/sq. in. at 7 days age and
4000 lb/sq. in. at 28 days age;

and the works cubes have to reach:

2000 lb/sq. in. at 7 days age and
3000 lb/sq. in. at 28 days age.

SPECIAL CONCRETE MIXES

On many works of good design where the control of materials, mixing, placing and supervision is of a suitably high standard, C.P. 114 (1957) permits the use of higher working stresses in special mixes, provided the cube strength tests are also appropriately high. This is a great step in the direction of progress. Such concrete mixes can make for considerable economies, allowing all members in a structure to be reduced in cross section, resulting in an overall saving in weight with consequent economies in the structure itself and in the foundations. The desirability of mechanical vibration for compacting concrete is referred to in many parts of this book. It is particularly worth while where advantage is to be sought from the use of special concrete mixes.

To reap the fullest benefits by getting the best strengths from the available aggregates and a given proportion of cement, a scientific approach needs to be adopted for the design of the concrete mix. A considerable step in mix design can now be reached directly by reference to tabulated results of innumerable tests made with different mix types. But no amount of calculation in the matter of mix design can replace the need for cube strength tests on mix trials in the laboratory; and, what is more important, cube strength tests on the site-concrete which is actually going to be used for the job.

Where the standard of control and technique in making the concrete is particularly high, C.P. 114 (1957) Amendment 1965 now refers specifically to *designed concrete mixes*, and allows appropriately higher working stresses in consequence.

CONCRETE MIX DESIGN

A direct method of concrete mix design is well described in detail both in Road Note No. 4[3] and in *The Vibration of Concrete*[4] from which Figs. 1.6 and 1.7 and Tables 1.5 to 1.7 have been taken. The following gives a summary of the steps involved in the method, and is limited to consideration of ¾ in. aggregates; but for a fuller appreciation of the method, reference should be made to the original works noted above.

Suppose it is required to design a mix to have some particular *minimum* cube crushing strength at 28 days. Clearly the *average* cube strength will be some value greater than the minimum, and the ratio minimum/average will vary depending on the degree of control of the concreting work on site. Values for this ratio, based on statistics, are expressed as a percentage in Table 1.5

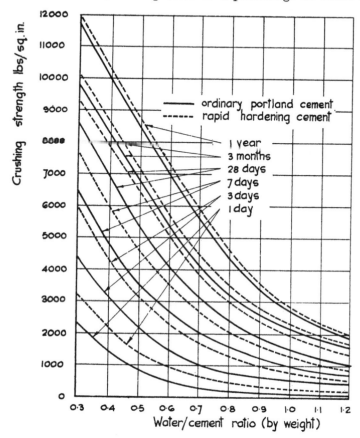

FIG. 1.6. RELATION BETWEEN AVERAGE CRUSHING STRENGTH AND WATER/CEMENT RATIO FOR 4 IN. CUBES OF FULLY COMPACTED CONCRETE FOR MIXES OF VARIOUS PROPORTIONS

for different degrees of control; and thus the required average cube strength may be determined.

Now in properly compacted concretes, provided the ingredients are up to the specifications referred to earlier, the strength of the concrete depends almost entirely on the water/cement ratio (by weight). The relation between crushing strength and water/cement ratio is given in Fig. 1.6 for different ages and different

TABLE 1.5. *Estimated relation between the minimum and average crushing strengths of works cubes for different degrees of control*

Degree of control	Minimum strength as percentage of average strength
Very good control with weigh-batching, moisture determinations on aggregates, etc., constant supervision	75
Fair control with weigh-batching	60
Moderate control; weigh-batching of cement, careful volume batching of aggregate	55
Poor control; volume batching of aggregates and cement	40

TABLE 1.6. *Uses of concrete of different degrees of workability*

Degree of workability	Slump: in.	Compacting factor	Use for which concrete is suitable
"Very low"	0 to 1	0·78	Vibrated or mechanically-compacted concrete in roads or other large sections.
"Low"	1 to 2	0·85	Road slabs without vibration or mass concrete foundations with or without vibration. Simple reinforced sections with vibration.
"Medium"	2 to 4	0·92	For normal reinforced work without vibration and heavily reinforced sections with vibration.
"High"	4 to 7	0·95	For sections with heavily congested reinforcement. Such concrete should only be vibrated to a limited extent under careful supervision.

cements, so that knowing the average crushing strength required, the necessary water/cement ratio can be determined. A first glance at Fig. 1.6 might suggest that for all situations it would be best to work to a very low water/cement ratio so as to get the highest values of crushing strength. But the fallacy of this is brought home by reference to Fig. 1.8 which shows that below some certain water/cement value the concrete becomes too dry to

TABLE 1.7. *Total aggregate/cement ratio (by weight) required to give four degrees of workability with different gradings and types of ¾ in. aggregates*

	Degree of workability	Very low			Low			Medium			High		
	Gradings of aggregate (Curve No. on Fig. 1.7)	1	2	3	1	2	3	1	2	3	1	2	3
ROUNDED	Water/cement ratio by weight 0·35	4·5	4·5	3·5	3·8	3·6	3·2	3·1	3·0	2·8	2·8	2·8	2·6
	0·40	6·6	6·3	5·3	5·3	5·1	4·5	4·2	4·2	3·9	3·6	3·7	3·5
	0·45	8·0	7·7	6·7	6·9	6·6	5·9	5·3	5·3	5·0	4·6	4·8	4·5
	0·50	—	—	8·0	8·2	8·0	7·0	6·3	6·3	5·9	5·5	5·7	5·8
	0·55	—	—	—	—	—	8·2	7·3	7·3	7·4	6·3	6·5	6·1
	0·60	—	—	—	—	—	—	—	—	8·0	×	7·2	6·8
	0·65	—	—	—	—	—	—	—	—	—	×	7·7	7·4
	0·70	—	—	—	—	—	—	—	—	—	×	—	7·9
	0·75										×	—	—
	0·80										×	—	—
	0·85										×	—	—
	0·90										×	—	—
IRREGULAR	Water/cement ratio by weight 0·35	3·7	3·7	3·5	3·0	3·0	3·0	2·6	2·6	2·7	2·4	2·5	2·5
	0·40	4·8	4·7	4·7	3·9	3·9	3·8	3·3	3·4	3·5	3·1	3·2	3·2
	0·45	6·0	5·8	5·7	4·8	4·8	4·6	4·0	4·1	4·2	×	3·9	3·9
	0·50	7·2	6·8	6·5	5·5	5·5	5·4	4·6	4·8	4·8	×	4·4	4·4
	0·55	8·3	7·8	7·3	6·2	6·2	6·0	×	5·4	5·4	×	4·8	4·9
	0·60	9·4	8·6	8·0	6·8	6·9	6·7	×	6·0	6·0	×	×	5·4
	0·65	—	—	—	7·4	7·5	7·3	×	×	6·4	×	×	5·8
	0·70	—	—	—	8·0	8·0	7·7	×	×	6·8	×	×	6·2
	0·75				—	—	—	×	×	7·2	×	×	6·6
	0·80				—	—	—	×	×	7·5	×	×	×
	0·85							×	×	7·8	×	×	×
	0·90							×	×	×	×	×	×
	0·95							×	×	×	×	×	×
	1·00										×	×	×
ANGULAR	Water/cement ratio by weight 0·35	3·2	9·0	2·9	2·7	2·7	2·5	2·4	2·4	2·3	2·2	2·3	2·1
	0·40	4·5	4·2	3·7	3·5	3·5	3·2	3·1	3·1	2·9	2·9	2·9	2·8
	0·45	5·5	5·0	4·6	4·3	4·2	3·9	3·7	3·7	3·4	3·5	3·5	3·2
	0·50	6·5	5·8	5·4	5·0	4·9	4·5	4·2	4·2	3·9	×	3·9	3·8
	0·55	7·2	6·6	6·0	5·7	5·4	5·0	4·7	4·7	4·5	×	×	4·3
	0·60	7·8	7·2	6·6	6·3	6·0	5·6	×	5·2	4·9	×	×	4·7
	0·65	8·3	7·8	7·2	6·9	6·5	6·1	×	5·7	5·4	×	×	5·1
	0·70	8·7	8·3	7·7	7·4	7·0	6·5	×	6·2	5·8	×	×	5·5
	0·75	—	—	8·2	7·9	7·5	7·0	×	×	6·2	×	×	5·8
	0·80	—	—	—	—	—	7·4	×	×	6·6	×	×	6·1
	0·85				—	—	7·8	×	×	7·1	×	×	6·4
	0·90				—	—	—	×	×	7·5	×	×	×
	0·95							×	×	8·0	×	×	×
	1·00							×	×	—	×	×	×

— Indicates that the mix was outside the range tested.
× Indicates that the mix would segregate.

be properly workable, depending on particle shape and grading of the aggregate, on the size of the member and congestion of reinforcement, and on whether the concrete is to be compacted by mechanical vibration or by hand tamping. Table 1.6 defines arbitrarily four degrees of workability required for different situations and methods of compaction, and Table 1.7 then gives

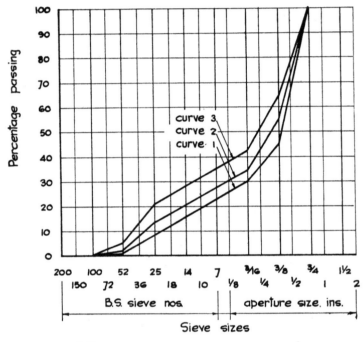

FIG. 1.7. CURVES FOR THREE GRADINGS OF $\frac{3}{4}$ IN.
AGGREGATE

the total dry aggregate/cement ratio (by weight) required for the different degrees of workability and shapes of the aggregate particles, for the water/cement ratio already determined.

In Table 1.7, each column for degree of workability refers to three gradings of the aggregate. These gradings are given in Fig. 1.7 and have been chosen so as to give good results with normal aggregates. Natural aggregate which does not follow any of the curves in Fig. 1.7 may well give good results, as referred to later. The curve No. 1 represents the grading with the least proportion of small material and accordingly is not suitable for situations where great workability is required: however, the reduced area of aggregate surface will make for a low water/ cement ratio, with consequent economy. Curve No. 3, on the other hand, has the highest proportion of small material, and is suitable for situations where the reinforcements are especially congested. Curve No. 2 is the happy medium.

An example of the use of the method is now given.

Suppose a mix is required to give a minimum cube strength of 4000 lb/sq. in. at 28 days using Ordinary Portland cement in a heavily reinforced section where vibrating tools will be used. The aggregates available locally are rounded, and the maximum nominal size will be ¾ in. All materials are to be measured by weigh-batching and supervision will be good.

From Table 1.5 it is seen that the average cube strength required is $4000 \times \dfrac{100}{60} = 6660$ lb/sq. in. at 28 days. From Fig. 1.6 this will require a water/cement ratio of 0·42. Table 1.6 shows the degree of workability of the mix required as "medium". By reference to Table 1.7, for a rounded aggregate, with medium workability, and with a water/cement ratio of 0·42 using grading No. 2 the total aggregate/cement ratio is seen to be 4·6. And from Fig. 1.7 the following proportions of true grading are seen to be required:

Aggregate size	Proportion of total mix
¾ in. to ⅜ in. ⅜ in. to ₃₆ in. Sand	45% 20% 35%

But in practice the aggregates available are not likely to be graded exactly as in Fig. 1.7, and accordingly some adjustment of the batching proportions will be necessary. This adjustment is made stage by stage working down from the largest aggregate sizes; and is described fully in *The Vibration of Concrete*. It is strongly emphasised that after a mix has been designed in the manner just described, trial mixes should be made up and test cubes crushed to give confirmation of the concrete strength.

QUANTITY OF WATER

With special concrete mixes it is essential that the water/cement ratio is maintained constant at its correct value, as determined in the previous section.

With *nominal concrete mixes* a little more latitude is intelligent, though a reasonable degree of control is still necessary. Clearly a curve relating strength to water/cement ratio must start from zero at one end (no water) and return to zero at the other (all water), and hence there will be an intermediate ratio for the practical maximum strength. This is indicated in Fig. 1.8 for a 1:2:4 mix at 7 days, though the exact characteristics of such a curve will depend on the shape and grading of the aggregate particles, and the method and degree of compaction achieved, particularly in the lower water/cement ratio range. The amount

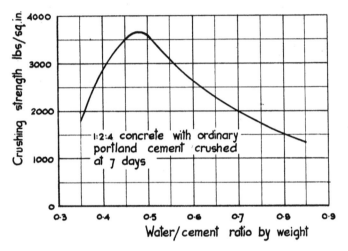

FIG. 1.8. PRACTICAL MAXIMUM CRUSHING STRENGTH
WITH DIFFERENT WATER/CEMENT RATIOS

of water corresponding to maximum strength usually results in a concrete too dry to be conveniently workable with the result that voids and honeycombing appear. Surely any concrete is better than none. Hence a compromise has to be struck between maximum strength and ready workability.

The degree of workability (*see* Table 1.6) is controlled on site by direct measurement of the water fed to the mix. A rough guide as to workability is given by the *slump test* (described fully in B.S. 1881 (1952)), where an open-ended metal former – this being a frustrum of a cone 12 in. high, 8 in. bottom diameter, 4 in. top diameter – is stood on a metal plate and filled in four stages with the concrete to be tested, the concrete at each stage being tamped with a steel rod. The cone is then carefully lifted and the reduction in height of the concrete is measured and known as the *slump*.

For laboratory work and high quality site work where near-zero slump concrete is required (as for example in prestressed work) the *compacting factor test* is now replacing the slump test. The compacting factor test is fully described in B.S. 1881 (1952). The apparatus is shown in Fig. 1.9 and consists of two conical hoppers set above a cylindrical mould. The bottoms of the hoppers are fitted with doors with quick-release catches. The upper hopper is filled with the concrete to be tested and the door opened allowing the concrete to fall into the lower hopper. The door of the lower hopper is then immediately opened allowing the concrete to fall into the bottom cylinder. The surplus concrete above the top of the cylinder is then cut off flush with a trowel and the weight of concrete remaining in the cylinder weighed and referred to as the "weight of partially compacted concrete". The same cylinder is then filled with concrete in layers rammed to achieve full compaction. This is then weighed and is known as the "weight of fully compacted concrete". The *compacting factor* is defined as the

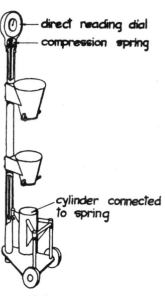

FIG. 1.9. APPARATUS FOR COMPACTING FACTOR TEST

ratio of the weight of partially compacted concrete to the weight of fully compacted concrete.

CUBE AND CYLINDER STRENGTHS

Reference is made in this article and elsewhere to the crushing strength of 6 in. concrete cubes in assessing the strength of a concrete. Cube testing is standard practice in Great Britain and Europe, but in the United States of America the ultimate strength of concrete is given by the strength of cylinders 6 in. diameter and 12 in. long. Cubes and cylinders of the same quality of concrete have different crushing strengths. The cylinder will fail at a stress of only about 0·8 times that required to crush the cube; and it gives a more realistic estimate of the strength of concrete in normal structures. The reason is that in the more slender cylinder, failure can take place by shear at a natural angle determined by the ratio of crushing to tensile strength (usually about 30° to the axis of the cylinder), whereas when a cube is tested the shape of the specimen interferes with the mode of failure and restricts this artificially to a plane at about 45°, bringing more friction into play.

In real structures, beams and columns are long enough for failure to occur at the natural angle, so that the test on cylinders corresponds more nearly to the way concrete will fail in practice. But there are practical advantages in testing cubes. They are loaded on their sides in the testing machine so that the top surface of the specimen as filled has little influence, and they can be stored and transported more easily. And since the ratio between cube and cylinder strengths is practically constant (1 to 0·78) there seems to be no objection to tests on cubes provided it is realised that the results they give are unnaturally enhanced. Thus a 1:2:4 nominal mix having a 28-day works cube strength of 3000 lb/sq. in. and a permissible working stress of 760 lb/sq. in. has not a factor of safety of $\dfrac{3000}{760} = 3.95$ but more nearly $\dfrac{3000 \times 0.78}{760} = 3.10$.

ART. 1.9. MIXING AND COMPACTING CONCRETE

All concrete today is mixed in mechanical mixers. Mixing continues until the various materials are uniformly distributed, and certainly for not less than two minutes. The cement must be measured by weight. The quantities of aggregates are also best measured by weight: this is essential when using special concrete mixes, though measurement by volume is permitted for the *nominal concrete mixes*. Immediately following mixing, the concrete has to be transported by trucks or pumping so that it can be placed in its final position in the work within 45 minutes of the water first coming into contact with the cement: otherwise the initial setting of the cement will be disturbed.

VIBRATED CONCRETE

It is normal now on all but the smallest works to compact the concrete mechanically using vibrating tools. This enables a low water/cement ratio to be used satisfactorily and achieves greater compaction. For in-situ structural work immersion type vibrators are generally used. They are effective for a radius of about 18 in. to 24 in. when allowed to sink under their own weight so as to be completely covered by concrete thus bringing into play the weight of the concrete above the vibrator head to assist in retaining within the concrete the energy of the vibrator. The vibrator has to be removed from the concrete slowly, while still acting, so as to allow the surrounding concrete to run in and take the position occupied by the departing vibrator. Each vibrator requires the full attention of one man who should keep it continually on the move from place to place and not leave it in one position for any length of time.

AIR-ENTRAINED CONCRETE

A number of additives are now on the market for assisting the compaction and handling of concrete. These materials are variously described as concrete plasticisers, wetting agents and air-entraining agents. They are fatty surface-active compounds which are fed into the mix in proportions up to about 0·1 per cent of the weight of the cement. In the mixing process they entrain air in the concrete up to about 5 per cent of the volume of concrete. The entrained air takes the form of numerous minute independent bubbles distributed uniformly throughout the mass, and has desirable qualities as described later. It is entirely different from the random porosity which results in normal concrete: and it is not to be confused with the much greater air inclusion achieved in the production of aerated concrete for purposes where light weight and low thermal conductivity is required.

The plasticisers reduce the surface tension of the water in the mix, increasing the likelihood of all particles of the aggregate and cement being properly wetted. Accordingly more efficient use is made of the cement, and there is less likelihood of parts of the aggregate being inadequately coated. The bubbles, acting as lubricated balls, together with the better wetting, increase the workability of the concrete so that better compaction is achieved, allowing of a lower water/cement ratio and consequently greater strength. Alternatively, for the same strength, the proportion of cement in the mix can be reduced with some economy. The better compaction, together with the restriction of entrained air into well defined bubbles, makes also for greater impermeability or waterproofness, and consequently in positions of exposure makes for greater permanence. Air-entrained concrete also has the property of being more cohesive than normal concrete, with the result that it can be transported and pumped considerable distances with less fear of segregation of the ingredients. All the special properties resulting from the use of plasticisers or air-entraining agents are due to the distribution of air in bubble form throughout the mass of the concrete, and not to any physical effect the additives have on the concrete material itself.

VACUUM CONCRETE

A more recent technique for assisting the compaction of concrete bears the misnomer "Vacuum Concrete". It has not yet widely caught the imagination of engineers, though there can be little doubt that the method will grow in favour, particularly for the manufacture of precast units. The system consists of covering the formwork internally with a filter cloth of wire mesh and gauze fabric; and as soon as the concrete has been placed in position and vibrated, a vacuum pump connected to the filter cloth reduces the pressure at the formwork from atmospheric to practically zero. This depression extends slowly across the full thickness of the concrete member, so that the internal pressure of the fresh concrete (which was made up originally of integranular pressure and fluid pressure) is drastically reduced and eventually becomes little more than the integranular contact pressure between the aggregate particles. Meanwhile the pressure of the atmosphere continues to act on the free surface of the concrete, with the result that the surplus water in the mix is expelled through the filter cloth, and the concrete is compressed in the formwork to a state of very high compaction.

The vacuum process is not to be confused with any idea of suction. It is the work done by the pressure of the atmosphere acting *against* vacuum that forces the aggregate particles together and consequently expels the surplus water. Within the mortar and the cement suspensoid the same process takes place, the water being expelled as the smaller grains are squeezed together. Thus there is practically no loss of cement from the mix, as indeed is shown by the clearness of water pumped out. But the reduced water/cement ratio gives the concrete a considerably enhanced strength, both in the immediate stage, and also finally. Alternatively the cement content of the mix can be reduced for a given strength with some economy.

A further advantage arising from the process is that when the vacuum pump is turned off, the atmosphere acting on the reduced internal fluid pressure of the concrete mass, sets up greater surface tension effects than would normally exist. Consequently the concrete immediately acquires some cohesive characteristics, and instead of slumping like a truly granular material, stands freely up to considerable heights when the side shutters are removed. Leviant[5] says this speeding up of striking the formwork can reduce the amount of shuttering in some cases by as much as 80 per cent, whilst the additional unit price by modifying the shuttering to make it "active" does not exceed about 25 per cent. The filter cloths can be used in some cases up to 200 times; and the cost of their replacement is compensated by the fact that their use avoids the need for greasing the formwork.

ART. 1.10. REINFORCEMENT

The reinforcement has normally to be one of the following:
   Rolled steel bars or hard drawn steel wire complying with B.S. 785 (1938);
   Cold twisted steel bars complying with B.S. 1144 (1943);
   Hot-rolled high-strength deformed bars;
   Steel fabric complying with B.S. 1221 (1964).

In all cases the steel is manufactured by the Open Hearth (Acid or Basic) process or by the Acid Bessemer process. The sulphur and phosphorous content is limited to 0·06 per cent for mild, medium tensile, and high tensile steels; and the carbon content in high tensile steel is limited to 0·30 per cent. Steels

TABLE 1.8. *Tensile test requirements for steel bars*

	Mild steel	Medium tensile	High tensile
ULTIMATE TENSILE STRESS (The maximum values not to apply to bars under ⅜ in.)	28–33	Tons/sq. in. 33–38	37–43
YIELD POINT (MIN) Up to and including 1 in. . . . Over 1 in. up to and including 1½ in. . Over 1½ in. up to and including 2 in. . Over 2 in. up to and including 2½ in. . Over 2½ in. up to and including 3 in. .	None specified	19·5 18·5 17·5 16·5 16·5	23·0 22·0 21·0 20·0 19·0
		Per cent	
ELONGATION (MIN) Under ⅜ in. (on 8 diameters) . . . Up to 1 in. (on 8 diameters) . . . Over 1 in. (on 4 diameters) . . .	16 20 24	14 18 22	14 18 22

prepared by the Basic Bessemer process (not permitted) tend to have a high nitrogen content. This produces embrittlement; so also does a high phosphorous content. The physical requirements of steel bars as rolled are that they shall have ultimate tensile strengths, yield points and elongations on test in accordance with the figures given in Table 1.8. These values are of course enhanced when the bars are cold-worked by twisting or by drawing to wire.

C.P. 114 (1957) sets out permissible working stresses in steel reinforcement as in Table 1.9.

TABLE 1.9. *Permissible stresses in steel reinforcement*

Type of stress	Permissible stress (lb/in.²)		
	Mild steel bars to B.S. 785 and all plain bars		All high-bond bars, and high-yield wire mesh, having a guaranteed yield or proof stress
	Effective diameter not exceeding 1¼ in.	Effective diameter exceeding 1½ in.	
Tensile stress other than in shear reinforcement	20,000	18,000	$0.55 f_y$, but not more than 33,000 for bars with an effective diameter not exceeding ⅞ in.
			$0.55 f_y$, but not more than 30,000 for bars with an effective diameter exceeding ⅞ in.
Tensile stress in shear reinforcement	20,000	18,000	$0.55 f_y$, but not more than 25,000
Compressive stress	18,000	16,000	$0.55 f_y$, but not more than 25,000

where f_y = the guaranteed yield or proof stress

These stresses are recommended having regard to the need for avoiding cracking of the concrete structure to a degree which would be undesirable for buildings. For liquid retaining structures and structures of a specialised character, such as silos for the storage of cement or wheat, considerably reduced values may be desirable. This is referred to in Article 2.2 and again in Chapters 10 and 12.

For all but the smallest and simplest work, descriptive lists are prepared in the design office showing the exact bending, shape, dimensions and diameter of every bar type, together with the number of each type of bar required. The bars are then bent cold, strictly to the details shown on the descriptive lists. Before placing in the forms, the reinforcements need to be brushed free from loose scale and rust, and must on no account be allowed to come into contact with mould oil or similar materials likely to lessen the bond strength between the steel and the concrete. When placed in the forms the reinforcements have to be firmly supported accurately in position to ensure the proper cover being obtained and to prevent the bars being dislodged while the concrete is placed and compacted. Binding wire is arranged with the ends bent away from the formwork so that the cover is not reduced by the wire by more than its own diameter.

Wires and cables for use in prestressed concrete work, together with end anchorages and grouting are discussed in Chapter 15.

REFERENCES

1. HAMILTON, S. B. A Note on the History of Reinforced Concrete in Buildings. *National Building Studies Special Report*, No. 24. H.M. Stationery Office (1956).
2. British Standard Code of Practice C.P. 114 (1957) Amended 1965, *The Structural Use of Reinforced Concrete in Buildings*. British Standards Institution (1965).
3. *Design of Concrete Mixes*. Road Note No. 4. Road Research Laboratory. H.M. Stationery Office (1950).
4. *The Vibration of Concrete*. Institution of Civil Engineers, and Institution of Structural Engineers (1956).
5. LEVIANT, I. Introduction to the Vacuum Concrete Processes. *The Structural Engineer*. Vol. 30. No. 11. (November, 1952).

CHAPTER 2

PROPERTIES OF REINFORCED CONCRETE

ART. 2.1. WEIGHT

OTHER things being equal, concretes made from aggregates of the same density and grading vary in weight according to the proportion of cement in the mix. With aggregates of normal density, contrary to popular belief, concretes rich in cement weigh less than concretes of weaker mixes, provided always there is enough cement in the mix to fill the voids. This follows from the fact that cement weighs less than concrete. In Fig. 2.1 the

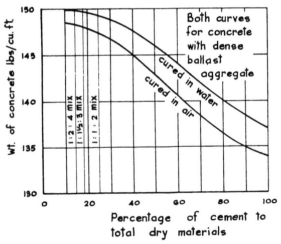

FIG. 2.1. REDUCTION OF WEIGHT OF CONCRETE WITH INCREASE OF CEMENT CONTENT

weights are shown of various concretes each made from the same dense-ballast aggregate but with different percentages of cement. One of the curves is for concrete cured in air: the other is for concrete cured by submersion in water. The cement percentages of the nominal mixes given in C.P. 114 (1957) are indicated. It will be seen from Fig. 2.1 that over normal ranges, concrete cured dry weighs about 2 lb/cu. ft. less than concrete cured wet.

When the amount of reinforcing steel is considerable, the weight of reinforced concrete increases rapidly. This is shown in Fig. 2.2 based on concrete weighing 150 lb/cu. ft. and steel at 490 lb/cu. ft. It will be seen that to take the weight of reinforced concrete as 150 lb/cu. ft., as is frequently done, is to underestimate, and that with high percentages of reinforcement an

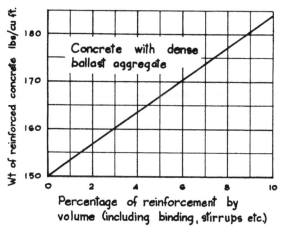

FIG. 2.2. EFFECT OF REINFORCEMENT PERCENTAGE
ON WEIGHT OF CONCRETE

adjustment is necessary. In the design of reinforced concrete ships and other structures where scantling sizes are cut down to a minimum, 10 per cent of reinforcement is neither uncommon nor uneconomic.

ART. 2.2. PERMANENCE

No structural material is everlasting in any absolute sense, and permanence is relative. Experience shows that long life for reinforced concrete depends on clean, impervious, chemically inert aggregates; a sufficiency of cement to produce a watertight concrete (not less than 1:2:4 mix); a sufficiency of water to enable the concrete to be thoroughly compacted, but no great excess which on drying out would leave the concrete porous; thorough compaction to ensure complete absence of voids; and ample cover of concrete to the reinforcements. The water content can be kept to a minimum and the best compaction achieved with rounded aggregates and when mechanical vibrators are used. It is too early for any pronouncement to be made on the degree of permanence of reinforced concrete, though enough experience has now been gained to show that with good design, good work-manship, good supervision and reasonable absence from corroding influences its life can certainly exceed sixty years. Furthermore, structures demolished after forty years' life have many times shown the reinforcements to be entirely free from corrosion and the concrete as sound as the day it was constructed : and this leads one to suppose that under good conditions reinforced concrete structures should last a hundred years or more, particularly if the concrete is mechanically vibrated.

The adverse conditions of exposure discussed below may reduce the life of reinforced concrete and require careful consideration. Wherever special cements are used to meet such adverse conditions, the use of those cements is no substitute for the necessity of providing a compact dense concrete free from voids or excess mixing water. Dense concrete into which deleterious solutions cannot enter or flow is the first essential measure towards prevention of attack. Precast concrete, made in factory conditions and well matured before exposure, is more resistant than concrete cast-in-situ and exposed while still green.

EXPOSURE TO ATMOSPHERIC CONDITIONS

Reinforced concrete stands exposure to weather relatively well : nevertheless it will last longer when protected as for example inside a dry building or, if used for outer walls, when protected by brickwork or stonework. When not so protected it is advisable to give additional concrete cover to reinforcements in the exposed portions of any structure. Rain is more corroding when containing carbonic, or sulphuric or sulphurous acids in solution, as occurs in industrial areas, or where sulphurous coal is burnt in numerous domestic grates. Rain containing salt blown in from the sea can also be harmful. Frost action – spalling – except in the uncured state, normally arises only with porous permeable concretes. The techniques necessary to guard against failure from this cause are therefore clear.

EXPOSURE TO DELETERIOUS SOIL CONDITIONS

Some soils contain sulphates in such form that ground water solutions can be strong enough to attack concrete made with Ordinary Portland cement.[1] In these cases adequate protection may be given by using Sulphate Resisting Portland cement : in more extreme cases it may be necessary to use High Alumina cement (see Arts. 1.3 and 1.5). Peaty soils sometimes contain organic acids in quantities sufficient to merit special precautions, such as the use of High Alumina cement.

EXPOSURE TO SEA WATER

Marine structures have sometimes suffered severe corrosion in about twenty years, requiring expensive repair works. The magnesium chloride in the sea-water acts on the free lime in the concrete and gradually dissolves it out. Cases arise where the use of Sulphate Resisting or High Alumina cements may be desirable. In general, marine structures suffer most between ordinary high and low tides, where subject to alternate wetting and drying.

EXPOSURE TO INDUSTRIAL SUBSTANCES

Certain vegetable oils or fats may become rancid and produce deleterious organic acids; and sugar solutions may ferment and become destructive when this fermentation occurs in contact with concrete. Certain flue gases and mineral acids are also destructive. In all such cases it is necessary to protect reinforced concrete with surface finishes of suitable tiles or bricks, laid in appropriate chemically-resistant jointing materials.

EFFECT OF HIGH STEEL STRESSES

It has long been established that in reinforced concrete construction the protection afforded to the reinforcement by the covering concrete is not impaired by the formation of very fine cracks in the concrete. However, where the reinforcing steel is highly stressed in tension, the normal hair-cracks and construction-joints open up to excessive widths, allowing air, moisture and other corroding influences to attack the reinforcements. Sometimes, for sake of commercial advantage or other reasons, very high steel stresses are adopted, resulting in too great strain or stretch. This may well be a short-sighted policy, especially in conditions of exposure to atmospheric or other corroding influences, resulting in considerable reduction to the life of the whole structure. Where complete weathertightness is required, as for example in silos for the storage of cement or wheat, the inconvenience and damage caused to the structure and to the stored materials by unwanted cracks forming for the total thickness of the external walls may be considerable; and any subsequent remedy is often difficult to achieve and generally expensive.

ART. 2.3. FIRE RESISTANCE

Fire resistance is not to be confused with *combustibility*. Uncased steel is incombustible at ordinary temperatures, but by no means fire-resisting. Fire resistance is the ability to resist fire for a certain time without serious loss of strength, distortion or collapse, and to withstand all the severe conditions arising from fire in modern civilisation, as for example the quenching effect of water from fire-fighting teams. It is clearly relative and not absolute, since all building materials would ultimately fail if the heat intensity and duration were sufficient.[2] The relative fire resistance of structural components of different sizes and materials is determined by practical tests at experimental stations, where the members to be tested are set up in specially constructed chambers where fires of standard intensities and duration are produced. The resistance is stated as the number of hours the component can properly carry its normal load at a specified temperature followed by watering from hoses without collapse. Apart from reducing financial risk or loss to the owner of the structure, high fire resistance is desirable to give the occupants time to escape,[3] to facilitate and make safer the work of the fire brigade, and to prevent damage and inconvenience to surrounding buildings and streets. Clearly then a higher standard of fire resistance is necessary for tall buildings than for low or single-storey buildings. The damage which would result if a skyscraper were to collapse has only to be considered to give point to this.

Reinforced concrete has about the best fire resistance of all materials normally used for structural work. Uncased structural steelwork has a relatively low fire resistance. It rapidly takes up high temperatures from direct contact with the fire, and since the strength of steel falls rapidly with increase of temperature, being quite low at red heat as evidenced by the blacksmiths' technique, the normal factor of safety disappears and collapse occurs. An uncased steel structure which has been subject to severe fire becomes a mass of twisted members difficult to believe unless witnessed. However, when steelwork is properly cased with brick or concrete the fire resistance can be as good as that of reinforced concrete. Care has then to be taken with the brick casing to ensure that it does not become detached in the fire and so cease to protect; this was shown up particularly in the disastrous fire following the San Francisco earthquake. The most secure casing is generally considered to be concrete held in place with steel binding-wire.

It appears that the high fire resistance of reinforced concrete is partly due to the steel being in numerous steel reinforcing rods instead of the larger flanges of structural steelwork, facilitating good adhesion and attachment of the concrete; and it is partly due to the poorer thermal conductivity of concrete compared with steel which prevents or retards the transmission of high temperatures across the individual members and from one part of a structure to another. For fire resistances of 2 hours and 4 hours, reasonable thicknesses of members of reinforced concrete using B.S. 882 aggregates are shown in Table 2.1.

TABLE 2.1. *Minimum dimensions of members to give 2-hour and 4-hour fire resistance*

Type of member	Period of fire	
	2 hours	4 hours
Walls	4 in.	7 in.
Floor slabs	5 in.	6 in.
Columns	12 in.	18 in.

Where light-mesh reinforcement is provided centrally in the concrete cover to the main reinforcement in columns, the thicknesses for 2 hours and 4 hours may be reduced to 9 in. and 12 in. respectively. It will be seen from the above that with reinforced concrete the requirements having regard to fire resistance will in many cases be otherwise met from structural considerations.

ART. 2.4. IMPERMEABILITY

Reinforced concrete can be made sufficiently impermeable for it to be readily suitable for the construction of reservoirs, water towers, and other liquid storage tanks. For such concrete, the aggregates must be impermeable ; and the mix must be sufficiently rich in cement and the aggregates well enough graded for the cement to fill the voids in the fine aggregate, and the resultant mortar to fill absolutely the voids in the large aggregate ; and consequently, it is normal to adopt a mix of 112 lb. cement to 2 cu. ft. fine aggregate and 4 cu. ft. coarse aggregate, being about 25 per cent richer than the nominal 1:2:4 mix. The mix requires sufficient water to make for good workability so that proper compaction and density are achieved ; on the other hand no undue excess of water is desirable since this later dries out leaving voids and accentuating shrinkage (*see* Art. 2.8). In other words a compromise has to be struck between high workability on the one hand and low water/cement ratio on the other ; and this normally requires the use of mechanical vibration, and where possible, evenly-graded rounded aggregates.

It is clear from the very make-up of concrete, a matrix of granular particles, that its impermeability cannot be absolute.

Consider six tennis balls as they lie in their rectangular box: whether they represent the large aggregate, or the fine aggregate, or the extremely fine particles of cement powder, there cannot be a hundred per cent dense solid resulting from the mixture of such ingredients. Accordingly the old *Code of Practice for the Design and Construction of Reinforced Concrete Structures for the Storage of Liquids* (*1950*) recommended a minimum thickness of concrete dependent upon the head of liquid retained. This minimum was 1 in. in excess of $\frac{1}{40}$ the depth below top water level, with a minimum value of 4 in. Curing of concrete is referred to in Article 2.5. It is of special consequence in water-retaining structures, the first seven days being critical. The damage done by ignoring proper water-curing during this period can never be repaired by subsequent efforts.

Some engineers carry out percolation tests on test slabs made of the concrete prior to the commencement of the work, and also from representative samples of the concrete as placed in the work. The test slabs may be about 12 in. square, and 3 or 4 in. thick, and they are subjected to a head of water of 40 to 50 ft. over a period of about a week. Such percolation tests will show up any blatant faults in the mix design or control of site concreting procedure: however, the evaporation throughout the test of percolated moisture from the face of the slab remote from the water is normally sufficient to give the impression of complete imperviousness. The test therefore may not be realistic. It is this combination of percolation and evaporation which leads to the popular comment that concrete floors are "cold", and the known phenomenon that concrete can be made waterproof but not damp-proof. Basements, in waterbearing ground, which have not been lined with asphalt or some other impervious membrane, may become musty and develop mould-growths.

As concrete dries out, it also shrinks, and unless it is free to shrink without restraint, tension stresses are set up, sometimes resulting in cracks. Accordingly a reservoir which is kept full of water is less likely to crack than one which is allowed to dry out. Tests have shown that the more thorough the curing of the concrete, and, up to a point, the longer the concrete has been kept saturated with water, the less permeable it is. Concretes made with Rapid-Hardening Portland cement are more likely to give shrinkage trouble than concretes using Ordinary Portland cement: and for this reason it is inadvisable to use Rapid-Hardening cement for liquid-retaining structures. It should also be noted that the enhanced mechanical bond offered by the use of high-bond bars may not become effective until after movements have occurred sufficient to allow minor cracks to form: thus there may be but little advantage gained from the use of high-bond bars in liquid-retaining structures, or indeed for silos or other storage structures requiring complete absence of cracking for reasons of weathertightness. Nor in such cases have steels with a high yield point any advantage over ordinary mild steel, since the eleastic moduli for the different steels is sensibly constant. To assist in controlling the effects arising from shrinkage, it is normal to provide partial or complete contraction joints at suitable intervals throughout all but the smallest liquid-retaining structures. The joints are later sealed with rubber/bitumen or other sealers, as further discussed in Chapters 12 and 14.

ART. 2.5. STRENGTH WITH AGE AND TEMPERATURE

The general characteristics of the strength of concrete are discussed in Chapter 1. The present article considers how the strength varies over the period from the date of mixing the concrete to its reaching maturity.

Figure 2.3 shows the crushing strengths of concrete cubes at

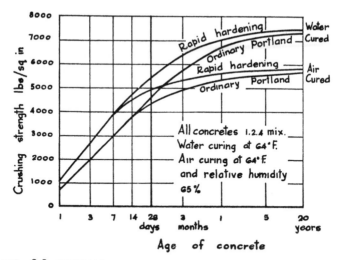

FIG. 2.3. INCREASE OF STRENGTH OF CONCRETE WITH AGE

different ages up to 20 years. Separate curves are given for Ordinary and Rapid-Hardening cements, and for the cubes stored in air or water, at a constant temperature of 64°F. In all cases the concrete mixture is 1:2:4 nominal. It will be seen from

Fig. 2.3 that concrete grows in strength with age, the rate of increase of strength being greatest during the first days of hardening, and becoming progressively less as the concrete gets older. Fig. 2.3 also shows that the Rapid-Hardening cement gives considerably enhanced strengths for the early life of the concrete, though relatively little advantage later on. This early strength is of advantage enabling the formwork to be struck earlier than would otherwise be safe: this has the double merit of saving time and enabling more re-uses to be made of the shuttering, thereby saving cost. Usually when referring to the strength of a concrete the 28-day strength is given. It will be seen from Fig. 2.3 that beyond this age the strength increase becomes small compared to the additional time it is necessary to wait for higher strength values. In C.P. 114 (1957), *age-factors* have now been introduced, permitting the use of compressive stresses higher than the 28-day values for members which will not receive their full design load until after the 28-day interval, as for example in foundations and the lower columns in multi-storey buildings. These age-factors are given in Table 2.2.

TABLE 2.2. *Age factors for permissible compressive stresses in concrete*

Minimum age of member when full design load is applied (months)	Age factor
1	1·0
2	1·10
3	1·16
6	1·20
12	1·24

From Fig. 2.3 it will have been observed that concrete cured in water achieves higher strengths than concrete cured in air. To understand this, one has to appreciate that the hardening of concrete is in no way associated with the water drying out, but results from a chemical reaction between the cement and water. However, when the concrete is wet at the time of mixing and placing, the cement particles are never hydralised throughout, the reaction being limited merely to the surface of the particles. The full amount of water added to the mix does not in fact take part in the reaction but is held within the concrete as moisture: and so long as this moisture is retained in the concrete it will act slowly on the part-hydrated cement, increasing the amount of chemical reaction and gradually enhancing the strength and density of the concrete. If the concrete is cured in water, there is always available a surplus of moisture to assist in this further hydration. But if the concrete is cured in air, moisture evaporates from the surface of the concrete leaving insufficient for completing the chemical reaction. Indeed, under circumstances of extreme sunshine or drying winds, the concrete may give up the whole of its moisture and the reaction virtually cease. Thorough water-curing is specially desirable in the first seven to ten days of the life of concrete. This enables hydration to continue further

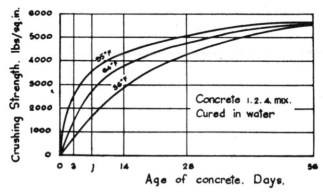

FIG. 2.4. EFFECT ON STRENGTH OF CURING ORDINARY PORT-LAND CEMENT CONCRETE AT DIFFERENT TEMPERATURES

into the cement particles before the outer crust of the particles becomes too dense.

It is for these reasons that proper curing of concrete is so important where high strengths and a high degree of impermeability are required. Water-curing is clearly the most efficient; if this is impracticable, the original moisture in the wet concrete can be retained by protecting the concrete from sun and wind by covering with burlap and hessian and spraying these continuously with water; or more recently it is becoming common practice where large areas of work are involved, as with roads and reservoirs, to apply by spray an impervious plastic-like film of selected resins and gums broken down with a volatile solvent, which serves to retain the moisture in the concrete for a period of about fourteen days before subsequently evaporating from the surface.

Figure 2.4 indicates the different rates of increase in strength of 1:2:4 nominal mix concretes cured in water at different temperatures, using Ordinary Portland cement. Fig. 2.5 gives similar information for 1:2:4 concretes using Rapid-Hardening cement. These figures show that at higher temperatures the growth of strength is more rapid, particularly in the earlier stages. Advantage may be taken of this, particularly in the early release of formwork. This applies especially in the manufacture of precast units, where by operating in artificial conditions of warmth and high relative humidity it is possible to re-use the moulds in only a fraction of the time required in cold dry outdoor conditions. Similar curves are not given for High Alumina cement concretes,

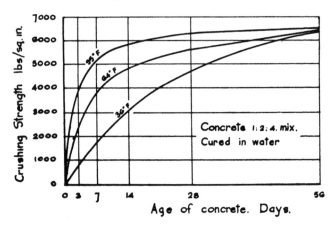

FIG. 2.5. EFFECT ON STRENGTH OF CURING RAPID-HARDEN-
ING PORTLAND CEMENT CONCRETE AT DIFFERENT
TEMPERATURES

because this cement is not suitable for high-temperature conditions, its strength tending to fall off rather than to increase.

ART. 2.6. ELASTICITY

The *modulus of elasticity* of any material is given by the ratio of stress/strain. Thus the modulus

$$E = \frac{\text{stress}}{\text{strain}} = \frac{\dfrac{P}{A}}{\dfrac{\delta}{L}} = \frac{PL}{A\delta} \qquad (2.1)$$

where P is applied load,
  A is area loaded,
  L is original length of specimen,
  δ is linear deformation of specimen.

With a truly *elastic* material the stress is directly proportional to the strain: and the modulus of elasticity of the material remains constant for all stresses. Concrete is not a truly elastic material. Figure 2.6 shows the stress/strain relationship for a 1:2:4 nominal mix concrete. It will be seen that although the curve starts roughly as a straight line at the origin, as the stresses increase, and long before failure, the strains increase disproportionately. Had the loading test indicated in Fig. 2.6 been spread over a greater

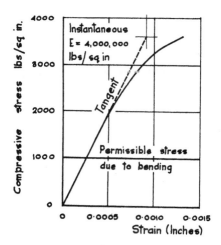

FIG. 2.6. TYPICAL STRESS/STRAIN CURVE FOR CONCRETE

period of time, the strains at all stresses would have been considerably increased, and the departure from the linear relationship more pronounced. This is associated with *creep* which is discussed more fully in Article 2.7: nevertheless an awareness of the effect of creep is necessary for any realistic appreciation of the elastic modulus of concrete, as will be shown.

In Fig. 2.6 the *instantaneous elastic modulus* of this concrete corresponds to the dotted tangent to the curve at the origin, thus

$$E_{c_{(\text{instantaneous})}} = \frac{\text{stress}}{\text{strain}} = \frac{3600}{0.0009} = 4,000,000 \text{ lb/sq. in.}$$

In practice the instantaneous elastic modulus is determined from tests on concrete cylinders 6 in. diameter and 12 in. long, the cylinders being stressed to only about one-third of the cube-crushing strength of the concrete in question. The details of these tests are given in B.S. 1881 (1952).

For different concretes the instantaneous elastic modulus varies roughly according to the square root of the ultimate concrete strength. This is indicated in Fig. 2.7, though the relationship is not exact. The elastic modulus for concrete features prominently in all reinforced concrete calculations, and from examination of Fig. 2.7 it would seem at first as though for calculation purposes different values should be taken according to the strength of concrete in question. Throughout this book, however, a constant value of E_c for concrete is adopted in all calculations: and for the following reason.

If a prism of concrete containing longitudinal reinforcing bars

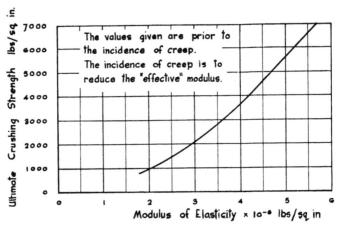

FIG. 2.7. MODULUS OF ELASTICITY OF CONCRETES
OF DIFFERENT ULTIMATE STRENGTHS

is stressed in compression within the range of permissible stresses, under circumstances in which there is no slipping of the steel in the concrete, the steel and the concrete must shorten by equal amounts. That is to say, the steel strain equals the concrete strain.

Now steel strain $= \dfrac{f_{sc}}{E_s}$ and concrete strain $= \dfrac{f_{cc}}{E_c}$,

so

$$\frac{f_{sc}}{E_s} = \frac{f_{cc}}{E_c}$$

or

$$\frac{f_{sc}}{f_{cc}} = \frac{E_s}{E_c} = m \ (modular \ ratio). \quad (2.2)$$

The *modular ratio* is therefore dependent on the steel and concrete elastic moduli, and determines the ratio of stresses as between the steel and concrete, provided the steel bars do not slip, and there has been no time for creep to act.

The value E_s may for all practical purposes be taken as 30,000,000 lb/sq. in. By reference to Fig. 2.7 a concrete of ultimate crushing strength 4000 lb/sq. in. has a value E_c of about 4,000,000 lb/sq. in. Hence the modular ratio is

$$m = \frac{E_s}{E_c} = \frac{30,000,000}{4,000,000} = 7\tfrac{1}{2}.$$

Yet within the range of permissible stresses it is common practice to take the modular ratio as 15. This is because of the influence of creep, which under persistent load reduces the *effective elastic modulus* of concrete as described in Article 2.7, so increasing m. Thus the modular ratio of any concrete mix is not precisely ascertainable, and the adoption of an approximate yet constant value of 15 in reinforced concrete design work is a reasonable compromise making for great convenience. Since E_s is roughly constant at 30,000,000 lb/sq. in., it follows that in our design calculations for reinforced concrete members we shall be considering an effective elastic modulus of concrete roughly constant, so that

$$E_{c_{(effective)}} = \frac{E_s}{m} = \frac{30,000,000}{15} = 2,000,000 \text{ lb/sq. in. approximately.}$$

ART. 2.7. CREEP

The stress/strain relationship for concrete is very closely dependent on the rate of loading of the specimen. Within the range of the permissible stresses given in C.P. 114 (1957), the *instantaneous* stress/strain curve is for all practical purposes a straight line (*see* Fig. 2.6): and if the stressing force is then *instantaneously* removed the specimen will recover almost to its original length with no permanent set. However, if the stressing force is applied slowly, or if the stress is allowed to persist, additional and permanent strains will develop over a period of years, the rate of development of these additional strains reducing with time. It is these strains additional to the instantaneous elastic strain which are known as *creep*. In Article 2.6 every endeavour is made to consider the *elasticity* as something separate from creep. However, little attempt is made in the present article to divorce *creep* from *shrinkage*; in practice the two are inter-related. This is because both occur over a considerable period of time, and although under laboratory conditions the two can be artificially separated, in the practical application of reinforced concrete this is not so. Creep is dependent on stress; shrinkage is not. The cause of shrinkage is dealt with in Article 2.8.

The significance of creep was not fully appreciated until Oscar Faber's paper[4] on the subject was published by the Institution of Civil Engineers in 1927. Later work by Dr. W. H. Glanville[5] showed that the substance of Faber's work was correct. Glanville's findings are available in Building Research Technical Paper No. 12. Faber's paper shows that:

(1) Structures of reinforced concrete, working at normal permissible stresses, continue to deform without change of load or temperature, a result due partly to shrinkage and partly to creep.

(2) The deformation under creep is about 3 times the deformation that occurs when the load is first applied.

(3) At normal working stresses, the deformation under creep is roughly proportional to the applied stress.

(4) The effect of creep and shrinkage is to produce a large though gradual redistribution of stress between the steel and the concrete in a reinforced concrete structure, the nature of this redistribution being such as to relieve the concrete and add to the steel stress.

(5) When the load is removed, only the instantaneous elastic deformation recovers, the deformation due to shrinkage and creep remaining as a permanent set.

(6) Reinforced concrete structures are not dangerous by reason of these phenomena when a knowledge of them is properly applied in the design, but special care is necessary in binding compression steel, whether in beams or columns.

(7) The modulus of elasticity of concrete is much higher than that specified in most regulations, but creep and shrinkage in some cases can produce results similar to those obtained with a very low modulus.

All the above findings still hold good today and are accepted by all reinforced-concrete engineers. What follows is merely elaboration of the fundamental principles set down by Faber and summarised above.

In his tests Faber separated the effects of *shrinkage* and *creep* by observing the shrinkage of unstressed specimens under the same conditions of temperature and humidity as his loaded specimens. Furthermore the elastic strain was readily calculable, since Faber had measured the instantaneous elastic modulus E_c; thus he was able to separate the *elastic strain* from the *creep*. From these results Fig. 2.8 has been prepared showing the proportion of total deformation due to elastic strain, creep and shrinkage, as took place over a period of 36 weeks for different concrete stresses. From Fig. 2.8 it can be seen that in round figures the creep was about twice the elastic strain; or in other words the original strain was roughly trebled.

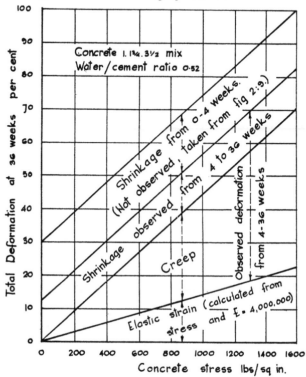

FIG. 2.8. COMBINED SHRINKAGE, CREEP AND ELASTIC STRAINS OF CONCRETE OVER 36 WEEK TEST. (AFTER OSCAR FABER.)

EFFECTIVE MODULAR RATIO

In the tests described above the measured elastic modulus E_c was just over 4,000,000 lb/sq. in., giving

$$m_e = \frac{E_s}{E_c} = \frac{30,000,000}{4,000,000} = 7\tfrac{1}{2} \text{ for elastic strain.}$$

But the strain after creep, being trebled, would reduce E_c to

$$\frac{4,000,000}{3},$$

giving $m_{e+c} = \dfrac{30,000,000}{4,000,000/3} = 22\tfrac{1}{2}$ for elastic strain + creep.

Thus the effect of creep, by causing a gradual increase in deformation, is to decrease the effective elastic modulus, with conse-
quent increase of m from $7\frac{1}{2}$ to $22\frac{1}{2}$. The value for m taken in our
calculations is a mean between these two values.

REDISTRIBUTION OF STRESS

From the above reasoning it is clear that when a reinforced
concrete column is first loaded, the concrete will be stressed more,
and the steel less, than in any calculation based on $m = 15$.
Nevertheless, as the load persists, the condition will be reversed,

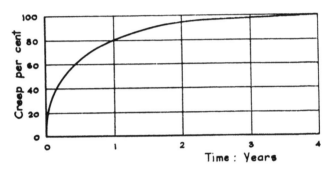

FIG. 2.9. TYPICAL INCREASE OF CREEP WITH TIME
UNDER CONSTANT LOAD

the concrete shedding its load to the steel which under certain
conditions may yield in compression and pass some of the load
back to the concrete. In this way there will be adjustments
made, both the concrete and steel relieving themselves as
necessary until, in the limit, an optimum distribution of stress is
achieved making for maximum ultimate safety.

With beams and slabs singly reinforced, i.e. reinforcement on
the tension face only, there will be but little redistribution of
stress; on the other hand the deflection of these members, being
very much greater than the axial strains, will increase progressively, the rate of increase being most pronounced over the first
few months (see Figs. 2.9 and 2.11). In the cases of beams and
slabs with double reinforcement, i.e. reinforcement on both the
tension and compression faces, the concrete on the compression
face will shed its load to the compression reinforcement (as with
columns), but the tension reinforcement will be but little affected.
As with singly reinforced beams and slabs, the deflections will
increase, but not to the same extent.

PRACTICAL EFFECT OF DEFORMATIONS

Although the ultimate shortening of columns and the deflection of beams may amount to about three or four times that which occurs when the load is first applied, these normally are sufficiently small to pass unnoticed by the naked eye, and are therefore generally of no consequence. Nevertheless there are cases where this is not so, for example where long lines of shafting, carefully levelled at the bearings, may get out of line. In any event it is important for engineers to be aware of creep and shrinkage deformations so as to be able to put a proper interpretation on observed phenomena; thus, when a certain cantilever grandstand was subjected to test load prior to acceptance it was found that the deflection gradually increased measurably day by day, and this was ascribed to some weakness in the construction and therefore created a feeling of alarm, whereas it was in fact what one would have expected from the normal behaviour of concrete displaying creep and shrinkage. On the other hand, the ability of concrete to relieve itself of high stresses by passing them on to the steel can, in skilled hands, result in enhanced safety. Though in pre-stressed concrete work, creep and shrinkage have an adverse effect, the concrete shying from the prestressing force which in consequence is reduced, and allowance has to be made for this as described in Chapter 15.

FACTORS INFLUENCING AMOUNT OF CREEP

For simplicity in the foregoing, the deformation ascribed to creep has been given in general terms as three times that due to

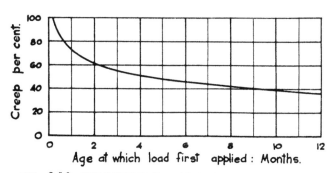

FIG. 2.10. REDUCTION OF TOTAL CREEP DEPENDING ON AGE WHEN CONCRETE FIRST LOADED

elastic deformation. This is as Faber originally found it, but in fact the deformations may vary slightly either way from this depending on a number of factors, including principally the following. The richer the concrete mix, the less the amount of creep; conversely weak concretes display greater creep. The amount of water in the mix is found to have but little influence. With concretes cured in air, the use of Rapid-Hardening cement gives less creep than Ordinary Portland cement. However, with water-cured concrete the difference is less marked. The creep displayed by concretes when loaded from an early age is considerably greater than concretes not loaded until later. This is indicated in Fig. 2.10 which is based on investigations carried out in Switzerland.

ART. 2.8. SHRINKAGE

When concrete sets and hardens in air it shrinks. This is due partly to the chemical action of the colloids produced by the water reacting on the cement, and partly to the physical drying out of the mass. The former action is irreversible: the drying-out action on the other hand is reversible. The amount of shrinkage is almost directly proportional to the amount of water contained in

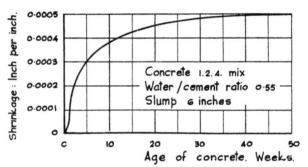

FIG. 2.11. AVERAGE SHRINKAGE OF CONCRETE WITH TIME

the mix. Therefore wet and sloppy mixes shrink more than low slump concretes; and also since for a given water/cement ratio the mixes rich in cement contain more water for a given volume of concrete, the richer the mix the more the shrinkage.

Figure 2.11 shows a typical curve of the shrinkage of a concrete over a 50-week period. The shrinkage of other concretes will vary from this curve according to the amounts of cement and water used in the mix, and can be estimated roughly from their water/cement ratios. Concrete in a humid atmosphere shrinks less than a concrete in very dry conditions, so that in practice some variations are to be expected. In general it can be said that to reduce shrinkage to a minimum, the amounts of water and cement in the mix should be kept to a minimum, necessitating mechanical vibration, and the concrete should be kept wet for as long as possible after it has been placed. From Fig. 2.11 it will be seen that shrinkage over a period of one year is about 0·0005 in./in. which is equal to about 0·6 in./100 ft. The effect of shrinkage is largely dealt with in Article 2.7 jointly with creep since frequently the effects of creep and shrinkage are inseparable in practice. But creep is caused solely by stress, whereas shrinkage is independent of stress.

MERIT OF SHRINKAGE

The advantage which stems from shrinkage is that, in shrinking, the concrete grips tightly on to embedded reinforcements and prevents slipping. This is fundamental to the full exploitation of reinforced concrete construction, enabling high stresses to be transmitted to and from the reinforcements. Thus shrinkage is the main cause of the *bond* between the concrete and the reinforcements, giving average values at the ultimate of about 400 lb/sq. in. of reinforcement surface in 1 : 2 : 4 concretes. Were it not for this gripping, bond would to a large extent be dependent on mechanical keying provided by specially-shaped bars with protrusions or indentations, or other devices for preventing slip.

DISADVANTAGES OF SHRINKAGE

One disadvantage of shrinkage is that unless a concrete structure is free to move and take up its deformations, the concrete will crack. Generally structures are restrained against such free movement. For example a road slab is restricted by friction on its underside where it rests on the ground, and retaining walls and most other structures are similarly restrained by nature of their anatomy. It is necessary then to construct in finite lengths (frequently somewhere between 15 ft. and 25 ft.) so that cracking occurs only along predetermined lines. It is sometimes necessary at some of these positions to interrupt the reinforcements also (*complete contraction joints*), though more frequently the reinforcements can be taken through (*partial contraction joints*). Proper attention to the design and spacing of contraction joints is especially important at reservoirs and in road construction : in such applications rubber/bitumen compounds are used to seal the joints against the passage of water. This is discussed more fully in Chapters 12 and 14.

Another disadvantage of shrinkage is the variation in deformations that exists between relatively moist and dry concrete. A length of road slab will dry out at its upper surface, exposed to sunshine and wind, much more rapidly than its under surface which remains moist in contact with the ground. Thus the upper surface shrinks causing the ends of the slab to curl upwards like a railway sandwich, leaving the slab free to rock and bounce as traffic passes over, and subject to considerable cantilever moments. This is one of the causes of cracked and broken corners of road slab panels, the beginnings of road failures.

REINFORCEMENT TO REDUCE SHRINKAGE CRACKS

If sufficient reinforcement is provided in a wall or slab (or other member) and there is no slip between the concrete and the steel, the shrinkage cracks will be so controlled as to be many in

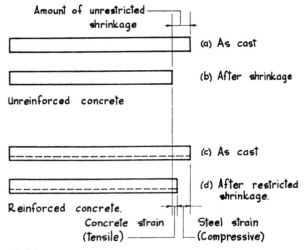

FIG. 2.12. DIAGRAM OF RELATIVE SHRINKAGE STRAINS OF CONCRETE WITH AND WITHOUT REINFORCEMENT

number but each negligible in width. If the percentage of steel is so small as to present no appreciable resistance to the overall shrinkage or shortening of the concrete, as it generally is, then since total tension = total compression,

$$A_c f_{ct} = A_s f_{sc} \tag{2.3}$$

where f_{ct} is the tensile stress in concrete,
             f_{sc} is the compressive stress in steel,
             A_c is the area of concrete,
             A_s is the area of steel.

And by reference to Fig. 2.12,

concrete strain + steel strain = unresisted shrinkage strain

or
$$\frac{f_{ct}}{E_c} + \frac{f_{st}}{E_s} = \varepsilon. \tag{2.4}$$

Substituting for f_{ct} and E_s in equation 2.4,

$$\frac{A_s f_{sc}}{A_c E_c} + \frac{f_{sc}}{m E_c} = \varepsilon$$

whence
$$f_{sc} = \frac{\varepsilon A_c E_s}{m A_s + A_c}.$$

If steel percentage $= r_p = \dfrac{A_s}{A_c} \cdot 100$,

then
$$f_{sc} = \frac{\varepsilon E_s}{1 + \dfrac{m r_p}{100}}. \tag{2.5}$$

f_{ct} can be determined from equation 2.3.

It is to be noted that to allow for the effect of creep, m is the *effective modular ratio* as described in Article 2.7, and to be consistent the *effective elastic modulus* is taken for E_c as described in Article 2.6. Taking then a percentage of steel $r_p = 1$ and shrinkage of 0·0005,

$$f_{sc} = \frac{0 \cdot 0005 \times 30{,}000{,}000}{1 + \dfrac{15 \times 1}{100}} = 13{,}000 \text{ lb/sq. in.}$$

and from equation 2.3, $f_{ct} = 130$ lb/sq. in.

The above form of calculation is not to be confused with the determination of reinforcement required in road slabs which is discussed in Chapter 14.

SHRINKAGE WHEN CONCRETE SETS AND HARDENS IN WATER

When concrete sets and hardens in water the shrinkage is very considerably reduced and may even be reversed, i.e. small expansion. Since the bond to reinforcements depends largely on the grip due to shrinkage, reinforced concrete which sets under water will develop lower bond stresses than concrete which sets in air. The provision of generous bond lengths and substantial hooks or other anchorages is therefore especially important. This does not apply to the same degree to reinforced concrete members which are part cured in air before being submerged, as the irreversible shrinkage on setting will then have taken place, and only the physical reversible action will occur on immersion.

ART. 2.9. THERMAL MOVEMENT

Table 2.3 gives the approximate coefficients of expansion of the basic components of reinforced concrete. Hence it is seen that concrete and steel expand and contract similarly, so that appreciable internal stresses will not be set up in reinforced concrete over reasonable temperature ranges.

TABLE 2.3. *Coefficients of expansion of concrete, cement and steel.*

	Ratio of cement to total aggregate	Aggregates	
		Gravel, granite, basalt	Limestone
Concrete	1 to 9 (1:3:6) 1 to 6 (1:2:4) 1 to 4½ (1:1½:2) 1 to 3 (1:1:2)	0·0000050 0·0000052 0·0000055 0·0000060	0·0000035 0·0000038 0·0000040 0·0000045
Neat cement		0·0000075	
Reinforcements		0·0000055	

It has been said that such similarity of expansion coefficients is a necessary condition for the success of reinforced concrete, but it is clear that this is an overstatement, when the greater relative movements of shrink and creep are considered, as well as the relief afforded by the latter.

The effects of expansion require consideration. Thus in roads, where the concrete may be subject to extremes of summer sun and winter frost, it is necessary to provide expansion joints at regular intervals. These same joints can be devised to function also as shrinkage joints, and a balance may be struck between the two contrary movements (*see* Chapter 14). On the other hand, tunnels and covered reservoirs may be sufficiently protected from temperature differences as to require very little provision for expansion, if any at all.

In some buildings, such as blocks of flats, where the internal temperature is kept fairly constant throughout the year, joints may not be required closer than about 500 ft. to 800 ft.; though special protection is then required to the roof construction by way of insulation to prevent thermal movements causing cracks in the upper floor height. Parapets should always be provided with expansion joints at frequent intervals. Buildings in tropical countries have a greater need for expansion joints than buildings in temperate climates, since although the temperature range may not be appreciably greater, the conditions of extreme temperature may persist by day and by night over many weeks resulting in a greater temperature build-up. Expansion joints in buildings are a great nuisance, and should be avoided wherever possible. Once provided, they will certainly act, generally with detrimental effect on the finishes and services. To be effective the joints have to give complete discontinuity, and should be provided in floors, roofs, walls and partitions.

ART. 2.10. SUMMARY OF LINEAR DEFORMATIONS

In order to appreciate the significance of the deformations referred to in Articles 2.6, 2.7, 2.8 and 2.9 the summary given in Table 2.4 is of interest. Since 0·0005 in./in. is equivalent to 0·6 in./100 ft., the figures in Table 2.4 give a graphic picture

TABLE 2.4. *Linear deformations*

Cause of deformation	Approx. deformation Inches/inch
Strain of steel stressed to 15,000 lb/sq. in.	0·0005
Instantaneous strain of 1:2:4 concrete stressed to 2000 lb/sq. in.	0·0005
Ultimate strain of 1:2:4 concrete stressed to 500 lb/sq. in. following creep	0·0005
Total shrinkage of concrete	0·0005
Thermal expansion of concrete due to 91°F temperature rise	0·0005

of the magnitude of the various deformations for the conditions described.

ART. 2.11. TEMPERATURE STRESSES

Apart from secondary stresses arising from thermal movements referred to in Article 2.9, an entirely separate problem arises when opposite faces of a concrete member suffer temperature changes of different amounts. The high-temperature face will then expand more than the other face causing curvature. If the member is free to curve without restraint, no stressing will result; but if the member is restrained, as for example the continuous wall of a silo, or the shell of a chimney, the resultant internal stresses can be appreciable, depending on the temperature gradient across the member, and the disposition of the reinforcements. This matter is discussed in greater detail in Chapter 13.

REFERENCES

1. *Concrete in Sulphate-bearing Clays and Ground Water.* Building Research Station Digest, No. 31. H.M. Stationery Office (June, 1951).
2. *Fire Grading of Buildings.* Part 1. Ministry of Works. H.M. Stationery Office (1946).
3. *Fire Grading of Buildings.* Parts 2, 3 and 4. Ministry of Works. H.M. Stationery Office (1946).
4. FABER, OSCAR. Plastic Yield, Shrinkage and other Problems of Concrete and their Effect on Design. *Proc. Inst. C.E.* (November, 1927).
5. GLANVILLE, W. H. *The Creep or Flow of Concrete under Load.* Building Research Technical Paper, No. 12. H.M. Stationery Office (1930).
6. MURDOCK, L. J. *Concrete Materials and Practice.* Arnold (1955).

CHAPTER 3

ELEMENTS OF REINFORCED CONCRETE DESIGN

ART. 3.1. INTRODUCTION

THE APPROACH TO DESIGN

BEFORE developing the basic equations for reinforced concrete under stress, it is perhaps appropriate to recapitulate here the physical characteristics of concrete described in Chapters 1 and 2 which influence its behaviour under load.

The water/cement ratio has a considerable effect on the strength of concrete, and on the amount of shrinkage. The elastic modulus varies according to the concrete strength; and changes with time due to creep, depending on the intensity of applied stress. The amount of creep depends also on the age of the concrete when it is first stressed. Further, the strength of concrete depends on the nature of the aggregates – their particle shape, surface texture and grading – and whether the concrete is cured in air or water, and at what temperatures. And stresses arise merely from the effects of shrinkage, and from temperature changes in restrained structures, without the application of any useful external loading: and variation of moisture content and temperature gradients across the thickness of a structural unit can cause considerable bending stresses.

Clearly then it falls outside practical limits to pursue, at the outset of design, all the variant complications that are known to exist. An intelligent stab has therefore to be made through a jungle of indeterminables, and later, when the worst of the bush has been cleared away and some sort of a design is seen to emerge, more precise attention can be given to the secondary effects of creep, shrinkage and so on in those circumstances where such further care is warranted. This is where the experience and skill of the designer come into play; knowing which secondary effects are relevant to the problem on hand. And here may lie the border between extravagance of design on the one hand, and inadequacy on the other.

But before the engineer starts his design calculations, he is faced with a greater problem – that of conception. What is he to

calculate? Is he to cover in a given space with a heavy slab
spanning directly the full width of the opening, or should he
provide beams spanning the opening with a thinner slab spanning
on to the beams? In the latter case the slab will be cheaper; but
will the cost of the beams outweigh the savings made on the
slab? Would pre-stressed beams show any advantage? This may
depend on the labour and supervision available, and on the value
set by enhanced headroom. Cases may arise where flat slab construction with intermediate columns would be best, requiring no
downstanding beam ribs at all thus enabling savings to be made
in the overall height and cost of the structure.

These matters of conception cannot be considered separately
from the problem of foundation design. For example, it would be
undesirable to collect all the loads from a superstructure to only
a few heavily loaded columns on a site of poor strata where a
raft foundation was indicated: the raft would then have to spread
the same loads back to perhaps roughly the original area of the
superstructure in order that the ground is not overstressed. Here
clearly, a close spacing of columns would show economy in the
foundation design. On the other hand, a foundation of long piles
end-bearing on rock would be better suited to accept large column
loads spaced greater distances apart.

The engineer has therefore to consider all aspects of structural
anatomy, and weigh up the relative costs before embarking on his
preliminary calculations. And here, more than anywhere else, is
the true key to economy of construction. It comes only from vivid
imagination, a wealth of experience, and properly digested statistics derived from detailed costings.

In the later chapters of this book examples are given of the
calculations of slabs, beams, columns, foundations, retaining walls,
silos, reservoirs, water-towers and so on; and from what has been
said above, it will be clear why these calculations are not taken
beyond slide-rule accuracy. The physical characteristics of concrete are liable to vary so widely that any attempts at closer
accuracy would be unrealistic.

The engineer's energy is better spent on giving fuller consideration to the anatomy of the structure, rather than calculating
meticulously the bending moments to see whether for example he
can save some small proportion of the reinforcement in a foundation raft. Perhaps if he stood back for a moment, where the wood
can be seen clear of the trees, he might find, under certain
conditions, that by doing away with the raft altogether and providing a piled foundation he could make greater reductions in
the overall cost.

THE APPROACH TO CALCULATION

Practical calculations for the purpose of achieving designs make bold and simplifying assumptions which lead to results that experience has shown are safe. This does not mean that such calculations give accurate information as to the actual stresses at any point within the materials; but despite phenomena such as creep, shrinkage, different coefficients of thermal expansion of steel and concrete, etc., the results are safe (whatever the actual stresses may be) and the methods are simple.

The assumptions are that at any cross-section plane sections remain plane, and all tensile stresses are taken by the reinforcement, except in the case of diagonal tensions arising from shear stresses.* Calculations are made either on the more usual Elastic Theory, assuming that stress is proportional to strain, and that the elastic modulus of concrete is effectively constant, so that m, the modular ratio, is equal to 15 (see Articles 2.6 and 2.7): alternatively calculations may be made on a Load-Factor basis which recognises that as failure is approached the compressive stresses adjust themselves to provide a greater total compressive resistance than would be indicated by the Elastic Theory.

(i) Bending

Consider an unreinforced concrete beam of rectangular cross-section supported on two end bearings and subjected to vertical loads as shown in Figure 3.1a. It is well known that due to the bending of the beam, the upper part will be in compression and the lower part in tension. This is clear from the fact that the top shortens and the bottom lengthens (Fig. 3.1b). If f is the fibre stress at the extreme top or bottom of the beam (see Fig. 3.2a), the resistance moment, according to the Elastic Theory, is given by

$$M_r = f \, \frac{bd^2}{6}$$
$$= fz$$

where z is the section modulus.

If the *ultimate crushing strength* of the concrete is say 3000 lb/sq. in., then the *ultimate tensile strength*, being only about one-tenth, will be say 300 lb/sq. in.; so that when the stress at the bottom of the beam reaches 300 lb/sq. in., the concrete will fail in tension, leaving a margin of compressive strength at the top of the beam of 3000 minus 300, equals 2700 lb/sq. in., of which no use is made.

* A further exception arises in checking against cracking of liquid-retaining structures, as referred to in Articles 12.2 and 12.3.

The purpose of reinforcing the concrete beam is to strengthen the tension part so as to bring into effective use the high compressive strength of the concrete in the other part. With f, the tensile stress of the concrete, limited to 300 lb/sq. in., M_r is limited to $300\,\dfrac{bd^2}{6} = 50\,bd^2$. However if the tension part were strengthened by the introduction of steel reinforcements so that the full strength

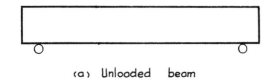

(a) Unlooded beam

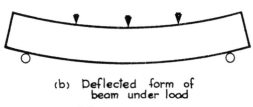

(b) Deflected form of beam under load

FIG. 3.1. BEAM BENDING

of the compression part could be developed, it would appear that M_r would be increased so as to depend only on the crushing strength of the concrete (3000 lb/sq. in.), and would become $3000\,\dfrac{bd^2}{6}$, equals $500\,bd^2$.

The principles given above are not greatly in error for stresses within the elastic range; but when the unreinforced concrete beam is actually tested to collapse, it is found that the resistance moment is something like 50 per cent greater than $50\,bd^2$. This is because, as ultimate conditions are approached, the concrete starts behaving plastically and the deformation increases disproportionately from the stress. The stress variation over the cross-section of the beam then departs from the linear relationship shown at Fig. 3.2a to a form more like that indicated at Fig. 3.2b. The neutral axis rises so as to give a greater proportion of the beam available for resisting tension, and the distribution of stress varies, increasing the average intensities of both tensile and compressive stresses. An unreinforced beam of 3000 lb/sq. in. compressive strength and 300 lb/sq. in. tensile strength will thus have an ultimate resistance moment on failure of more like 75 bd^2. The apparently greater tensile strength to produce this enhanced resistance moment (assuming the stress relationship had remained linear), would be

$$R = \frac{M_{r(ult)}}{z} = \frac{M_{r(ult)}6}{bd^2}$$

and at the moment of collapse is defined as the *modulus of rupture*. It is not to be confused as any true value of stress.

Referring again to Fig. 3.2, the stress distribution shown at (*a*) is approximately true within the range of permissible working stresses of the materials, which for calculation purposes is regarded as being within the *elastic* range of the materials. On the other hand, the *plastic* distribution shown at (*b*) approximates to conditions as collapse is approached: and this consideration is not

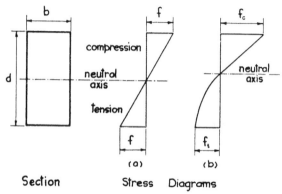

FIG. 3.2. STRESS DIAGRAMS FOR UNREINFORCED BEAM

dissimilar from the Load-Factor method of design described in detail in Article 3.3, where safe loads are based on collapse conditions moderated by the application of a suitable Load Factor.

(ii) *Compression*

Consider a different case. A simple prism of concrete of cross-sectional area A will safely carry an axial load of $p_{cc} \times A$ where p_{cc} is the permissible concrete stress. If however the concrete is reinforced with longitudinal steel bars, as shown in Fig. 3.3, the calculated safe axial load on the prism will depend on whether the materials are considered as behaving elastically or plastically.

From *elastic* considerations, the stress in the steel will be m times the concrete stress, so that the total contribution by the steel is $m \times p_{cc} \times A_{sc}$, where A_{sc} is the cross-sectional area of the steel. Thus the safe load on the reinforced prism is

$$p_{cc}(A - A_{sc}) + mp_{cc}A_{sc}$$

or $\qquad\qquad p_{cc}A + (m - 1)p_{cc}A_{sc}. \qquad\qquad (3.1)$

Here the steel stress is clearly limited to m times the permissible concrete stress. With a 1:2:4 nominal mix this is $m \times 760$, or 15×760 equals 11,400 lb/sq. in., whereas steel of the quality

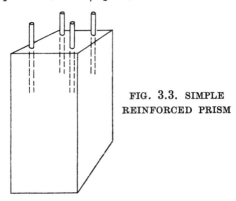

FIG. 3.3. SIMPLE
REINFORCED PRISM

used in reinforced concrete work is capable of sustaining safely much higher stresses than this.

Now note the *plastic* behaviour of the same reinforced prism. If the applied load is increased so that conditions of collapse are approached, the rate of strain of the concrete increases disproportionately whereas the stress/strain relationship for the steel remains linear. Thus the steel carries an increased proportion of the load; and the steel stress increases rapidly until the yield point is reached. Accordingly, just prior to collapse, both the concrete and the steel stresses are at maximum values. The application of a load factor to these values clearly determines a working stress for the concrete based on its ultimate strength, and a working stress for the steel based on its yield strength. The safe working load for the reinforced prism may thus be expressed in the form

$$P_0 = p_{cc}A_c + p_{sc}A_{sc} \qquad\qquad (3.2)$$

where A_c is the net area of concrete, and p_{sc} is the permissible stress in the steel. This is the formula given in C.P. 114 (1957)[1].

With a 1:2:4 nominal mix, an unreinforced prism 12 in. $\times$ 12 in. would safely carry

$$760 \times 12 \times 12 = 109,000 \text{ lb.}$$

With four 1 in. reinforcements, the strength is increased, according to the expression 3.1, to

$$(760 \times 12 \times 12) + (14 \times 760 \times 3.14) = 142,000 \text{ lb.}$$

And according to formula 3.2, the strength is increased to

$$(760 \times 140.86) + (18,000 \times 3.14) = 163,500 \text{ lb.}$$

In practice, simple cases of axially loaded columns seldom arise. The matter becomes more complicated when direct load is combined with bending. This is dealt with later in the present chapter.

(iii) *Tension*

With members subjected to direct tension forces, conditions are different. Where precautions have to be taken to prevent the concrete cracking, as in liquid retaining structures, or structures required to be weathertight, elastic considerations apply, and a check has to be made that the tensile working stress in the concrete is restricted to some suitable value, say 175 lb/sq. in. This of course restricts the tensile working stress in the steel (according to the modular ratio) to $m \times 175 = 3060$ lb/sq. in., so that in such cases the use of the steel becomes very ineffective. (This is dealt with more fully in Chapters 10 and 12.)

However, in circumstances where hair cracks in tension members would not be objectionable, as for example internal members not subject to contact with liquids, it is satisfactory to ignore the tensile strength of the concrete by allowing it to be loaded beyond even any plastic state, so that it actually fails by cracking. It is then necessary to provide sufficient steel to take the whole of the tension force at the full permissible working stress of 18,000 lb/sq. in. or more. The assumption that the concrete is cracked does not mean that the cracks will necessarily be visible to the naked eye; and where the stress is limited to 18,000 or 20,000 lb/sq. in. frequently the cracks are not visible. With higher stresses this may not be so. It is important that cracks are not wide enough to permit corrosion of the steel by exposure to atmospheric or other corrosive influences, and for this reason C.P. 114 (1957) sets a maximum limit on permissible steel stresses no matter what the guaranteed yield stress for the steel may be. This applies equally for direct tension members and for tension forces in members subject to bending.

CHOICE OF ELASTIC-THEORY OR LOAD-FACTOR METHOD OF DESIGN

In dealing with direct stresses, C.P. 114 (1957) permits the use of both methods of design – the Elastic-Theory or the Load-Factor Method. Both have their uses.

The full theory for both Elastic-Theory and Load-Factor Method follows in the present chapter. In later chapters, examples of practical applications are given and fully worked out. Normally, to avoid duplication in the examples, only one of the two methods is demonstrated for any one problem. Often it will be quite obvious why one method of design has been used in preference to the other. Elsewhere, where clearly either method would be readily applicable, the choice has been made to demonstrate the particular method rather than for any other reason: it does not follow that a more economical design could not be achieved using the alternative method.

ART. 3.2. ELASTIC THEORY: (WITH FACTORS OF SAFETY)

PRINCIPLE OF ELASTIC THEORY

In the Elastic Theory, the stresses in the members are considered for the normal working-load condition. That is to say no attention is given to the conditions that arise at the time of structural collapse: and all interest is centred on the state of stress as it is assumed to exist when the member is functioning within some acceptable margin of safety.

Experience has shown that when due allowance is made for faulty materials and errors of workmanship, a suitable margin of safety entails restricting the concrete and steel stresses to limits below which it so happens that the materials behave reasonably elastically. Thus the stresses in the concrete and steel are roughly proportional to their strains. For steel, which departs abruptly from elastic behaviour at its yield stress, the permissible working stress is taken at about half the yield stress. For concrete, which has no well-defined yield point, the permissible working stress is determined perhaps a little more arbitrarily from the crushing strength of 28-day test cubes. With Portland cement concrete and Portland Blastfurnace cement concrete where aggregates comply with B.S. 882, for *nominal concrete mixes* the permissible compressive stress due to bending is one-third the Works cube strength or one-quarter the Preliminary cube strength, and the permissible stress due to direct compression is 76 per cent of this value: whereas for *designed concrete mixes* CP. 114 (1957) Amendment 1965 recommends working 10 per cent nearer the cube strength results.

There are many complications which prevent the Elastic Theory from being precise. Creep and shrinkage of course head the list, and these have been discussed in Articles 2.7 and 2.8. They prevent there being any true constant value for the modular ratio – the ratio between the elastic moduli of steel and concrete. However, for convenience, and as a sweeping approximation, it is normally reasonable to work to a modular ratio m equal to 15.

BENDING

When a reinforced concrete member is subjected to bending, the assumptions made in the theory are as follows:

At any cross-section plane sections remain plane.
All tensile stresses are taken by the reinforcement.

(i) *Singly reinforced rectangular beam*

Figure 3.4 represents the simplest example of a reinforced concrete member designed to resist bending. It is a beam reinforced on the tension side only. In the upper part of the beam

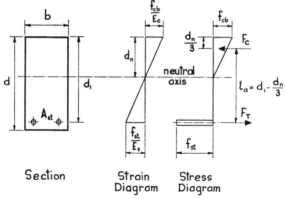

Section   Strain   Stress
                 Diagram   Diagram

FIG. 3.4. STRESS AND STRAIN DIAGRAMS FOR SINGLY
REINFORCED BEAM. ELASTIC THEORY

the whole of the compression is taken by the concrete, but in the lower part the concrete is assumed to have cracked leaving the reinforcements to take the whole of the tension. The section is therefore clearly unsymmetrical so that d_n, the depth to the neutral axis, is not known, and has to be determined.

If f_{cb} is the compressive stress in the concrete (outer fibre),
  f_{st} is the tension stress in the steel,
  b is the breadth of the beam,
  d_1 is the depth to the tension steel ("effective" depth),
  A_{st} is the area of the tension steel,

then from the properties of similar triangles in the strain diagram

$$\frac{\dfrac{f_{cb}}{E_c}}{\dfrac{f_{st}}{E_s}} = \frac{d_n}{d_1 - d_n}, \tag{3.3}$$

and since the total compression force across the section must balance the total tension force,

$$\frac{bd_n f_{cb}}{2} = A_{st} f_{st}. \tag{3.4}$$

If the area of tension steel is r_{p1} per cent of the effective area of the beam, i.e. $r_{p1} = \dfrac{A_{st}}{bd_1} \times 100$;

and if we put $\quad \dfrac{f_{st}}{f_{cb}} = t_1$ and $\dfrac{d_n}{d_1} = n_1,$

then since $\dfrac{E_s}{E_c} = m$, we have from equation 3.3

$$n_1 = \frac{m}{t_1 + m} \tag{3.5}$$

and from equation 3.4

$$t_1 = \frac{50 n_1}{r_{p1}}. \tag{3.6}$$

Eliminating t_1 from equations 3.5 and 3.6,

$$n_1 = \sqrt{(0\cdot 01 m r_{p1})^2 + 0\cdot 02 m r_{p1}} - 0\cdot 01 m r_{p1} \tag{3.7}$$

which for the value $m = 15$ gives

$$n_1 = \sqrt{0\cdot 0225 r_{p1}^2 + 0\cdot 3 r_{p1}} - 0\cdot 15 r_{p1} \tag{3.8}$$

which is the equation for the curve for n_1 given at Fig. 3.5. Alternatively, eliminating n_1 from the equations 3.5 and 3.6, the curve for t_1 in terms of r_{p1} is derived, as shown at Fig. 3.6. This curve shows the ratio of stresses f_{st} and f_{cb} existing simultaneously in any singly reinforced member for various percentages of steel; so that if for example the steel stress is known, then the concrete stress can be immediately determined.

Now from Fig. 3.4 it is clear that the resistance moment lever arm l_a is equal to $d_1 - \dfrac{d_n}{3}$. This may be expressed as a proportion of the effective depth so that by definition

$$a_1 = \frac{l_a}{d_1} = 1 - \frac{n_1}{3}. \tag{3.9}$$

The curve for a_1 shown in Fig. 3.5 is obtained by solving equations 3.8 and 3.9. This curve demonstrates that for normal percentages of steel a_1 does not vary very greatly and may reasonably be taken as about 0·86.

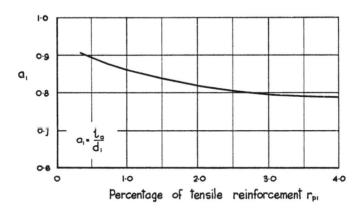

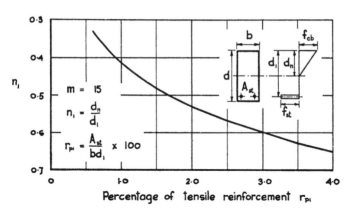

FIG. 3.5. CURVES OF a_1 AND n_1 PLOTTED AGAINST r_{p1} FOR SINGLY REINFORCED MEMBER. ELASTIC THEORY

Clearly the resistance moment of the section may be written either

$$M_r = \text{total tension force} \times \text{lever arm} = A_{st}f_{st}l_a \quad (3.10)$$

or

$$M_r = \text{total compression force} \times \text{lever arm} = \frac{bd_nf_{cb}}{2} l_a. \quad (3.11)$$

These two values for M_r are of course equal, and the ratio between the stresses f_{st} and f_{cb} will be as shown in Fig. 3.6 for any given percentage of tension steel. However, if f_{st} and f_{cb} are to be maximum *permissible* stresses, the two expressions for M_r (3.10 and 3.11) will only agree if the percentage of steel happens to be such as to coincide exactly with the ratio of permissible stresses under consideration. This is an unlikely chance; and if the steel percentage is less than shown in Fig. 3.6, the resistance moment

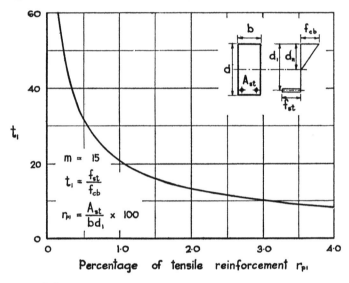

FIG. 3.6. CURVE OF t_1 PLOTTED AGAINST r_{p1} FOR SINGLY REINFORCED MEMBER. ELASTIC THEORY

of the section will be limited by the strength of the steel, and equation 3.10 will dominate; whereas if the steel percentage is greater than shown in Fig. 3.6, the resistance moment will be determined by the strength of the concrete, and given by equation 3.11. Since the ratio t_1 is constant for any two given permissible values of stress f_{st} and f_{cb}, so also is the percentage of steel which will enable both the steel and the concrete to be fully stressed simultaneously. This percentage has become known as the "economic percentage"; an unfortunate misnomer since many other considerations can affect the economy, as for example the relative costs of the concrete and steel and the formwork.

From equations 3.5, 3.9 and 3.11 we have

$$\frac{M_r}{bd_1{}^2} = \frac{mf_{cb}(3t_1 + 2m)}{6(t_1 + m)^2}. \tag{3.12}$$

Now m is taken as constant at 15. So clearly if the maximum permissible values of stress f_{st} and f_{cb} are defined, $\dfrac{M_r}{bd_1^2}$ may be determined. Thus for any particular design problem the permissible stresses may be specified, as for example 20,000 lb/sq. in. for the steel, and 1000 lb/sq. in. for the concrete. Then

$$t_1 = \frac{20,000}{1000} = 20$$

and

$$\frac{M_r}{bd_1^2} = \frac{15 \times 1000(3 \times 20 + 2 \times 15)}{6(20 + 15)^2} = 184.$$

For a known value of M_r in any member it is then possible to choose suitable dimensions b and d_1 knowing that the compressive

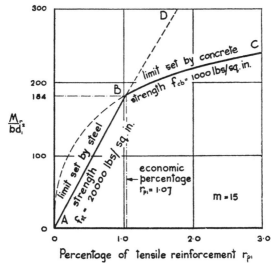

FIG. 3.7. SAFE $\dfrac{M_r}{bd_1^2}$ VALUES FOR SINGLY REINFORCED MEMBERS.

ELASTIC THEORY

(f_{st} = 20,000 lb/sq. in. : f_{cb} = 1000 lb/sq. in.)

stress in the concrete will be restricted to the maximum value permitted. The area of steel reinforcement required to develop the necessary tension force across the section may then be found from equation 3.10, where normally it is sufficiently accurate to take $l_a = 0.86$. Worked examples are included in later chapters.

In Fig. 3.7 two intersecting curves are shown plotted for

$\dfrac{M_r}{bd_1^2}$ against the steel percentage r_{p1}. The steeper curve is derived from equations 3.8, 3.9 and 3.10 with f_{st} as 20,000 lb/sq. in. and shows the limit of $\dfrac{M_r}{bd_1^2}$ set by the steel strength: the flatter curve is derived from equations 3.8, 3.9 and 3.11 with f_{cb} as 1000 lb/sq. in

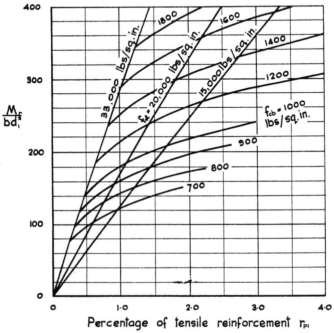

FIG. 3.8. SAFE $\dfrac{M_r}{bd_1^2}$ VALUES FOR SINGLY REINFORCED MEMBERS (VARIOUS STRESSES). ELASTIC THEORY

and shows the limit of $\dfrac{M_r}{bd_1^2}$ set by the concrete strength. For any value of r_{p1} there are clearly two values given for M_r, but only the lower of these is allowable since otherwise the other material would be overstressed. Accordingly for the stresses of 20,000 lb/sq. in. in the steel and 1000 lb/sq. in. in the concrete the allowable values for M_r are obtained from the heavy kinked line ABC.

From Fig. 3.7 it is clear that for low percentages of steel, the resistance moment of a member increases up to B roughly *pro rata* as the steel percentage is increased; and as the steel normally represents only a small part of the total cost of a member, it seldom pays to reinforce with a percentage less than that corresponding to B. On the other hand, beyond B the benefit of increasing the percentage of reinforcement tails off markedly. The point B where the two curves intersect indicates where the steel and concrete are both fully stressed simultaneously. This corresponds to the "economic percentage" referred to previously; and it certainly appears that with singly reinforced rectangular beams any steel greatly in excess of this percentage would not be very economical.

Figure 3.8 is derived in the same way as Fig. 3.7 but gives curves for a variety of permissible stresses in both the steel and the concrete.

(ii) *Doubly reinforced rectangular beam*

Figure 3.7 shows that in a singly reinforced beam it is not possible to make effective use of the tension reinforcement if this exceeds the so-called "economic percentage." This is because of the limit set by the compressive strength of the concrete in the upper part of the beam – the compression zone. However, if the

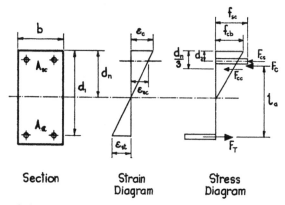

Section   Strain Diagram   Stress Diagram

FIG. 3.9. STRESS AND STRAIN DIAGRAMS FOR DOUBLY REINFORCED BEAM. ELASTIC THEORY

compression zone is strengthened or reinforced in some suitable way, the moment of resistance of the beam can be increased up to the limit set by the curve ABD. Since steel is much stronger in compression than concrete, it is possible without increasing the size of the beam to increase its strength by reinforcing the compression zone as well as the tension zone. In this way the size and weight of a beam to perform a certain duty may be minimised, with attendant savings in cost arising from the reduced quantities of concrete and formwork.

Figure 3.9 shows a rectangular beam which has in addition to the usual tension steel an area of compression steel A_{sc} arranged at a depth d_2 below the top of the beam.

If f_{sc} is the compression stress in the steel, and r_s is the ratio of areas of compression to tension steel, so that $r_s = \dfrac{A_{sc}}{A_{st}}$, then, using where applicable the same notation as before, it follows from similar triangles in the strain diagram that

$$\frac{\dfrac{f_{cb}}{E_c}}{\dfrac{f_{sc}}{E_s}} = \frac{d_n}{d_n - d_2}$$

or
$$f_{sc} = m f_{cb} \frac{(d_n - d_2)}{d_n}, \tag{3.13}$$

and since the total compression force across the section must balance the total tension force,

$$F_C = F_T. \tag{3.14}$$

Now F_C is made up of a compression force in the concrete and a compression force in the top steel, so that

$$F_C = F_{cc} + F_{sc}$$
$$= \left(\frac{b d_n f_{cb}}{2} - A_{sc} f_{cb} \frac{(d_n - d_2)}{d_n} \right) + A_{sc} f_{sc}. \tag{3.15}$$

Substituting from equation 3.13 in equation 3.15, we can now rewrite equation 3.14 as

$$\frac{b d_n f_{cb}}{2} + A_{sc}(m - 1) f_{cb} \frac{(d_n - d_2)}{d_n} = A_{st} f_{st}$$

which, by rearrangement and substitution of r_s for $\dfrac{A_{sc}}{A_{st}}$, gives

$$t_1 = \frac{50 n_1}{r_{p1}} + r_s(m - 1) \frac{\left(n_1 - \dfrac{d_2}{d_1} \right)}{n_1}. \tag{3.16}$$

Furthermore, in the same way that equation 3.3 led to equation 3.5, we have

$$t_1 = \frac{m}{n_1} (1 - n_1). \tag{3.17}$$

Equating the equations 3.16 and 3.17, we have

$$n_1 = \sqrt{\{0{\cdot}01r_{\mathrm{p1}}[r_{\mathrm{s}}(m-1)+m]\}^2 + 0{\cdot}02r_{\mathrm{p1}}\left(r_{\mathrm{s}}\frac{d_2}{d_1}(m-1)+m\right)} - 0{\cdot}01r_{\mathrm{p1}}[r_{\mathrm{s}}(m-1)+m] \quad (3.18)$$

which for the value $m = 15$ gives

$$n_1 = \sqrt{[r_{\mathrm{p1}}(0{\cdot}14r_{\mathrm{s}}+0{\cdot}15)]^2 + 2r_{\mathrm{p1}}\left(0{\cdot}14r_{\mathrm{s}}\frac{d_2}{d_1}+0{\cdot}15\right)} - r_{\mathrm{p1}}(0{\cdot}14r_{\mathrm{s}}+0{\cdot}15).$$

This equation gives the curves at Fig. 3.10 showing how the ratio of compression steel to tension steel affects the position of the neutral axis in a beam for different percentages of tension steel.

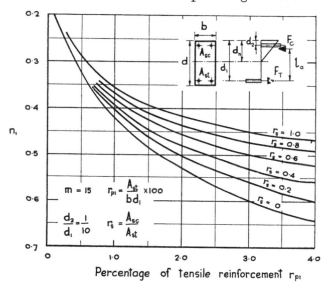

FIG. 3.10. CURVES OF n_1 PLOTTED AGAINST r_{p1} FOR DOUBLY REINFORCED MEMBER. ELASTIC THEORY

Now taking moments about the tension steel, we may equate the total compression F_{C} times the lever arm, to the moments arising from the concrete and the compression steel considered separately: so that

$$\left(\frac{bd_{\mathrm{n}}f_{\mathrm{cb}}}{2} + A_{\mathrm{sc}}f_{\mathrm{cb}}(m-1)\frac{(d_{\mathrm{n}}-d_2)}{d_{\mathrm{n}}}\right)l_a$$

$$= \frac{bd_{\mathrm{n}}f_{\mathrm{cb}}}{2}\left(d_1-\frac{d_{\mathrm{n}}}{3}\right) + A_{\mathrm{sc}}f_{\mathrm{cb}}(m-1)\frac{(d_{\mathrm{n}}-d_2)}{d_{\mathrm{n}}}(d_1-d_2)$$

which reduces to

$$l_a = d_1 - \frac{bd_n{}^3 + 6A_{sc}(m-1)(d_n - d_2)d_2}{3bd_n{}^2 + 6A_{sc}(m-1)(d_n - d_2)}.$$

Substituting for $A_{sc} = r_s A_{st} = r_s r_{p1} \cdot \dfrac{bd_1}{100}$, the equation becomes

$$a_1 = 1 - \frac{50n_1{}^3 + 3r_s r_{p1}(m-1)\left(n_1 - \dfrac{d_2}{d_1}\right)\dfrac{d_2}{d_1}}{3\left\{50n_1{}^2 + r_s r_{p1}(m-1)\left(n_1 - \dfrac{d_2}{d_1}\right)\right\}}. \qquad (3.19)$$

From this equation the family of curves for a_1 given in Fig. 3.11 has been obtained.

From Figs. 3.10 and 3.11 it will be seen that the effect of

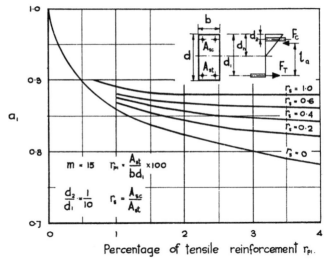

FIG. 3.11. CURVES OF a_1 PLOTTED AGAINST r_{p1} FOR DOUBLY REINFORCED MEMBER. ELASTIC THEORY

introducing compression steel is to raise the neutral axis and to increase the lever arm. When the area of compression steel is equal to the area of tension steel, the lever arm becomes practically independent of the percentage of tension steel and approximates very closely to the actual distance between the compression and tension reinforcements. This indicates that with such a high proportion of compression steel, the contribution from the concrete is only small: and this is the basis of the so-called "steel-beam theory" where equal steel is provided in the top and bottom of the beam, and the resistance moment is calculated on the assumption that all internal bending stress is taken by the top and bottom reinforcements acting at a lever arm equal to the distance apart of the reinforcements. The strength of the concrete in compression is then ignored.

Clearly, by considering the tension steel, the resistance moment of the section shown in Fig. 3.9 may be written

$$M_r = A_{st}f_{st}l_a. \tag{3.20}$$

Or similarly, from consideration of the internal compression forces given at equation 3.15 multiplied by their appropriate lever arms about the tension steel, we can write

$$M_r = \frac{bd_n f_{cb}}{2}\left(d_1 - \frac{d_n}{3}\right) + A_{sc}f_{cb}(m-1)\frac{(d_n - d_2)}{d_n}(d_1 - d_2) \tag{3.21}$$

which can be rearranged to read

$$\frac{M_r}{bd_1^2} = \frac{mf_{cb}(3t_1 + 2m)}{6(t_1 + m)^2} + \\ + 0{\cdot}01r_s r_{p1}f_{cb}\frac{(m-1)}{m}\left(m - \frac{d_2}{d_1}.(t_1 + m)\right)\left(1 - \frac{d_2}{d_1}\right). \tag{3.22}$$

This equation is similar in form to equation 3.12 for a singly reinforced beam. Here $\frac{M_r}{bd_1^2}$ is not independent of the percentage of tension steel (as might first appear) since t_1 is related to both r_{p1} and r_s by equations 3.17 and 3.18. Further, it should be noted that the second term of the equation does not represent simply the increase in $\frac{M_r}{bd_1^2}$ of the doubly reinforced beam over the singly reinforced beam; this again is due to the influence of the introduction of compression steel on t_1.

From equations 3.17, 3.18 and 3.22, Fig. 3.12 has been prepared giving values for $\frac{M_r}{bd_1^2}$ plotted against the percentage of tension steel in doubly reinforced beams with different proportions of compression steel. The permissible concrete stress has been taken as 1000 lb/sq. in. and for a permissible stress in the tension steel of 20,000 lb/sq. in. the figure is seen to resemble closely Fig. 3.7 for a singly reinforced beam; the difference being that the introduction of various proportions of compression steel enables enhanced values of $\frac{M_r}{bd_1^2}$ to be developed. The figure is convenient for use in design work, enabling the required proportion of compression steel to be read off directly, once the percentage of tension steel has been determined by calculation. Alternatively,

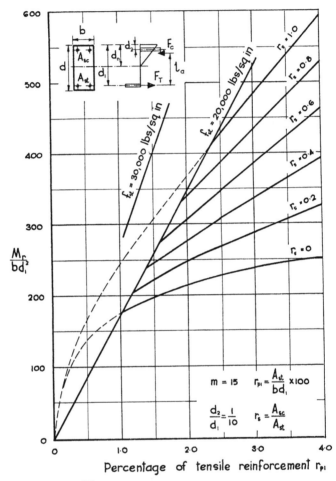

FIG. 3.12. SAFE $\dfrac{M_r}{bd_1^2}$ VALUES FOR DOUBLY REINFORCED MEMBERS.

ELASTIC THEORY

($f_{st} = 20{,}000$ lb/sq. in.: $f_{cb} = 1000$ lb/sq. in.)

in analysing members where the reinforcements have been predetermined, the actual stress in the steel can be gauged by inspection relative to the 20,000 lb/sq. in. line.

Similar figures can be prepared for different values of p_{cb}, p_{st} and $\frac{d_2}{d_1}$.

Practical examples of beams with compression reinforcement are given in Chapter 4 with full calculations.

(iii) T-*beams and* L-*beams*

When reinforced concrete beams are constructed monolithically with the floor slabs they support, the beam rib can only act in bending by bringing the slab into play as a compression flange. This interaction between the rib and slab gives rise to beam sections of T and L shapes as indicated in Fig. 3.13. Appreciation

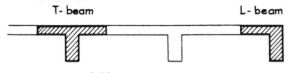

FIG. 3.13. T-BEAM AND L-BEAM

of T- and L-beam action enables a high compression force to be developed in the beam without the need for any compression steel: and the compression force being centred well up from the tension steel gives an increased lever arm and enhanced resistance moment. The width of slab likely to act in conjunction with the rib will depend on many considerations. For example, it will clearly be greater for T-beams than for L-beams. C.P. 114 (1957) gives empirical values for the different slab breadths to be taken into account in calculations of T- and L-beams, and these are referred to later in Chapter 4.

There is no other difference between the theory for bending in T-beams and L-beams; and the approach to the problem is governed only by the thickness of the top flange in relation to the depth of the neutral axis. Clearly when the neutral axis lies within the flange the theory will follow exactly the line already given for simple rectangular beams, taking the breadth of the beam in this case as the effective breadth of the T- or L-beam flange. However, when the neutral axis lies below the underside of the flange as shown in Fig. 3.14 (as is usual), the breadth of the compression zone of the beam is not constant, and the true theory becomes complex. For simplicity, in what follows the effect of the small compression in the rib between the neutral axis and the underside of the flange is ignored. And later, yet further simplifications will be described.

If b_r is the breadth of the beam rib,
   b is the breadth of the flange,
   d_s is the depth of the flange (or slab),
   d_1 is the depth to the tension steel ("effective" depth),
and f'_{cb} is the compressive stress in the concrete at the underside
   of the flange,

then, as before from equation 3.5,

$$t_1 = \frac{m}{n_1}(1 - n_1) \qquad (3.23)$$

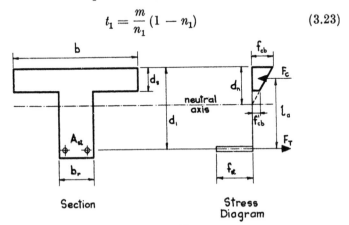

Section   Stress
                               Diagram

FIG. 3.14. STRESS DIAGRAM FOR T-BEAM. ELASTIC THEORY

and since the total compression force across the section must
balance the total tension force,

$$bd_s f_{cb} \frac{\left(d_n - \dfrac{d_s}{2}\right)}{d_n} = A_{st} f_{st}$$

or
$$t_1 = \frac{bd_s}{2A_{st}} \cdot \frac{(2d_n - d_s)}{d_n}. \qquad (3.23a)$$

Putting $r_{p1} = \dfrac{A_{st}}{bd_1} \times 100$ and $n_1 = \dfrac{d_n}{d_1}$, this becomes

$$t_1 = \frac{50}{r_{p1}n_1} \cdot \frac{d_s}{d_1}\left(2n_1 - \frac{d_s}{d_1}\right) \qquad (3.24)$$

and eliminating t_1 from equations 3.23 and 3.24,

$$n_1 = \frac{mr_{p1} + 50\left(\dfrac{d_s}{d_1}\right)^2}{mr_{p1} + 100\dfrac{d_s}{d_1}}. \qquad (3.25)$$

From equations 3.24 and 3.25 it can be seen that the position of the neutral axis, and the ratio of stresses are dependent not only on the percentage of steel in the member (as was the case of the simple rectangular beam – equations 3.6 and 3.7) but also on the ratio of flange depth to effective depth at rib.

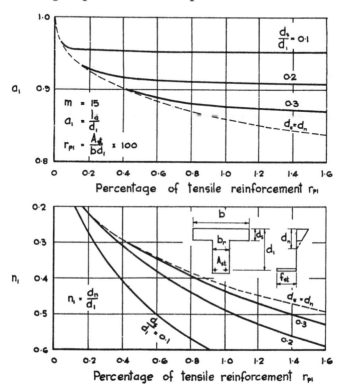

FIG. 3.15. CURVES OF a_1 AND n_1 PLOTTED AGAINST r_{p1} FOR T- AND L-BEAMS. ELASTIC THEORY

Now taking moments about the tension steel,

$$d_s b \frac{(f_{cb} + f'_{cb})}{2} \cdot l_a = d_s b f'_{cb} \left(d_1 - \frac{d_s}{2} \right) + d_s b \frac{(f_{cb} - f'_{cb})}{2} \left(d_1 - \frac{d_s}{3} \right),$$

then, since
$$f'_{cb} = f_{cb} \frac{(d_n - d_s)}{d_n}, \tag{3.26}$$

we have
$$l_a = d_1 - \frac{d_s}{3} \left(\frac{3d_n - 2d_s}{2d_n - d_s} \right)$$

or
$$a_1 = 1 - \frac{1}{3} \cdot \frac{d_s}{d_1} \frac{\left(3n_1 - 2\frac{d_s}{d_1}\right)}{\left(2n_1 - \frac{d_s}{d_1}\right)}. \tag{3.27}$$

From equations 3.25 and 3.27 the two families of curves given in Fig. 3.15 for n_1 and a_1 have been obtained for different ratios of flange to rib depth.

Clearly by considering the tension steel, the resistance moment of the section shown in Fig. 3.14 may be written

$$M_r = A_{st} f_{st} l_a. \tag{3.28}$$

Alternatively, considering the concrete,

$$M_r = d_s b f'_{cb} \left(d_1 - \frac{d_s}{2}\right) + d_s b \frac{(f_{cb} - f'_{cb})}{2} \left(d_1 - \frac{d_s}{3}\right). \tag{3.29}$$

And from equations 3.23, 3.26 and 3.29 an expression for $\frac{M_r}{bd_1^2}$ can be obtained similar to those for the rectangular beams. Thus

$$\frac{M_r}{bd_1^2} = \frac{f_{cb}}{6m} \cdot \frac{d_s}{d_1} \left[3m\left(2 - \frac{d_s}{d_1}\right) - \frac{d_s}{d_1}\left(3 - 2\frac{d_s}{d_1}\right)(t_1 + m)\right]. \tag{3.30}$$

Approximate method

To avoid using all the complicated formulae given above, it is sufficiently accurate in most cases to consider that the centre of the compression acts at the centre of the flange.

Then
$$l_a = d_1 - \frac{d_s}{2} \tag{3.31}$$

and the area of tension steel is given by

$$A_{st} = \frac{M}{\left(d_1 - \frac{d_s}{2}\right) f_{st}}. \tag{3.32}$$

The average stress in the flange is then $\frac{M}{l_a b d_s}$, and from equation 3.31 and by similar triangles in the strain diagram

$$f_{cb} = \frac{M}{\left(d_1 - \frac{d_s}{2}\right) b d_s} \cdot \frac{n_1}{\left(n_1 - \frac{d_s}{2d_1}\right)}. \tag{3.33}$$

For this calculation n_1 can be taken from Fig. 3.15. The use of equation 3.32 shows slightly more reinforcement to be required than does the more exact method. And equation 3.33 gives a concrete stress higher than the true value. But the method is simple, and errs on the side of safety.

In Chapter 4 worked examples are given for the design of a T-beam, both by the approximate method and also by the more exact method. A comparison is made of the results given by the two methods.

BENDING AND DIRECT COMPRESSION COMBINED

Many applications arise in practice where members are subjected simultaneously to bending moments and direct forces. Obvious examples are a column carrying a crane gantry on an eccentric bracket, or a column in a rigid frame subjected to bending moments from incoming beams. Where the members are of one material such as timber, or steel, the stress analysis presents no special problem and in all cases the fibre stresses due to bending and axial load can be determined separately and summated algebraically. But with a composite member like reinforced concrete, the matter is more complicated, and once the concrete has cracked on the tension side the section becomes unsymmetrical so that the position of the neutral axis is not immediately known.

With the elastic theory there are two separate cases to be considered; first where the eccentricity of the moment is so small that no tensile stress is developed in the member, and secondly where the eccentricity produces tension on one side of the member, and the concrete is assumed to have cracked.

If P is the direct load on the column, and M is the applied bending moment, it is convenient throughout to express these in the form that P is acting at an eccentricity e from the centre or centroid of the member such that $e = \dfrac{M}{P}$.

The assumptions made in the elastic theory are as before, namely:

At any cross-section plane sections remain plane.
All tensile stresses are taken by the reinforcement.

Where applicable the same symbols are used as before, and as shown in Fig. 3.16. Additional symbols are as follows:

d is the overall depth of the member (between concrete faces),
A_c is the area of concrete,
A_s is the total area of steel,

A_{st} is the area of tension steel $= \dfrac{A_s}{2}$,

A_{sc} is the area of compression steel $= \dfrac{A_s}{2}$,

A_E is the equivalent area of concrete,

I_C is the second moment of area of the section (ignoring the steel),

I_E is the second moment of area of a member of equivalent area A_E.

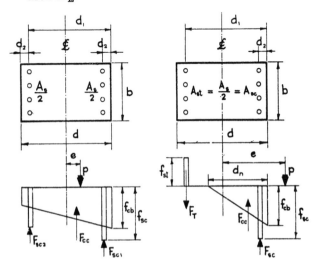

Stress Diagram   Stress Diagram

(a) Tensile stress not   (b) Tensile stress developed
    developed

FIG. 3.16. STRESS DIAGRAMS FOR MEMBERS SUBJECT TO
BENDING AND DIRECT COMPRESSION COMBINED.
ELASTIC THEORY

(i) *Case where tension does not develop*

Figure 3.16a shows a symmetrical reinforced concrete section carrying a load P at an eccentricity e from its centre so small that none of the section develops tensile stress.

Now the strain in the steel is equal to the strain in the concrete immediately adjacent to it; so the stress in the steel is m times the stress in the concrete. For convenience therefore we may consider the steel as being replaced by additional concrete of area $(m - 1)$ times the area of the steel. The equivalent area of concrete is then

$$A_E = bd + (m - 1)A_s \qquad (3.34)$$

and its *second moment of area* from elementary principles is given as

$$I_E = \frac{bd^3}{12} + (m - 1)A_s \left(\frac{d}{2} - d_2\right)^2. \qquad (3.35)$$

Because the section is uncracked, the simple theory of combined bending and direct load will apply, so that

$$f = \frac{P}{A} \pm \frac{My}{I}$$

or $$f = P\left(\frac{1}{A} \pm \frac{ey}{I}\right). \qquad (3.36)$$

Substituting in this equation the expressions of A_E for A, I_E for I, and $\frac{d}{2}$ for y, we have, on simplifying,

$$f_{cb} = P\left[\frac{1}{bd + (m - 1)A_s} \pm \frac{6ed}{bd^3 + 12(m - 1)A_s(0{\cdot}5d - d_2)^2}\right].$$

If now we define r_p as $\frac{A_s}{bd} \times 100$, the equation becomes

$$f_{cb} = \frac{P}{bd}\left[\frac{1}{1 + 0{\cdot}01(m - 1)r_p} \pm \frac{6ed}{d^2 + 0{\cdot}12(m - 1)r_p(0{\cdot}5d - d_2)^2}\right]. \qquad (3.37)$$

The positive sign in this expression gives the maximum concrete stress on the eccentrically loaded side of the member: the negative sign gives the minimum stress on the opposite side of the member.

In the limiting condition when the minimum stress is zero

$$\frac{1}{1 + 0{\cdot}01(m - 1)r_p} = \frac{6ed}{d^2 + 0{\cdot}12(m - 1)r_p(0{\cdot}5d - d_2)^2},$$

so that $$e = \frac{d^2 + 0{\cdot}12(m - 1)r_p(0{\cdot}5d - d_2)^2}{6(1 + 0{\cdot}01(m - 1)r_p)d}. \qquad (3.38)$$

From this equation it can be seen that the limiting value of e for no tension to develop depends only on the dimensions of the concrete section and the amount and position of the reinforcement. The magnitude of the direct load has no influence.

In design work it is not practical to calculate directly the necessary size and reinforcements for a member subject to combined bending and compression. The method adopted is to select a suitable section and analyse this to see whether the stresses in the chosen section are reasonable. This requires experience, if the number of trials and errors is not to be wearisome; but in any event it will be found easier to work this way about, rather than attempt a direct calculation for the section required. To assist in the analysis of selected sections, Fig. 3.17 has been prepared. Knowing the eccentricity of the load, and the proposed percentage of steel in the member, the value $\dfrac{P}{bdf_{cb}}$ can be read directly from the graph. Then knowing P, b and d, the maximum stress in the concrete can be determined. Alternatively the graph can be used the other way round, and by fixing p_{cb} for f_{cb}, and knowing P, n d and e, the necessary percentage of steel can be read off. A b, example of the use of Fig. 3.17 is given in Chapter 6.

While Fig. 3.17 is convenient for use where it so happens that none of the section develops tensile stress, it should be noted that there is no special merit in design in aiming to achieve this condition. More often it happens that tensile stresses do develop in the member, and in many cases it is more economical for this to be so. The theory for such cases now follows.

(ii) *Case where tension develops*

Figure 3.16b shows a symmetrical section carrying a load P at such an eccentricity e that part of the section develops tensile stress.

The total compression force across the section must balance the total applied load, so

$$P = F_{cc} + F_{cs} - F_T$$

$$= \left(\frac{f_{cb}d_n b}{2} - A_{sc} \frac{(d_n - d_2)}{d_n} f_{cb} \right) + A_{sc}f_{sc} - A_{st}f_{st}.$$

But $f_{sc} = \dfrac{(d_n - d_2)}{d_n} m . f_{cb}$

and $\quad f_{st} = \dfrac{(d_1 - d_n)}{d_n} m . f_{cb},$ $\qquad\qquad$ (3.39)

and since $\qquad\qquad r_p = \dfrac{A_s}{bd} \times 100,$ and $n_1 = \dfrac{d_n}{d_1},$

we have

$$P = \frac{f_{cb}bd_n}{2} + A_{sc}f_{cb}(m-1)\frac{(d_n - d_2)}{d_n} - A_{st}f_{cb}m\frac{(d_1 - d_n)}{d_n}$$

$$= \frac{f_{cb}bd}{2}\left\{ n_1\frac{d_1}{d} + \frac{0\cdot01r_p}{n_1}\left[\left(n_1 - \frac{d_2}{d_1}\right)(m-1) - (1-n_1)m \right] \right\}. \quad (3.40)$$

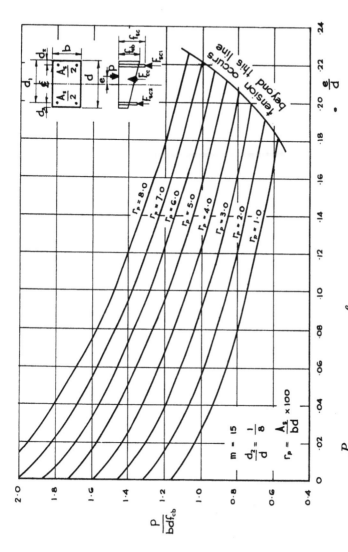

FIG. 3.17. $\dfrac{P}{bd f_{cb}}$ PLOTTED AGAINST $\dfrac{e}{d}$ FOR COMBINED BENDING AND COMPRESSION WHERE NO TENSION DEVELOPS. ELASTIC THEORY

89

Now the resistance moment of the section must balance the applied bending moment, so that

$$P \cdot e = \frac{f_{cb}bd_n}{2}\left(\frac{d}{2} - \frac{d_n}{3}\right) + A_{sc}f_{cb}(m-1)\frac{(d_n - d_2)}{d_n}\left(\frac{d}{2} - d_2\right) +$$

$$+ A_{st}f_{cb}m\frac{(d_1 - d_n)}{d_n}\left(\frac{d}{2} - d_2\right).$$

Substituting as before and simplifying, we have

$$P \cdot e = \frac{f_{cb}bd^2}{12}\left\{n_1\frac{d_1}{d}\left(3 - 2n_1\frac{d_1}{d}\right) + \right.$$

$$\left. + \frac{0 \cdot 03r_p}{n_1}\left(1 - 2\frac{d_2}{d}\right)\left[\left(n_1 - \frac{d_2}{d_1}\right)(m-1) + (1-n_1)m\right]\right\}. \quad (3.41)$$

It will be seen that equations 3.40 and 3.41 both contain the unknowns f_{cb} and n_1.

Eliminating f_{cb}, we have

$$n_1{}^2\left(\frac{d_1}{d}\right)^2\left(3 - 2n_1\frac{d_1}{d} - 6\frac{e}{d}\right) - 0 \cdot 03r_p\left[(m-1)\left(n_1\frac{d_1}{d} - \frac{d_2}{d}\right)\right.$$

$$\left(2\frac{d_2}{d} + 2\frac{e}{d} - 1\right) + m\frac{d_1}{d}(1-n_1)\left(2\frac{d_2}{d} - 2\frac{e}{d} - 1\right)\right] = 0 \quad (3.42)$$

which is a cubic in n_1. Then if r_p and $\dfrac{e}{d}$ are known, equation 3.42 can be solved to give n_1 but not without considerable labour. f_{cb} can then be found by substitution in equation 3.40. As with the previous case of combined bending and direct compression, the practical method of design here is again to select by experience an appropriate section and analyse this to see whether the choice has been suitable. To assist in this, Fig. 3.18 has been prepared from equations 3.40, 3.41 and 3.42, and gives values for $\dfrac{M}{bd^2f_{cb}}$ plotted against $\dfrac{e}{d}$ for various percentages of steel in the member. The graph is used in much the same way as previously described for Fig. 3.17. However, an additional check has then to be made that the tension steel is not overstressed, and this is done by reading from the graph the appropriate value for n_1 and substituting this with f_{cb} in formula 3.43 which has been derived from formula 3.39:

$$f_{st} = \frac{(1 - n_1)}{n_1}mf_{cb}. \quad (3.43)$$

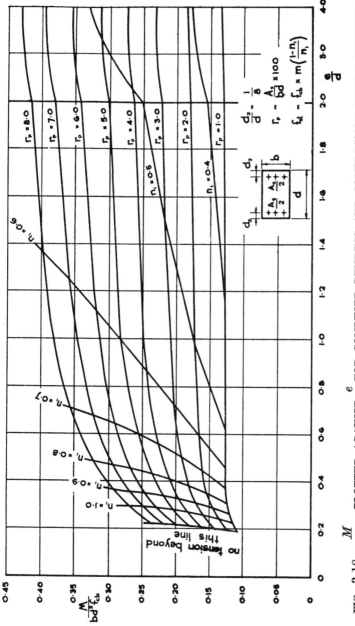

FIG. 3.18. $\dfrac{M}{bd^2f_{cb}}$ PLOTTED AGAINST $\dfrac{e}{d}$ FOR COMBINED BENDING AND COMPRESSION WHERE TENSION

DEVELOPS. ELASTIC THEORY

91

Examples of the design of columns carrying loads at considerable eccentricities are included in Chapter 6, using Fig. 3.18.

ART. 3.3. PLASTIC THEORY: (WITH LOAD FACTOR)

FUNDAMENTALS OF PLASTIC BEHAVIOUR

True plastic theory involves the study of internal stresses in
a member when loaded up to the condition where a section of the
member actually fails. The materials have then ceased to behave
elastically: the strains have increased many times beyond those
considered in the elastic theory, and the stress distribution across
the section has adjusted itself in such a way as to enable the
member to develop its maximum capacity for resisting the condition of applied loading.

In a member supported in a redundant manner, if the applied
load is increased until the material at some point along the
member becomes plastic, this point will then act as a hinge
enabling the moments in the member to readjust themselves
in the most favourable way. This normally has the effect of
increasing the load-carrying capacity of the member before
collapse occurs. But the behaviour of members beyond the stage
where plastic hinges form is not dealt with here, as this has no
influence on the study of stress distribution at any section under
consideration, and is only a matter of more complex structural
analysis of the member as a whole.

Ignoring the question of hinge formation, strictly speaking in
any true load factor method of design the basis should be to
calculate the maximum load the member can carry immediately
prior to collapse (allowing for the plastic behaviour of the
materials) and then divide the value for this load by a suitable
load factor which will constitute an insurance that under working conditions there is adequate margin against risk of collapse.
However, as will be seen at the end of this introduction to the
present article, the practical approach to design given in C.P. 114
(1957) departs fundamentally from this basic concept: nevertheless
the method propounded makes for simple procedure in engineers'
design offices, if not for quite the greatest economy. In other words
it represents a good rough tool, though free from refinements and
free from the fullest efficiency.

But before dealing with the simplified method given in C.P.
114 (1957), a further introduction to the true features of plastic
behaviour will not be out of the way here. Consider a beam.
Unless the percentage of steel is unusually small, ultimate collapse
will always occur as a result of crushing of the concrete. This is

because it is in the nature of the steel that once it has yielded it
is still able to provide the same tension force while yet it continues
to stretch; and in the meantime the centre of compression moves
up the beam nearer and nearer the top face until in the ultimate
condition the depth to the neutral axis becomes so small, and the
concrete stress becomes so great, that failure of the concrete occurs
at the top edge of the beam.

(i) *Stress and Strain*

Tests show that the stress/strain relationships of concrete in
bending up to failure is of the form given in Fig. 3.19. It will

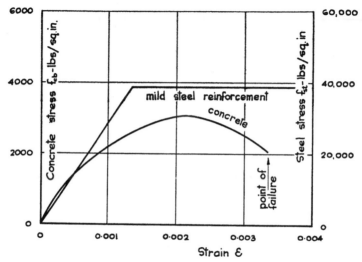

FIG. 3.19. TYPICAL STRESS/STRAIN CURVES FOR MILD STEEL
AND CONCRETE IN BENDING THROUGH PLASTIC RANGE

be seen that as the stress is increased so also does the strain
increase, though not linearly, until a maximum stress value is
reached. Beyond this point, as the strain increases the concrete
becomes capable of carrying less stress.

Now Fig. 3.20 shows the distribution of stress and strain in a
beam at the moment prior to collapse due to concrete failure.
Experiments show that even approaching the condition of failure
plane sections remain plane, so that strain at all points remains
everywhere proportional to the distance from the neutral axis.
Therefore the curve for the stress distribution in the concrete in
the compression zone is identical to the stress/strain curve given

in Fig. 3.19. The area of this compressive stress distribution above
the neutral axis has come to be known as the "compressive stress
block".

It has alas been found that the curve for the stress distribution
varies according to the cube strength of the concrete. Before an
exact study can be made of the resistance moment of a member
it is necessary to know the total compressive resistance of the
stress block and the position of its resultant F_C. This has recently

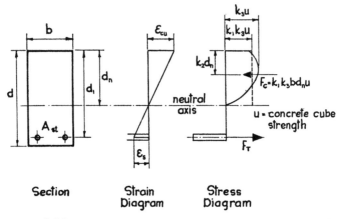

Section   **Strain**   **Stress**
                   Diagram   **Diagram**

FIG. 3.20. STRESS AND STRAIN DIAGRAMS FOR SINGLY
REINFORCED BEAM AT COLLAPSE

been the subject of much research. In Fig. 3.20 the maximum
and average concrete stresses at failure have been related to the
cube strength of the concrete, so that:

the maximum compressive stress is $k_3 u$,
the average compressive stress is $k_1 k_3 u$,
and the depth to the centre of compression is $k_2 d_n$,

where u is the cube strength of the concrete and k_1, k_2, and k_3 are
factors which vary according to the cube strength of the concrete.
Figure 3.21 gives values for k_1, k_2 and $k_1 k_3$ and ultimate strains for
concretes of different cube strengths. These values are based on
the research results of Hognestad, Hanson and McHenry[2] which
were originally published in terms of cylinder strengths: the
values in Fig. 3.21 have been converted from cylinder strength
to cube strength values using the relationship that cylinder
strength is 0·78 times the cube strength.

The stress/strain relationship for ordinary mild steel reinforcement is near enough linear up to the yield point of the steel;

and the modulus of elasticity can be regarded as constant at 30×10^6 lb/sq. in. When the load in the steel persists after the yield point has been reached, the steel continues to act at the yield stress while yet the strain increases very considerably. This is indicated in Fig. 3.19. Reinforcements other than mild steel may behave less simply and require special consideration when used in work designed by the plastic theory. What follows in this chapter applies to the use of mild steel: but other steels may be dealt with by similar methods.

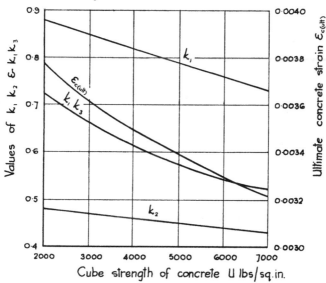

FIG. 3.21. CURVES OF k_1, k_2, k_1k_3 AND $\varepsilon_{c(ult)}$ PLOTTED
AGAINST CUBE STRENGTH u

(ii) *True plastic behaviour*

The true plastic behaviour at a section of a member subject to bending is now described. Later it will be shown how simplified load factor formulae are derived as used more commonly in design work. But the significance of the simplified formulae will be better appreciated after considering the true behaviour of a member suffering plastic deformation.

In what follows the symbols used have the same meaning as in the earlier part of this chapter; and are contained in the Glossary of Symbols at the beginning of the book. New symbols are:

M_r for the working moment of resistance of a section; and
$M_{r(ult)}$ for the ultimate moment of resistance of a section.

Consider the bending of a singly reinforced beam. For any state of stress in the beam the resistance moment developed by the concrete is

$$M_r = bd_n f_{c(av)} l_a$$
$$= bd_1{}^2 n_1 a_1 f_{c(av)} \tag{3.44}$$

where $f_{c(av)}$ is the average compressive stress in the concrete; and the resistance moment developed by the steel is

$$M_r = A_{st} f_{st} l_a$$
$$= 0.01 r_{p1} bd_1{}^2 a_1 f_{st}. \tag{3.45}$$

Three conditions arise.

(a) The first condition is the case of a beam in which the reinforcement yields before the concrete is stressed to the ultimate. Such a beam is known as "under-reinforced". With such a beam, even after the reinforcement has yielded, it is nevertheless possible for the resistance moment to increase. This is achieved by the neutral axis rising which gives an increased lever arm. Because the compression zone is now reduced in size, the concrete stress increases; and if additional moments are applied, the concrete stress will continue to increase until the outer fibre stress reaches the maximum value the concrete is capable of resisting. Further applied moments will then lead to plastic deformation with the outer fibre stress falling off until ultimately the stress distribution shown in Fig. 3.20 is reached.

The condition at failure of an "under-reinforced" beam is shown in Fig. 3.22a. Since at failure $f_{c(av)} = k_1 k_3 u$, we have, by equating equations 3.44 and 3.45,

$$k_1 k_3 u n_1 = 0.01 r_{p1} f_{sy}$$

where f_{sy} is the stress of the reinforcement at yield.

This gives
$$n_1 = \frac{0.01 r_{p1} f_{sy}}{k_1 k_3 u}. \tag{3.46}$$

Now the depth of the centre of compression is $k_2 d_n$, i.e. $k_2 n_1 d_1$, so that at failure we can express equation 3.45 as

$$M_{(ult)} = 0.01 r_{p1} bd_1{}^2 (1 - k_2 n_1) f_{sy}.$$

Substituting for n_1, this becomes

$$M_{(ult)} = 0.01 r_{p1} bd_1{}^2 \left(1 - \frac{k_2}{k_1 k_3} \cdot \frac{0.01 r_{p1} f_{sy}}{u} \right) f_{sy}. \tag{3.47}$$

(*b*) The second condition to be considered is that of the beam which fails before the tension reinforcements yield. Such a beam is known as "over-reinforced". Here the mechanism leading to failure is the very opposite of that for the under-reinforced beam. As the applied moment is increased and failure approaches, the

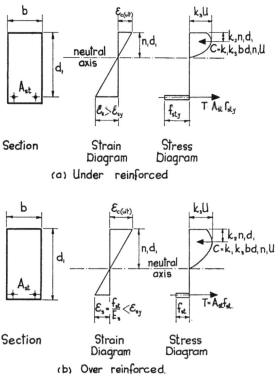

FIG. 3.22. STRESS AND STRAIN DIAGRAMS FOR "UNDER-" AND "OVER-REINFORCED" BEAMS AT COLLAPSE

neutral axis drops and the lever arm is reduced. This causes the steel stress to increase. Then as the extent of plasticity increases, the lever arm continues to reduce and very little gain comes from the increase in size of the compressive stress block. Failure occurs when the concrete reaches its ultimate strain at the top, and the small gain in compressive resistance is totally offset by the reduced lever arm.

The condition at failure of an over-reinforced beam is shown in

Fig. 3.22b. If $\varepsilon_{c(ult)}$ is the ultimate strain in the concrete and ε_s is the elastic strain in the steel, then since in this case the steel has not yielded but is still within the elastic range, we have from Fig. 3.22b

$$n_1 = \frac{\varepsilon_{c(ult)}}{\varepsilon_{c(ult)} + \varepsilon_s}. \qquad (3.48)$$

Since the total compression force across the section must balance the total tension force,

$$0 \cdot 01 r_{p1} b d_1 f_{st} = k_1 k_3 u b n_1 d_1.$$

Therefore

$$f_{st} = \frac{k_1 k_3 u n_1}{0 \cdot 01 r_{p1}} = E_s \varepsilon_s. \qquad (3.49)$$

Substituting in equation 3.49 the expression for ε_s from equation 3.48, we have

$$\frac{k_1 k_3 u n_1{}^2}{0 \cdot 01 r_{p1}} + E_s \varepsilon_{c(ult)} n_1 - E_s \varepsilon_{c(ult)} = 0 \qquad (3.50)$$

which is a quadratic in n_1 readily solvable once a value for the concrete strain has been chosen. The ultimate resistance moment of the beam can then be found from the equation

$$M_{(ult)} = b d_1{}^2 n_1 k_1 k_3 u (1 - k_2 n_1). \qquad (3.51)$$

Equation 3.51 is developed from equation 3.44.

(c) The third and last condition to be considered arises when the ultimate resistance of the concrete and the yield stress of the steel are reached simultaneously. This is known as "balanced" design. It is the basis normally used when the reinforcements are of mild steel. In this case the depth of the neutral axis may be derived from equation 3.48 which becomes

$$n_1 = \frac{\varepsilon_{c(ult)}}{\varepsilon_{c(ult)} + \varepsilon_{sy}}. \qquad (3.52)$$

The ultimate resistance moment for this value of n_1 may then be found from equation 3.51 and the percentage of steel from equation 3.46.

With mild steel reinforcements and normal grades of concrete the value of n_1 in balanced design approximates to 0·75. The exact figure, of course, depends on the cube strength of the concrete and the yield point of the steel. In under-reinforced beams the value of n_1 will be smaller, and for over-reinforced beams it will be greater than 0·75. Where higher yield point steels are used the depth of the compressive stress block becomes less as the yield stress and strain increase.

To arrive at the *working moment* a beam can safely resist, a suitable load factor has to be applied to the *ultimate resistance moments* given by equations 3.47 and 3.51. It should be noted that the collapse of an over-reinforced beam will occur suddenly without warning; whereas the rapidly increasing deflection of an under-reinforced beam as it approaches failure gives some previous warning. Accordingly there may be a case for preferring the under-reinforced, rather than the over-reinforced beam.

(iii) *Simplified approach to load factor design*

The theory set out above gives the true plastic behaviour at failure of a beam. However, in practice it is more usual to adopt a simplified approach when designing on a Load Factor basis. Instead of working to a curved stress distribution for the compressive stress block, it is assumed that the distribution of compressive stress is uniform. Such a distribution of stress has become known as a "rectangular stress block". Then so long as the assumed uniform stress at ultimate load is kept down to the average stress value given by k_1k_3u in Fig. 3.21, it is clear from the k_2 values in Fig. 3.21 that the assumed rectangular stress block will give a reduced imaginary lever arm, making for a low estimate of the ultimate strength of the member, with a consequent hidden margin of safety.

If the ultimate resistance moment of the beam were calculated on the basis of the simple rectangular stress block, and a load factor then applied, one could fairly say one was designing to a simplified Load Factor Method. However, as will be shown in the remainder of this article, the simplified method propounded by CP. 114 (1957) appears not to be this at all. Instead of applying a load factor to the ultimate carrying capacity of the member at its *failing load*, what is done is to calculate its capacity at *working load* using working stresses but a stress distribution approximating to that which occurs at failure. Thus this procedure in design is really to follow the old-established sequence of events which was appropriate to elastic design methods, but which is in a manner of speaking back to front for the Load Factor Method. Nevertheless one can say in favour of the simplified system that it makes for directness, if not for the strictest accuracy or economy.

In the whole of the formulae for Load Factor design developed in the remainder of this article no account is taken of the loss of concrete area arising from the area of reinforcements in the compression block. The errors arising from this are small.

BENDING

C.P. 114 (1957) requires the resistance moments of beams and slabs to have a load factor generally of 1·8, and the stresses at working loads to be such as not to cause excessive cracking. For the tension steel, the resistance moment corresponding to working loads (M_r) is to be calculated at a stress not exceeding the permissible stress appropriate to the particular steel; in which case the load factor on ordinary mild steel, being $\dfrac{\text{yield stress}}{\text{permissible stress}}$, appears to be about 1·8.

For the concrete, the resistance moment corresponding to working loads is to be calculated at a permissible stress over the rectangular stress block equal to two-thirds of the permissible compressive stress of the concrete in bending. From Tables 1.3 and 1.4 it is clear that for *nominal concrete mixes* the permissible stress of the concrete in bending is limited to one-third of the Works cube strength, so that at working loads we have the permissible stress over the rectangular stress block equal to $\frac{2}{3}p_{cb}$ is $\frac{2}{3} \cdot \frac{1}{3} u$ equals $\frac{2}{9} u$. With a load factor of 1·8 this would appear to correspond to an average compressive stress at ultimate conditions of $1·8 \cdot \frac{2}{9} u$ equals $0·4 u$; and by reference to the ratio of average to maximum stress values given in Fig. 3.21 it will be seen that $0·4 u$ provides quite a hidden margin of safety. With *designed concrete mixes* the permissible stress over the rectangular stress block is taken as 10 per cent higher.

By restricting the depth of the compressive stress block to half the effective depth, C.P. 114 (1957) confines the design of members by the simplified formulae to sections under-reinforced on the tension side.

(i) *Singly reinforced rectangular beam*

Consider the singly reinforced beam shown in Fig. 3.23. In the under-reinforced condition we have from equation 3.45

$$M_{r(ult)} = A_{st}f_{sy}l_a$$

and with a load factor of about 1·8 at working loads this becomes

$$M_r = A_{st}p_{st}l_a. \tag{3.53}$$

Now for a rectangular stress block form we have

$$l_a = d_1 - \frac{d_n}{2} = d_1 - \frac{n_1 d_1}{2}.$$

And from equation 3.46

$$n_1 = \frac{A_{st}}{bd_1} \cdot \frac{1\cdot 8p_{st}}{1\cdot 8 \times \frac{2}{3}p_{cb}} = \frac{3A_{st}p_{st}}{2bd_1p_{cb}},$$

so that

$$l_a = d_1 - \frac{3A_{st}p_{st}}{4bp_{cb}}. \tag{3.54}$$

These equations 3.53 and 3.54 are the formulae given in C.P. 114 (1957) for the working resistance moment of a beam based on the tensile reinforcement.

Now in equation 3.54 substituting r_{p1} for $\frac{A_{st}}{bd_1} \times 100$, t_1 for $\frac{p_{st}}{p_{cb}}$, and a_1d_1 for l_a,

we have

$$a_1 = 1 - \frac{0\cdot 03r_{p1}t_1}{4}.$$

But since

$$a_1 = 1 - \frac{n_1}{2},$$

we have

$$n_1 = \frac{0\cdot 03r_{p1}t_1}{2}. \tag{3.55}$$

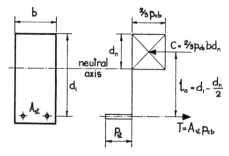

FIG. 3.23. STRESS DIAGRAM FOR SINGLY REINFORCED BEAM. SIMPLIFIED LOAD FACTOR METHOD

And substituting similarly in equation 3.53,

we have

$$M_r = 0\cdot 01r_{p1}bd_1{}^2 \left(1 - \frac{n_1}{2}\right) p_{st}$$

which from equation 3.55 becomes

$$M_r = \tfrac{2}{3}bd_1{}^2n_1 \left(1 - \frac{n_1}{2}\right) p_{cb}, \tag{3.56}$$

and since d_n is limited to $\frac{d_1}{2}$, so that $n_1 = 0\cdot 5$, the maximum value for M_r is given by

$$M_r = bd_1{}^2 \frac{p_{cb}}{4}. \tag{3.57}$$

This equation is the formula given in C.P. 114 (1957) for the working resistance moment of a beam based on the strength of the concrete in compression. It demonstrates as one might expect that for any section M_r is limited by the size of the section and the permissible concrete stress. The corresponding steel percentage to provide the maximum permissible resistance moment is obtained from equation 3.55 putting $n_1 = 0\cdot5$, so that

$$r_{p1} = \frac{1}{0\cdot03t_1}. \tag{3.58}$$

Alternatively, analysing any given section, the working resistance moment is the lesser of the values obtained from equations 3.53 and 3.57.

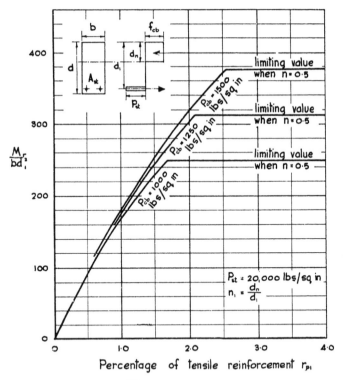

FIG. 3.24. WORKING $\dfrac{M_r}{bd_1{}^2}$ VALUES FOR SINGLY REINFORCED MEMBERS (VARIOUS STRESSES). SIMPLIFIED LOAD FACTOR METHOD

Figure 3.24 has been prepared from equations 3.55, 3.56 and 3.57. It gives values for $\frac{M_r}{bd_1^2}$ plotted against the percentage of steel for three different permissible concrete stresses. The permissible steel stress is taken as 20,000 lb/sq. in. throughout.

From the foregoing it is clear that with singly reinforced beams the simplified load-factor method of design gives higher values of $\frac{M_r}{bd_1^2}$ for a given quality of concrete than does the elastic-theory method: whereas with the optimum section corresponding to $\frac{M_r}{bd_1^2} = \frac{p_{cb}}{4}$, in order to develop the full value M_r the load-factor method requires a greater percentage of steel than does the elastic-theory method. Or expressing this another way, if the breadth of the beam is constant, the depth as given by load-factor design is less than by the elastic-theory, showing a saving in concrete, formwork and weight: whereas if the size of the beam is settled by other considerations such as shear or deflection, then unless $\frac{M_r}{bd_1^2}$ exceeds the value given by equation 3.12, the elastic-theory will demand a smaller percentage of reinforcement than will the load-factor method.

(ii) *Doubly reinforced rectangular beam*

When the bending moment applied to a beam exceeds the value for M_r given by equation 3.57, it becomes necessary to provide compression reinforcement as already described at part (ii) of

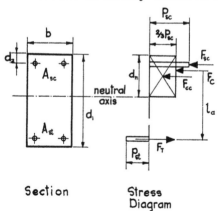

Section   Stress
                        Diagram

FIG. 3.25. STRESS DIAGRAM FOR DOUBLY REINFORCED
BEAM. SIMPLIFIED LOAD FACTOR METHOD

Article 3.2. The stress diagram is then as shown in Fig. 3.25. Since normally the strains at failure are such that the compression steel will have yielded before the concrete fails, the resistance moment, as limited by the compression, will be enhanced by the compression steel to the tune of $A_{sc}f_{sy}(d_1 - d_2)$, or at working loads

the added compressive $M_r = A_{sc}p_{sc}(d_1 - d_2)$,   (3.59)

so that the total resistance moment due to compression is

$$M_r = \frac{p_{cb}}{4}\,bd_1{}^2 + A_{sc}p_{sc}(d_1 - d_2)$$ (3.60)

which is the formula given in C.P. 114 (1957).

It also then becomes necessary to increase the amount of tension steel. And since equation 3.59 given in C.P. 114 (1957) must have been based-on the assumption that the compressive force due to the concrete is not affected by the introduction of compression steel, it is consistent to assume that the depth of the neutral axis is not affected either. And we can by the same token assume also that an increase in the percentage of tension steel (over and above that given by equation 3.58) will contribute directly to an increase of resistance moment, without adjustment to the depth of the neutral axis.

Now if r'_{p1} is the percentage of additional tension steel and A'_{st} is the area of the additional tension steel so that $r'_{p1} = \dfrac{A'_{st}}{bd_1} \times 100$, then, due to the additional tension steel,

the added tension $M_r = A'_{st}p_{st}(d_1 - d_2)$

$$= 0\!\cdot\!01r'_{p1}bd_1{}^2p_{st}\left(1 - \frac{d_2}{d_1}\right).$$ (3.61)

And if now the *total* percentage of tension steel is expressed as $r_{p1} = \dfrac{A_{st}}{bd_1} \times 100$, and we put $r_s = \dfrac{A_{sc}}{A_{st}}$ (the familiar symbols without the dashes), then, due to the compression steel,

the added compression $M_r = 0\!\cdot\!01r_sr_{p1}bd_1{}^2p_{sc}\left(1 - \dfrac{d_2}{d_1}\right),$ (3.62)

always providing r_{p1} is not less than for the case of the singly reinforced beam.

Now we require the increase in tension to equal the increase in compression, so equating formulae 3.61 and 3.62, we have

$$r_sr_{p1}p_{sc} = r'_{p1}p_{st},$$

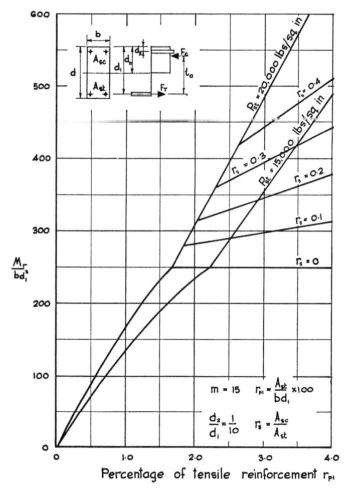

FIG. 3.26. WORKING $\dfrac{M_r}{bd_1^2}$ VALUES FOR DOUBLY

REINFORCED MEMBERS

($p_{sc} = 18{,}000$ lb/sq. in.: $p_{cb} = 1000$ lb/sq. in.)

SIMPLIFIED LOAD FACTOR METHOD

105

and since the *total* tension steel is equal to the original plus the additional tension steel, we have from the result of equation 3.58

$$\text{total } r_{p1} = \frac{1}{0 \cdot 03 t_1} + r'_{p1}$$

which enables us to rewrite the previous equation as

$$r_s = \frac{p_{st}}{p_{sc}} \left(1 - \frac{1}{0 \cdot 03 t_1 r_{p1}} \right). \tag{3.63}$$

Then, for example, if $p_{st} = 20{,}000$ lb/sq. in., $p_{sc} = 18{,}000$ lb/sq. in. and $p_{cb} = 1000$ lb/sq. in., we can determine the required ratio of compression to tension steel as

$$r_s = 1 \cdot 11 - \frac{1 \cdot 85}{r_{p1}}.$$

If in equation 3.60 which gives the compression resistance moment, we substitute the value r_s from equation 3.63, we derive

$$r_{p1} = \frac{1}{0 \cdot 01 p_{st} \left(1 - \dfrac{d_2}{d_1} \right)} \left[\frac{M_r}{b d_1{}^2} + \frac{p_{cb}}{12} \left(1 - 4 \frac{d_2}{d_1} \right) \right]. \tag{3.64}$$

For any required resistance moment, this equation will give the percentage of tension steel; and the amount of compression steel can then be found by determining r_s from equation 3.63.

Figure 3.26 has been prepared from equations 3.62, 3.63 and 3.64, and gives curves for $\dfrac{M_r}{b d_1{}^2}$ plotted against percentage of tension steel for various amounts of compression steel. The figure is based on $p_{sc} = 18{,}000$ lb/sq. in., $p_{cb} = 1000$ lb/sq. in., and $\dfrac{d_2}{d_1} = \dfrac{1}{10}$. It is seen to be similar in form to Fig. 3.12 (prepared from the elastic theory), and may be used for design work in just the same way.

(iii) T-*beams and* L-*beams*

The principles of T- and L-beam action have been described in part (iii) of Article 3.2. For the same reasons as given there, the following study is confined to consideration of T-beams, though is equally applicable to L-beams. And as before, if the neutral axis lies within the depth of the flange, the same theory will apply as for a simple rectangular beam, but of enhanced breadth. In all other cases the analysis is as follows.

Consider the simplified stress distribution at working loads as shown in Fig. 3.27. Owing to the supposed uniform distribution of concrete stress, the contribution of the compressive force provided by the rib cannot be ignored, and we have

$$M_r = \frac{2}{3} p_{cb} \left[b_r d_n \left(d_1 - \frac{d_n}{2} \right) + b d_s \left(d_1 - \frac{d_s}{2} \right) - b_r d_s \left(d_1 - \frac{d_s}{2} \right) \right]$$

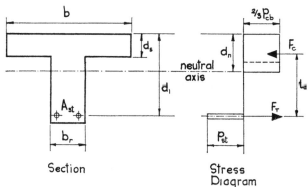

Section   Stress Diagram

FIG. 3.27. STRESS DIAGRAM FOR T-BEAM. SIMPLIFIED
LOAD FACTOR METHOD

which may conveniently be rearranged to give

$$M_r = \frac{2}{3} p_{cb} b d_1^2 \left[\frac{b_r}{b} n_1 \left(1 - \frac{n_1}{2} \right) + \left(1 - \frac{b_r}{b} \right) \frac{d_s}{d_1} \left(1 - \frac{d_s}{2d_1} \right) \right]. \quad (3.65)$$

Now the maximum allowable value for M_r arises when $d_n = \dfrac{d_1}{2}$, i.e. when $n_1 = \frac{1}{2}$.

Equation 3.65 then reduces to

$$M_r = p_{cb} b d_1^2 \left[\frac{b_r}{4b} + \frac{d_s}{3d_1} \left(1 - \frac{b_r}{b} \right) \left(2 - \frac{d_s}{d_1} \right) \right] \quad (3.66)$$

$$= \gamma p_{cb} b d_1^2 \quad (3.67)$$

where
$$\gamma = \frac{b_r}{4b} + \frac{d_s}{3d_1} \left(1 - \frac{b_r}{b} \right) \left(2 - \frac{d_s}{d_1} \right). \quad (3.68)$$

These are the equations given in C.P. 114 (1957) as the simplified formulae for the design of T- and L-beams.

The total compressive force is clearly given by

$$F_c = \frac{2}{3} p_{cb} \left(b_r \frac{d_1}{2} + b d_s - b_r d_s \right), \quad (3.69)$$

so that from equations 3.66 and 3.69 we can express the lever arm as

$$l_a = \frac{M_r}{F_c} = \frac{d_1 \left[\dfrac{b_r}{4b} + \dfrac{d_s}{3d_1} \left(1 - \dfrac{b_r}{b} \right) \left(2 - \dfrac{d_s}{d_1} \right) \right]}{\dfrac{2}{3} \left[\dfrac{b_r}{2b} + \dfrac{d_s}{d_1} \left(1 - \dfrac{b_r}{b} \right) \right]}$$

which on reduction gives

$$a_1 = \frac{l_a}{d_1} = \left(1 - \frac{d_s}{2d_1} \right) + \frac{b_r}{4b} \frac{\left[3 - 2 \left(2 - \dfrac{d_s}{d_1} \right) \right]}{\left[\dfrac{b_r}{b} + \dfrac{2d_s}{d_1} \left(1 - \dfrac{b_r}{b} \right) \right]}. \tag{3.70}$$

This full expression is rather complicated and depends on the ratios $\dfrac{b_r}{b}$ and $\dfrac{d_s}{d_1}$. However, in most practical examples the second term of the expression can be safely ignored, giving an error of about 4 per cent or less. But where the flange is unusually thin and the rib unusually broad, say for example $\dfrac{d_s}{d_1} = \dfrac{1}{8}$ and $\dfrac{b_r}{b} = \dfrac{1}{3}$, then the error of omitting the second term becomes as much as 13 per cent.

For normal cases we shall assume

$$a_1 = 1 - \frac{d_s}{2d_1} \tag{3.71}$$

which gives the resistance moment based on tension as

$$M_r = A_{st} p_{st} \left(1 - \frac{d_s}{2d_1} \right) d_1, \tag{3.72}$$

as given in C.P. 114 (1957).

Substituting $0 \cdot 01 r_{p1} b d_1$ for A_s in equation 3.72, we have

$$M_r = 0 \cdot 01 r_{p1} p_{st} b d_1{}^2 \left(1 - \frac{d_s}{2d_1} \right). \tag{3.73}$$

And in order to reinforce so as to develop the maximum resistance moment of the section (when $n_1 = \frac{1}{2}$), we have to equate the values for M_r in equations 3.67 an 3.73. This simplifies to give

$$r_{p1} = \frac{\gamma}{0 \cdot 02 t_1 \left(2 - \dfrac{d_s}{d_1} \right)} \tag{3.74}*$$

In designing T-beams the normal procedure is to select suitable

* Where $\gamma = \dfrac{b_r}{4b} + \dfrac{1}{3} \left[1 - \dfrac{b_r}{b} \right] \left[2 \dfrac{d_s}{d_1} - \left(\dfrac{d_s}{d_1} \right)^2 \right]$.

concrete dimensions, and calculate the area of tension steel from equation 3.72. The concrete stress is then checked using equation 3.67: and in the rare event of this proving excessive, compression steel has to be added as described in part (ii) of this article so that

$$M_r = \gamma p_{cb} b d_1{}^2 + A_{sc} p_{sc} (d_1 - d_2), \qquad (3.75)*$$

as given in C.P. 114 (1957). A further check on the tension steel must then be made.

A variety of worked examples of beam design are given in Chapter 4.

BENDING AND DIRECT COMPRESSION COMBINED

As in the previous study on beams, the present theory for combined bending and direct compression is developed on the simplified assumption of a rectangular compressive stress block, with the analyses made at working load conditions and the stresses appropriately restricted.

As with beams, C.P. 114 (1957) requires columns in bending to have a load-factor generally of 1·8, and the stresses at working loads to be such as not to cause excessive cracking. For the steel, the stress at working loads is again to be restricted to the permissible stresses appropriate to the particular steel.

For the concrete, the maximum strain in compression is to be assumed not to exceed 0·33 per cent at failure, and the maximum stress at failure is not to exceed two-thirds of the cube strength. Further, in strength calculations the cube strength for nominal concrete mixes is to be taken as only 68 per cent of the actual cube strength. Thus the maximum allowable stress in the concrete at failure is $\frac{2}{3} . u . \frac{68}{100}$ so that at working loads the maximum permissible stress with a load factor of 1·8 is $\frac{1}{1·8} . \frac{2}{3} . u . \frac{68}{100}$ equals $\frac{76}{100} . \frac{u}{3}$. Reference to Tables 1.3 and 1.4 will show that for Portland cement concretes with B.S. 882 aggregates this recommendation of C.P. 114 (1957) has the effect that the working stress in the concrete is equal to the permissible stress for concrete in direct compression. With *designed concrete mixes* the permissible stress over the rectangular compressive stress block is taken as 10 per cent higher. In addition, it will be seen later, at part (i) of this section, that in deriving the formulae in C.P. 114 (1957) it must have been the intention to restrict the extent of the compressive stress block to $0·85 \times d_n$.

* See footnote, p. 108.

There are with this problem three possible modes of failure as follows. Each requires separate study.

(a) The ultimate resistance of the concrete can be reached simultaneously with the yielding of the tension steel.

(b) The concrete can reach failure before the tension steel yields.

(c) The tension steel can yield before the concrete is stressed to the ultimate.

In studying each of these conditions, the member together with its reinforcements is taken as symmetrical. The dimensions are as shown in the appropriate figures referred to; and the notation used is the same as before, and as given in the Glossary of Symbols at the beginning of the book. New symbols are:

ε_s for the strain of tension reinforcement,

ε_c for the outer fibre strain of the concrete,

r for the ratio $\dfrac{A_s}{bd}$.

(i) Balanced Design

Figure 3.28a gives the diagrams for ultimate strain and working stress in the case where the maximum permissible stresses are reached simultaneously in the concrete and in the tension steel. In this special case the applied load is denoted by P_b and the eccentricity at which it acts as e_b. The extent of the compression block Xd_1 is shown restricted to $0.85d_n$: otherwise equations 3.76, 3.77 and 3.78 will not fall in with the recommendations given in C.P. 114 (1957).

From the strain diagram we have

$$\frac{\varepsilon_s}{\varepsilon_c} = \frac{d_1 - d_n}{d_n}$$

or, with a load factor of 1.8 on the steel stress, and ε_c limited to $\dfrac{1}{300}$,

$$1.8 p_{st} = \frac{d_1 - d_n}{d_n} \cdot \frac{E_s}{300},$$

so that

$$d_n = n_1 d_1 = \frac{E_s}{E_s + 540 p_{st}} \cdot d_1.$$

With mild steel, for which E_s equals 30×10^6 lb/sq. in. the depth of the stressed section is given by

$$Xd_1 = 0.85 d_n = \frac{0.85 \times 30 \times 10^6}{30 \times 10^6 + 540 p_{st}} \cdot d_1$$

$$= \frac{85{,}000}{100{,}000 + 1.8 p_{st}} \cdot d_1. \tag{3.76}$$

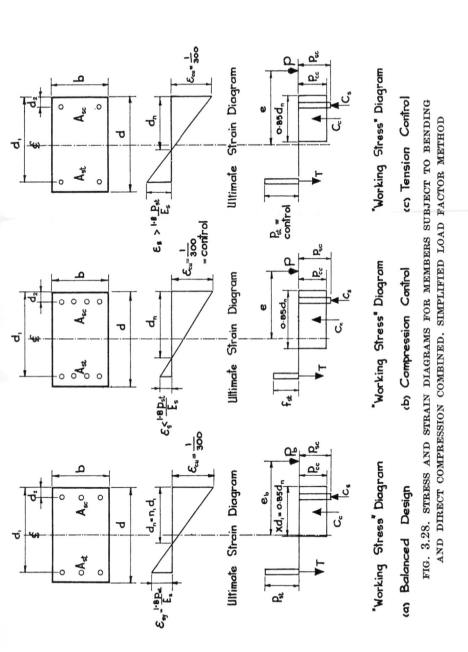

"Working Stress" Diagram

(a) Balanced Design

Ultimate Strain Diagram

"Working Stress" Diagram

(b) Compression Control

Ultimate Strain Diagram

"Working Stress" Diagram

(c) Tension Control

Ultimate Strain Diagram

FIG. 3.28. STRESS AND STRAIN DIAGRAMS FOR MEMBERS SUBJECT TO BENDING AND DIRECT COMPRESSION COMBINED. SIMPLIFIED LOAD FACTOR METHOD

111

Now still with the restricted extent of stress block as referred to earlier, since the total compression force across the section must balance the total applied force, we have

$$P_b = bd_1 X p_{cc} - A_{st} p_{st} + A_{sc} p_{sc}$$

and putting $A_{st} = A_{sc}$, we have

$$P_b = bd_1 X p_{cc} - A_{sc}(p_{st} - p_{sc}), \qquad (3.77)$$

and similarly, since the resistance moment of the section must balance the applied bending moment,

$$M_b = P_b e_b = bd_1 X p_{cc} \left(\frac{d - Xd_1}{2}\right) + A_{sc}(p_{sc} + p_{st}) \left(\frac{d}{2} - d_2\right). \quad (3.78)$$

These equations 3.77 and 3.78 are as given in C.P. 114 (1957) except that for convenience equation 3.78 has here been expressed in a slightly different form.

Thus, for a selected section, the working load and moment under balanced-design conditions are determined. And from the point of view of stressing all the materials to their permissible stress values simultaneously, the balanced design is the ideal solution. However, other physical requirements and the economics of the matter do not necessarily lead to conditions which give a balanced design.

Useful constants to have available may be determined from equation 3.76; thus when $p_{st} = 20{,}000$ lb/sq. in., then

$$n_1 = 0 \cdot 736 \text{ and } X = 0 \cdot 625.$$

(ii) *Design Controlled by Compression*

Figure 3.28b gives the diagrams of ultimate strain and working stress in the case where the design is controlled by compression. At the working load the concrete and compression steel are both at their maximum permissible stresses, whereas the tension steel is not. The applied load is P acting at an eccentricity e.

If f_{st} is the stress in the tension steel at working load, then from the ultimate strain diagram using a load factor of $1 \cdot 8$, we have

$$\varepsilon_s = \frac{1 \cdot 8 f_{st}}{E_s} = \frac{(d_1 - d_n)}{d_n} \cdot \varepsilon_c.$$

With ε_c limited to $\dfrac{1}{300}$, and taking E_s as 30×10^6 lb/sq. in., and substituting n_1 for $\dfrac{d_n}{d_1}$, we have

$$f_{st} = 55{,}500 \frac{(1 - n_1)}{n_1}. \qquad (3.79)$$

Equating forces, we have

$$P = bd_1 \, 0.85 n_1 p_{cc} + A_{sc} p_{sc} - A_{st} f_{st}$$

and substituting for f_{st} from equation 3.79, and r for $\dfrac{A_s}{bd}$ equals $\dfrac{2A_{st}}{bd}$, we can rearrange the equation as

$$P = p_{cc} bd \left\{ 0.85 n_1 \frac{d_1}{d} + \frac{r}{2 p_{cc}} \left[p_{sc} - 55{,}500 \left(\frac{1 - n_1}{n_1} \right) \right] \right\}. \quad (3.80)$$

Taking moments about the centre of the section, we have

$$M = Pe = bd_1 \, 0.85 n_1 p_{cc} \left(\frac{d}{2} - \frac{0.85 n_1 d_1}{2} \right) + A_{sc} p_{sc} \left(\frac{d}{2} - d_2 \right) +$$
$$+ A_{st} \, 55{,}500 \left(\frac{1 - n_1}{n_1} \right) \left(\frac{d}{2} - d_2 \right)$$

and, substituting as before,

$$M = Pe = bd^2 p_{cc} \left\{ 0.425 \frac{n_1 d_1}{d} \left(1 - 0.85 n_1 \frac{d_1}{d} \right) + \right.$$
$$\left. + \frac{0.25 r}{p_{cc}} \left(1 - 2\frac{d_2}{d} \right) \left[p_{sc} + 55{,}500 \left(\frac{1 - n_1}{n_1} \right) \right] \right\}. \quad (3.81)$$

It is evident that the number of solutions to equations 3.80 and 3.81 is infinite, since for given conditions of applied loading and permissible concrete stress, the values for r, b, d, d_1 and n_1 can be varied at will to give solutions satisfying both equations simultaneously. The trying labour of this process can be obviated by working from a suitable design graph. Fig. 3.29 has been prepared for this purpose, and gives values for $\dfrac{M}{bd^2 p_{cc}}$ plotted against $\dfrac{e}{d}$ for various ratios of $\dfrac{r_p}{p_{cc}}$. The curves of this graph *above* the "balanced design line" have been derived from equations 3.80 and 3.81 with p_{sc} taken as 18,000 lb/sq. in., and apply to cases where the design is controlled by compression. (The curves for values of n_1 less than 0.736 are of course not derived in the manner described above, but come from the theory in part (iii) of this section.) The purpose of giving curves for $\dfrac{r}{p_{cc}}$ (rather than for r alone) is to make the graph equally applicable to all concrete cube strengths.

It is to be noted that, because the graph is based on $\dfrac{d_2}{d} = \dfrac{1}{8}$, it follows that when n_1 is equal to $\dfrac{1}{0.875}$ (i.e. 1.145) or greater,

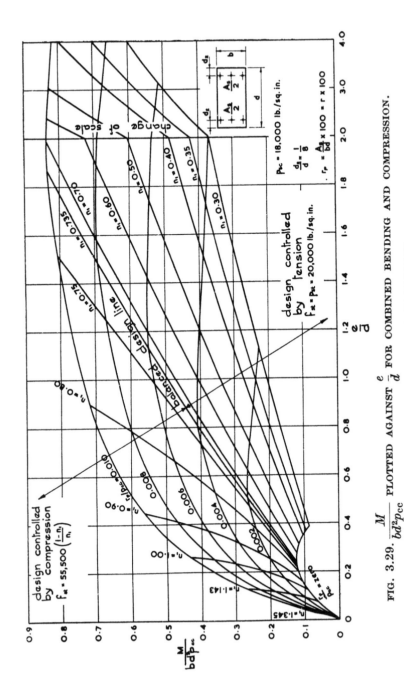

FIG. 3.29. $\dfrac{M}{bd^2 p_{cc}}$ PLOTTED AGAINST $\dfrac{e}{d}$ FOR COMBINED BENDING AND COMPRESSION.

SIMPLIFIED LOAD FACTOR METHOD

the section is uncracked. And also n_1 cannot exceed $\dfrac{1}{0\cdot 875 \times 0\cdot 85}$ (i.e. $1\cdot 345$), because the stress block would then come outside the section, and the equations would not apply.

For a section controlled by compression, C.P. 114 (1957) relates the permissible load and eccentricity (P and e) to the permissible load P_0 for an axially loaded column (as given by equation 3.2) in accordance with equation 3.82 as follows:

$$P = \frac{P_0}{1 + \left(\dfrac{P_0}{P_b} - 1\right)\dfrac{e}{e_b}}. \tag{3.82}$$

This equation appears to have been derived from the study given above, but with additional approximations, and gives values of P slightly lower than given by equations 3.80 and 3.81, particularly where $\dfrac{e}{d}$ approaches zero and the percentage of reinforcement is small. Nevertheless equation 3.82 does conform to experimental findings.

(iii) *Design Controlled by Tension Steel Yielding*

Figure 3.28c gives the diagrams of ultimate strain and working stress in the case where the design is controlled by the yielding of the tension steel. In the ultimate condition the yield of the tension steel reduces the extent of the stress block until the concrete reaches its full plastic state which leads in the end to collapse. Thus at working load all materials are at their maximum permissible stresses but the stress block is of less extent than for the case of "balanced design".

Equating forces, we have

$$P = p_{cc}b\,0\cdot 85d_n + A_{sc}(p_{sc} - p_{st})$$
$$= bdp_{cc}\left[0\cdot 85n_1\frac{d_1}{d} + \frac{r}{2p_{cc}}\left(p_{sc} - p_{st}\right)\right]. \tag{3.83}$$

And taking moments about the centre of the section,

$$M = Pe = p_{cc}b\,0\cdot 85n_1d_1\left(\frac{d}{2} - \frac{0\cdot 85n_1d_1}{2}\right) +$$
$$+ A_{sc}p_{sc}\left(\frac{d}{2} - d_2\right) + A_{st}p_{st}\left(\frac{d}{2} - d_2\right)$$
$$= bd^2p_{cc}\left[0\cdot 425n_1\frac{d_1}{d}\left(1 - 0\cdot 85n_1\frac{d_1}{d}\right) +$$
$$+ \frac{0\cdot 25r}{p_{cc}}\left(1 - \frac{2d_2}{d}\right)\left(p_{sc} + p_{st}\right)\right]. \tag{3.84}$$

Here again the labour of solving the two equations simultaneously can be wearisome, and a design graph is welcome. And here the curves of Fig. 3.29 *below* the "balanced design line" have been derived from equations 3.83 and 3.84, with p_{sc} taken as 18,000 lb/sq. in., and p_{st} as 20,000 lb/sq. in.

As a point of interest, equation 3.83 may be rewritten as

$$0.85n_1 = \frac{P}{p_{cc}bd_1} - \frac{r}{2p_{cc}d_1}(p_{sc} - p_{st})$$

and if the substitution of $0.85n_1$ is then made in equation 3.84, and the expression rearranged, we have

$$P = p_{cc}bd \left\{ \left(0.5 - \frac{e}{d} - Y \right) + \right.$$
$$\left. + \sqrt{\left(0.5 - \frac{e}{d} - Y \right)^2 + r \cdot \frac{p_{sc}}{p_{cc}} \frac{(d_1 - d_2)}{d} + Y \left(2\frac{d_1}{d} - Y \right)} \right\} \quad (3.84a)$$

where
$$Y = \frac{r}{2} \left(\frac{p_{st} - p_{sc}}{p_{cc}} \right).$$

This is the equation given in C.P. 114 (1957) for the case where the control is by tension.

A variety of worked examples of column design are given in Chapter 6.

ART. 3.4. SIMPLE SHEAR

When a load is applied to a member transversely, as for example the load on a beam, failure may occur due to shear even though the compressive and tensile bending stresses already discussed in Articles 3.2 and 3.3 are well within the permissible values. This matter was considered at length by Oscar Faber in his book *Reinforced Concrete Beams in Bending and Shear*[3] published in 1924; and the views he expressed 35 years ago still hold good today, and form the basis of all shear calculations for reinforced concrete.

Shear failure is most likely to occur where the shear forces are a maximum. This is generally near the ends of the member. Shear failure first appears by the formation of diagonal cracks, as indicated in Fig. 3.30.

For a homogeneous rectangular beam the distribution of bending and shear stresses, according to elastic theory, is as shown in Fig. 3.31a. The bending stresses vary uniformly from a maximum compression at the top to a maximum tension at the bottom Now the vertical distribution of transverse shear is such that the shear at any level is equal to the summation of axial forces arising from bending stress, compressive or tensile, above or below the level considered: thus for the homogeneous member the shear stress distribution is of parabolic form being a maximum at the neutral axis and tailing off to zero at the top and bottom fibres.

For a singly reinforced concrete beam the distribution of bending and shear stresses is as shown in Fig. 3.31b. As in the previous case the shear stress distribution above the neutral axis will be parabolic; but below the neutral axis, the maximum value of shear stress is maintained down to the level of the tension steel because there is no change in the axial bending force down to this

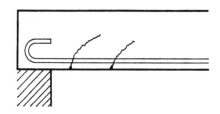

FIG. 3.30. SHEAR FAILURE OF BEAM, WHERE NO
SHEAR REINFORCEMENT PROVIDED

point. Below the tension steel the shear stress is of course zero. Now the total area of the shear stress diagram must equal the total applied shear acting across the section, so that if q is the maximum transverse shear stress at the section, then since the area within a parabola is equal to two-thirds the area of the enclosing rectangle, and using the same notation as before, we have

$$Q = qb(d_1 - d_n) + qb\tfrac{2}{3}d_n,$$

that is to say

$$Q = qb\left(d_1 - \frac{d_n}{3}\right)$$

and since

$$l_a = d_1 - \frac{d_n}{3},$$

we have

$$Q = qbl_a$$

or

$$q = \frac{Q}{bl_a}. \tag{3.85}$$

For a doubly reinforced concrete beam the distribution of transverse shear stress is not greatly different, and the form is shown in Fig. 3.31c. Here again it can be demonstrated that q has the value given by equation 3.85.

DIAGONAL TENSION

Consider a small square element of unit dimensions and unit thickness near the neutral axis of a beam where bending stresses are negligible. Due to vertical shear, there will be an upward shear stress q_1 on the left-hand face of the element and a balancing

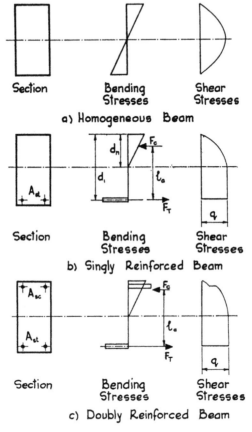

a) Homogeneous Beam

b) Singly Reinforced Beam

c) Doubly Reinforced Beam

FIG. 3.31. DISTRIBUTION OF SHEAR STRESS IN BEAMS

downward shear stress q_1 on the right-hand face of the element as shown in Fig. 3.32a. These stresses form a couple tending to rotate the element in a clockwise sense. Equilibrium can only be obtained by a similar system of complementary stresses acting horizontally. There are therefore horizontal shear stresses q_2, complementary to the vertical shear stresses q_1 as shown in Fig. 3.32b such that $q_1 = q_2 = q$.

If now the adjacent stresses q_1 and q_2 at any corner of the element are combined vectorially, since $q_1 = q_2 = q$, the resultant elemental forces are inclined at 45° and of magnitude $q\sqrt{2}$, in the one case compressive and in the other case tensile as shown in Fig. 3.32c; and since the areas on which these resultant forces

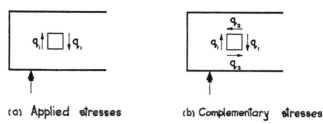

(a) Applied stresses   (b) Complementary stresses

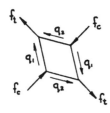

(c) Resultant forces   (d) Resultant stresses

FIG. 3.32. DIAGRAM OF UNIT SHEAR STRESSES IN BEAM

act are $\sqrt{2}$, being the diagonal of a unit square, the diagonal compressive and tensile stresses are $\dfrac{q\sqrt{2}}{\sqrt{2}} = q$, so that from equation 3.85 we see that

the diagonal compressive and tensile stresses $= \dfrac{Q}{bl_a}$.   (3.86)

Since concrete is strong in compression but weak in tension, the first indications of distress will be due to the diagonal tensile stress and will appear in the form of inclined tension cracks as shown in Fig. 3.30.

CP. 114 (1957) Amendment 1965 generally requires all shear forces to be carried by stirrups or bent-up bars as described later. However for slabs, footings, and members of secondary importance the shear strength of the concrete may be relied upon to the values given in Table 3.1. As shown above these permissible shear stresses are in fact nothing more than the safe tensile stress of the various concrete mixes: and this is the only instance in reinforced concrete design where structural reliance is placed upon the tensile strength of concrete. Even so, the Code recommends a conservative approach in calculating the shear resistance of members where no shear reinforcement is provided: otherwise the margin of safety may be less than desirable.

TABLE 3.1. *Permissible shear stress* (q)

Nominal mix	Safe shear stress
1:1:2	130 lb/sq. in.
1:1½:3	115 lb/sq. in.
1:2:4	110 lb/sq. in.

In slabs, where the shear per unit width is limited by considerations of bending, and b is relatively great, it is unusual for $\dfrac{Q}{bl_a}$ to exceed the permissible values for q given in Table 3.1. However, with beams, the end shear is generally very much greater, yet for reasons of economy the beam rib is kept as narrow as possible, generally only wide enough to accommodate the main longitudinal reinforcements: and under these circumstances $\dfrac{Q}{bl_a}$ can be very much greater than the permissible values, and it then becomes necessary to reinforce the member appropriately. The reinforcement usually takes the form either of stirrups (Fig. 3.33a), or bent-up bars (Fig. 3.34a), or a combination of both.

STIRRUPS

In Fig. 3.33b an arrangement is indicated diagrammatically to illustrate how concrete and stirrups act together to resist transverse shear. The concrete, by virtue of its high compressive strength, acts as the diagonal compression members of a lattice girder system where the stirrups act as vertical tension members. The diagonal compression force is such that its vertical component is equal to the tension force in the stirrup.

If the spacing or pitch of stirrups s is equal to the lever arm of the beam l_a it is clear that the tension per stirrup will be equal to the total shear Q across the section, since any diagonal tension crack across the beam at 45° to its axis must cut one stirrup (at this spacing) and cannot cut two. Similarly if the stirrups are spaced twice as close so that s equals $\dfrac{l_a}{2}$ then a diagonal crack will cut two stirrups and the shear strength of the member will be doubled. Thus

$$Q = \frac{p_{st} A_w l_a}{s} \qquad (3.87)$$

where p_{st} is the permissible tensile stress in the stirrups, A_w is the total cross-sectional area of the stirrup. This is the formula given in C.P. 114 (1957).

Stirrups must not be spaced further apart than the lever arm of

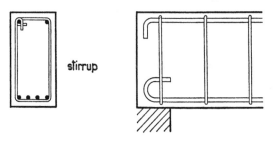

stirrup

(a) Simple stirrups.

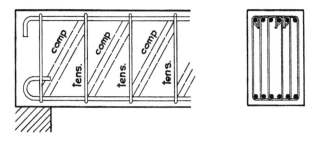

(b) Stirrup action.   (c) Compound stirrups.

FIG. 3.33. STIRRUPS

the member, otherwise a diagonal crack could form entirely clear of any two consecutive stirrups. For practical reasons stirrups should never be less than $\frac{5}{16}$ in. diameter or they are likely to be deformed in the concreting operations and finish with inadequate side cover: on the other hand stirrups greater than $\frac{1}{2}$ in. diameter are troublesome to fix except in special circumstances.

The cross-sectional area of a stirrup A_w is the total area of all strands at any one position along the beam, and whereas Fig. 3.33a shows only two strands, cases arise where 4,6 or more strands become necessary as shown in Fig. 3.33c.

BENT-UP BARS

Near the ends of a beam the positive bending moment decreases, with a consequent decrease in the amount of main tension steel required at the bottom face. Part of the main steel then becomes available to be bent up in to the top of the beam where it can assist in resisting any negative moment at the support as shown in

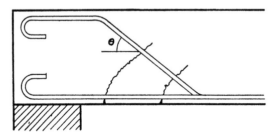

(a) Bar cutting planes of diagonal tension

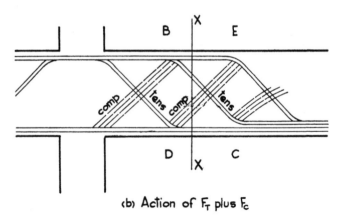

(b) Action of F_T plus F_c

FIG. 3.34. BENT-UP BARS

Fig. 3.34a. Reinforcement bent up in this way crosses the planes of diagonal tension and is thus available to contribute towards the shear strength of the member.

If A_{st} is the area of bent-up bar, or bars, θ is the inclination of the rod to the horizontal, and p_{st} is the permissible stress in the steel, then the vertical component of tension in the bar available for resisting vertical shear is given by

$$F_T = A_{st} p_{st} \sin \theta. \qquad (3.88)$$

Now it is clear from Figs. 3.32c and d that diagonal tensions and compressions are complementary to one another. Therefore any enhanced diagonal tension provided by the bent-up reinforcement must be associated with a complementary diagonal compression. This is indicated diagrammatically by the dotted bands of compression in Fig. 3.34b for a system involving three panels of bent-up bars. It will be seen that the shear plane XX is crossed both by the diagonal tension BC and also the diagonal compression DE, and the vertical components of both these are available for resisting shear. Thus the arrangement in Fig. 3.34b is known as a Double System and gives a shear resistance of

$$Q = F_T + F_C$$
$$= 2F_T$$
$$= 2A_{st}p_{st} \sin \theta. \tag{3.89}$$

If the bent-up bars were spaced further apart, so that B came to the right of D, and E to the right of C, etc., the addition of the effects of the diagonal tensions and compressions would no longer be admissible, and consequently the arrangement would act as a Single System. Single Systems are generally less efficient in use of reinforcement; nor do they make maximum use of the available compressive strength of the concrete.

In any system of bent-up bars, if the arguments given above are to be valid, the stress in the bars must be developed over the whole of their inclined lengths. Thus in bar BC there must be sufficient anchorage at the upper end by continuing the bar horizontally along the top of the beam and providing a substantial hook. Similar consideration is required at the bottom end of BC, though the stress in the horizontal part of the bar (here due to its action in resisting the positive bending moment) will in many cases alleviate the problem of anchorage. Consideration should be given to the total tensile stress in the bar due to the summation of its two duties – shear resistance and bending resistance.

DIRECT INCLINED COMPRESSION

Figure 3.35 illustrates diagrammatically the action of direct inclined compression separately for the cases of a uniformly distributed load and a central point load. For the uniform load, the system acts as a parabolic arch, with the bottom reinforcement, in conjunction with the vertical end reactions, providing the end thrusts. For the point load the system acts as a triangulated truss with the reinforcement acting as the bottom cord. Other systems of loading would effect a compromise between these two.

As with other arches or trusses, the ratio of rise to span influences the shear resistance: clearly, in the limit, a flat arch or a flat truss have no resistance as such. Similarly with a reinforced concrete beam, the contribution to shear strength arising from direct inclined compression varies according to the depth/span ratio of the member.

It is uncertain from C.P. 114 (1957) whether it is intended that

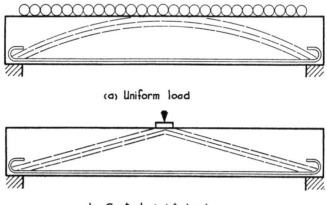

(a) Uniform load

(b) Central point load

FIG. 3.35. DIRECT INCLINED COMPRESSION

any reliance should be placed upon the effect of direct inclined compression in determining the shear resistance of a member; though provided the end anchorage of the bottom reinforcement is efficient there is no doubt as to the usefulness of such internal action: and shear resistance by direct inclined compression places no reliance on the tensile strength of the concrete, and can accordingly be added to the shear resistance from diagonal tension, or from any system of reinforcements.

COMBINED EFFECT OF VARIOUS SYSTEMS

As previously explained at part (iii) of Article 3.1, if reliance is put on the tensile strength of concrete, the corresponding stress in the reinforcements has to be so low as to make the action of the reinforcements extremely inefficient: otherwise the concrete will crack and can no longer display even its limited tensile strength. Accordingly if the permissible diagonal tension resistance of the concrete (as given in Table 3.1) is to be exceeded, it follows that it has to be entirely ignored. Reliance has then to be placed on the systems of shear reinforcement and direct inclined compression.

From Fig. 3.36 it can be seen that the arrangement of stirrups and bent-up bars can function simultaneously. The diagonal tension cracks indicated will in no way interfere with the diagonal compression bands indicated in Fig. 3.33b, while the stirrups will act satisfactorily in tension; and the tension cracks are a necessary condition for the bent-up bars to function at full stress. And as previously indicated, since direct inclined compression places no reliance on the tensile strength of the concrete, it is admissible to consider resistance by this action as additional to the resistance

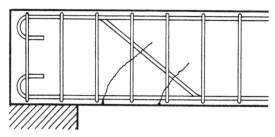

FIG. 3.36. COMBINED SYSTEM OF SHEAR REINFORCEMENT

provided by stirrups and bent-up bars. There can be no doubt that the action of direct inclined compression will influence beneficially the over-simplified lattice girder arrangement indicated diagrammatically in Fig. 3.34, but this view is apparently not yet generally accepted or taken into account in design methods.

MAXIMUM SHEAR RESISTANCE OF REINFORCED MEMBER

Having exceeded and consequently ignored the diagonal tension strength of the concrete, and having introduced reinforcements to cope with all internal tension forces, there must ultimately arise a limit for the total shear resistance of the member when the diagonal compressive stresses in the concrete reach the maximum value permissible.

The maximum resistance will occur when the compressions are at 45° to the axis of the member, and when the bands shown diagrammatically in Figs. 3.33b and 3.34b are spread along the length of the beam to give a continuous zoning of diagonal compressions. The compressions should also then be the full width of the member, which of course would mean considerably greater stresses in the immediate proximity of the shear reinforcements. In view of the uncertainties which may arise in this condition C.P. 114 (1957) limits the maximum stress given by equation 3.85 to four times the permissible shear stress for the concrete alone.

Many examples of calculations for shear strength are included later in the book.

ART. 3.5. TORSION

Torsion arises in most reinforced concrete members though frequently as a secondary effect. In columns it may be the result of unbalanced movements due to shrinkage or temperature. In beams it arises from secondary beams coming in one one side only, or from loads or spans on either side of the beam being unbalanced, a condition certain to occur at all edge beams of a floor or roof. Often the stresses from torsion may be of little consequence requiring no special calculation, and indeed in T- or L-beam

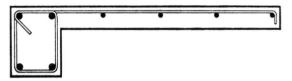

FIG. 3.37. BEAM SUBJECTED TO TORSION

construction, the slab may provide sufficient lateral or restraining influence on the beam to practically nullify the twisting actions. However, examples do arise of secondary torsion where careful investigation is required: and of course primary torsion occurs in cases such as where columns are subject to eccentric lateral thrusts and where single beams are used to support eccentric loads like a walkway or cantilevered canopy as indicated in Fig. 3.37.

Concrete being a brittle material is ill-suited to accommodate the effects of torsion, and failure occurs by tension on a helical plane at 45° to the axis of the member. The failure is abrupt and occurs without warning. In view of the inherent brittleness of concrete and its proneness when unreinforced to crack under the effects of shrinkage or sudden jarring, it is generally imprudent to rely on the torsional strength of any concrete member which does not have a reasonable minimum percentage of reinforcement.

Numerous torsion tests have been carried out in many parts of the world, though unfortunately many of these have been on specimens rather too small for the purpose. All tests have shown, as one might expect, that the form of reinforcement most effective for resisting torsion is a 45° spiral arranged so as to cut the helical planes of potential failure at right angles. However, except for the limited application of such reinforcement to circular members, a system of spiral reinforcement is difficult to achieve in practice, and would be required as two series of counter-spirals in order to cater for stress reversals.

For practical reasons reinforcement is provided in the form of longitudinal bars situated close to the outer faces of the member, and transverse stirrups. The longitudinal bars are best uniformly distributed about the perimeter of the member and not necessarily limited to the four corners of a rectangular section. The applied torque is frequently a maximum at the supported ends of the member and it is important here for the longitudinal bars to be well anchored and provided with substantial hooks (in much the same way as good end fixity is an essential requirement in supporting a simple cantilever). The stirrups are best arranged as close together as possible and certainly not further apart than half the minimum dimension of the enclosed concrete core. The stirrups should be well anchored by effectively lapping the strands and hooking over the longitudinal bars.

Tests show that members reinforced in this way carry very much greater torques than unreinforced members, possibly as a result of the inner core benefiting from the corseting action of the reinforcements. Failure is no longer abrupt, and twisting strains of considerable magnitude can be accommodated without disintegration of the concrete. However, all tests have shown that whatever the amount of reinforcement, initial cracking of the member occurs at the same torque as with the same sized member unreinforced.

It is clear that in any reinforced concrete member with reinforcements arranged longitudinally and transversely, in order to influence economically the strength of a member liable to suffer tension failure on planes at 45° to the axis, the required percentages of longitudinal and transverse reinforcements should be equal. If the transverse reinforcements exceed the longitudinal reinforcements they will be in excess and understressed; and vice versa.

Many workers have sought a rational basis for designing reinforced concrete members subject to torsion. A review of these works appears to indicate that the more involved theoretical studies are founded on an incomplete recognition of the internal straining actions. There seems little doubt that concrete in torsion at high stresses behaves inelastically; and indeed tests to failure indicate that the torsional resistance of concrete reinforced with a reasonable minimum percentage of steel conforms more nearly with that of a fully plastic section, in accordance with the formula

$$T = \frac{b^2}{2}\left(a - \frac{b}{3}\right)f_s \qquad (3.90)$$

where T is the applied torque, a and b are respectively the longer and shorter sides of the core enclosed within the stirrups, and f_s is the shear stress in the concrete across the section. As already indicated in Article 3.4 such a shear stress gives rise to complementary diagonal tensile and compressive stresses of intensity equal to that of the shear stress, and since the resistance of the member will be limited by the tensile strength of the concrete, we have, for the safe torsional resistance,

$$T = \frac{b^2}{2}\left(a - \frac{b}{3}\right)q \qquad (3.91)$$

where q is the permissible tensile or shear stress for the particular concrete mix as given in Table 3.1. Tests have shown that initial cracks of reinforced members subject to torsion, whatever the percentage of reinforcement, follows very closely on formula 3.91 above, but substituting for q the ultimate tensile strength of the concrete (or approximately 1/10th the compressive cube strength).

The best shape for any member in resisting torsion is obviously a circle, and the best rectangular shape is the square. It is doubtful whether formula 3.91 should be applied where the ratio $\frac{a}{b}$ exceeds 2·0; certainly there is no indication of its suitability where $\frac{a}{b}$ exceeds 3·0. With T-beams and L-beams very high stress intensities develop at the re-entrant angles under the flange and these angles should be eased with generous haunches. The flanges of T- or L-beams are shown by experiment generally to give no increase in torsional resistance over and above that of the basic rectangle.

L. Turner and V. C. Davies[4], in a Selected Engineering Paper published by the Institution of Civil Engineers in 1934, give careful consideration to the influence of reinforcements in rectangular members subject to torsion. In their Paper they analyse the findings of various experimenters in Germany, America and Japan as well as the results of their own tests in London. They conclude that if $\frac{1}{2}p$ is the percentage by volume of longitudinal steel, and $\frac{1}{2}p$ is also the percentage by volume of transverse steel, so that p is the total percentage of steel, then the reasonable minimum percentage of reinforcement p for a member subject to torsion is 1 per cent. Further that the effect of the reinforcement is to increase the torsional resistance given by the concrete acting on its own in accordance with the formula

$$T_{c+s} = T_c(1 + 0.25p) \qquad (3.92)$$

where T_{c+s} is the total safe effective torsional resistance of concrete plus steel, and T_c is the safe torsional resistance of the concrete acting on its own. Turner and Davies considered that equation 3.92 would be valid up to values of 2·0 for p.

Later tests carried out at the University of Nebraska and reported by G. C. Ernst[5] in the Journal of the American Concrete Institute, October 1957, confirm consistently the suitability of formula 3.91. These tests also indicate that formula 3.92 gives conservative results on the ultimate torsional resistance even with values of p up to 3·2 per cent giving a factor against failure of about 5 or 6 times. It must be appreciated however that the torque to produce initial cracking is sensibly constant whatever the percentage of reinforcement, and any enhanced values arising from the use of formula 3.92 will give a lower factor of safety against initial cracking. However, the test results reported by Ernst indicate that for values of p up to 3·2 per cent the factor of safety against initial cracking (not failure) appears to be well in excess of 2·0; and this may be regarded as giving a reasonable margin.

With a beam, the vertical shear stress across the concrete due to torsion will be upwards on one side of the beam and downwards on the other. On one side, therefore, the stress due to torsion will act in the same sense as the simple shear arising from normal transverse loading, and the arithmetical sum of these two shearing stresses must not exceed the permissible shear stress q given in C.P. 114 (1957) for the particular concrete mix. As the study given before for the calculation of torsional resistance depends on the concrete not cracking, it is not admissible when considering the simple shear strength of a beam subject to combined torsion and normal loading to allow anything for the effect of stirrups or bent-up bars, since neither of these can act without the concrete first cracking.

While formula 3.92 for the effect of reinforcements might perhaps be criticised for its empirical nature, it can be gratefully accepted as essentially practical and giving assuredly safe results within the limits for which it is intended. On the other hand, more elegant expressions derived by Rausch and by Cowan[6, 7] need great caution in application since although by coincidence there may be some evidence to substantiate the results of these formulae at ultimate conditions when the behaviour of the concrete is plastic, it must be appreciated that the theories are based on stress conditions which are essentially elastic and the true mathematical solution to the problem of torsion in rectangular sections requires correction for the warping deformation of cross-sections originally plane. Furthermore the admissibility of adding the ultimate resistances of the steel and the concrete as calculated separately appears dubious. Accordingly unless the designer has considerable experience of reinforced concrete members subject to torsion, he would be wise in the present state of knowledge to restrict himself to the use of formulae 3.91 and 3.92. Chapter 4 includes a worked example for a beam subject to torsion.

ART. 3.6. BOND

When a straight steel bar of round section is embedded in a substantial block of concrete as shown in Fig. 3.38, it is found that

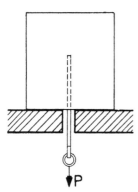

a considerable force P is required before the bar will pull out of the concrete even though the bar be perfectly plain and not provided with serrations or other intentional roughnesses. Indeed, if the length of embedment exceeds forty times the diameter of the bar, or thereabouts, depending on the quality of the steel and the strength of the concrete, the bar may fail in tension leaving the embedded length behind in the concrete block. The frictional force between the steel and the concrete is known as *bond*. And if A is the surface area of steel embedded, and P is the pull at failure resulting solely from the bar pulling out of the concrete, then by definition

FIG. 3.38. PULL-OUT
TEST FOR BOND

$$\frac{P}{A} = \text{ultimate average } bond \text{ stress.} \qquad (3.93)$$

Bond is caused by two phenomena. The first of these is *adhesion*, as evidenced by the difficulty experienced in cleaning a shovel that has been used for mixing concrete and left a month or so before cleaning. The second is the *grip* of the concrete on the steel resulting from the shrinkage of the concrete in setting, in just the same way as an iron tyre grips on a wooden wheel if applied hot and allowed to cool. And the better the grip the better will be the adhesion, just as when gluing together two pieces of wood a carpenter clamps them firmly in position until the glue has hardened. Thus

$$\text{bond} = \text{adhesion} + \text{grip.}$$

Tests show that if the block in Fig. 3.38 is not of sufficient size to allow very ample side cover to the bar, failure occurs by partial disruption of the block before the full ultimate bond stress has had a chance to develop. Fig. 3.39 gives the results of a series of

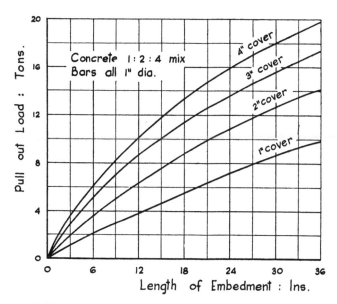

FIG. 3.39. EFFECT OF EMBEDMENT LENGTH ON PULL-OUT
LOADS

tests and indicates how the ultimate loads varied with different lengths of embedment and different side covers of concrete.

Bond then is influenced mainly by four factors as follows:

(i) *Roughness of steel surface*

Rusty steel can give bond strengths about one and a half times the values of new steel as rolled, whereas polished steel shafting gives values very much less (only about half). C.P. 114 (1957) allows bond stresses to be increased by 40 per cent for high-bond bars of suitable pattern.

(ii) *Concrete mix*

Rich concrete mixes have greater adhesion than weaker mixes. And rich mixes, by virtue of the greater water content required for a given water/cement ratio, have high shrinkage characteristics and consequently give better grip.

(iii) *Shrinkage*

Beneficial shrinkage arises from rich and moderately wet mixes of concrete. However, exaggerated shrinkage from the use of excessively wet and sloppy mixes is to be avoided, since the tensile strength, and thus the shear strength of the concrete suffers. It is important to note that when concrete sets and hardens in water the shrinkage is very considerably reduced (*see* Article 2.8).

(iv) *Cover of concrete*

The need for adequate cover is indicated by Fig. 3.39. Due to the strain of a bar in tension, the concrete immediately adjacent must be stressed in tension well beyond failure, so that no reaction can be provided to the tension without there being other concrete remote from the bar itself capable of acting in compression. Further, if a bar is too close to the surface of the concrete, the concrete in gripping the bar by shrinkage will fail circumferentially with the formation of radial cracks. This is one of the reasons why C.P. 114 (1957) requires the cover of concrete to any bar to be never less than the diameter of the bar itself, and frequently more. Cracking is controlled to some extent by the use of small diameter binding reinforcements such as stirrups, links, etc.

AVERAGE BOND

In pull-out tests as indicated by Fig. 3.38 it has been found that the unit bond resistance depends on the length of embedment of the bar. Thus in the ultimate condition the average bond stress developed in a bar of short embedment is greater than for a bar of longer embedment, as shown in Fig. 3.40. This indicates that at first a greater unit bond resistance is developed nearer to the surface of the concrete, than deeper in; which is exactly as one would expect since not only will the unit stretch of the bar at the surface of the concrete be very much greater than deeper in where the stress will have reduced, but the axial movement of the bar at the surface will be the summation of all the stretching strains suffered within the block. Thus while the stress given by formula 3.93 is an average value, locally at the surface of the concrete the actual stress is considerably greater.

The bond stress has been shown by pull-out tests[8] to vary along the embedded length of the bar according to an exponential law. And when the maximum bond stress near the concrete surface reaches the ultimate bond stress, slip occurs here, and the position of maximum bond stress is transferred progressively further along the bar, deeper into the concrete: but although there has been local slip near the concrete surface, a residual adhesion persists here, contributing to an increased pull-out resistance.

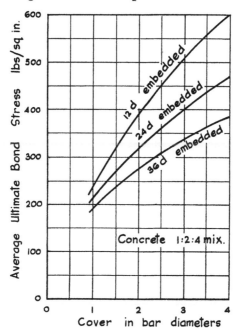

FIG. 3.40. EFFECT OF CONCRETE COVER ON
AVERAGE BOND STRESS

C.P. 114 (1957) gives values as shown in Table 3.2 for permissible bond stresses using ordinary rods as rolled for the nominal concrete mixes shown, provided adequate covers are given.

TABLE 3.2. *Permissible bond stresses*

Nominal concrete mix	Average bond	Local bond
1:1:2	150 lb/sq. in.	220 lb/sq. in.
1:1½:3	135 lb/sq. in.	200 lb/sq. in.
1:2:4	120 lb/sq. in.	180 lb/sq. in.

In the case of high-bond bars, the bond stresses may be increased by 40 per cent. Caution is necessary when dealing with concrete which is to set and harden in water, and it may then be desirable to work to very much lower bond stresses.

(i) *Bars in tension*

If the anchorage of a tension bar in concrete is to be such that the permissible average bond stress is reached at the same time as the permissible tensile stress in the bar, and if the bar diameter is D and the length of embedment is l, then

$$\text{bond stress} \times \pi Dl = \frac{\pi D^2}{4} \cdot f_{st},$$

so

$$l = \frac{D \times f_{st}}{4 \times \text{permissible average bond stress}}. \qquad (3.94)$$

This is the minimum anchorage length required by C.P. 114 (1957). Furthermore a bar should always extend at least $12D$ beyond the point at which it is no longer required to resist stress.

From formula 3.94 it is possible to calculate the minimum embedment lengths necessary to restrict the bond to permissible stresses, using different concrete mixes and different steels. Such results are given in Table 3.3. Hence for a particular nominal concrete mix with the steel stressed at any section as tabulated, the bar is to be extended beyond that section by the length indicated.

TABLE 3.3. *Minimum bond lengths for round bars in tension and compression*

Nominal concrete mix	Average bond stress	Steel in tension			Steel in compression		
		Stress 20,000 lb/sq. in.	Stress 25,000 lb/sq. in.	Stress 30,000 lb/sq. in.	Stress 16,000 lb/sq. in.	Stress 18,000 lb/sq. in.	Stress 25,000 lb/sq. in.
1:1:2	150	$33.3D$	$41.7D$	$50.0D$	$21.4D$	$24.0D$	$33.4D$
1:1½:3	135	$37.0D$	$46.3D$	$55.5D$	$23.7D$	$26.7D$	$37.0D$
1:2:4	120	$41.7D$	$52.1D$	$62.5D$	$26.7D$	$30.0D$	$41.6D$

However, if a bar stressed in tension is mechanically anchored by having its end bent to a hooked shape, the bond length required may be reduced by the anchorage value of the hook. The dimensions and anchorage values for suitable hooked ends are given in Fig. 3.41. The internal radii of $2d$ are appropriate for mild steel bars only: with high-yield bars an internal radius of at least $3d$ is necessary. Usually the U-type of hook is to be preferred. Where the hook of a mild steel bar fits over to grasp a main reinforcement or other adequate bar, the radius of the bend may be reduced to that of the grasped bar.

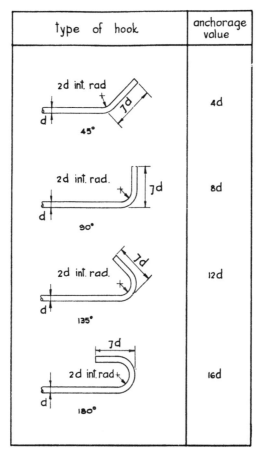

type of hook	anchorage value
2d int. rad / 7d / d / 45°	4d
2d int. rad. / 7d / d / 90°	8d
2d int. rad. / 7d / d / 135°	12d
7d / 2d int. rad / d / 180°	16d

FIG. 3.41. DIMENSIONS AND ANCHORAGE VALUES FOR
HOOKED ENDS OF MILD STEEL BARS

Figure 3.42 illustrates the eccentric nature of anchorage provided by an end-hook, and the potential deformations α, β and γ. If these deformations are not adequately resisted, the bar may burst outwards, spalling the concrete. The deformations α and γ are frequently catered for by applied column loads or other reactions: nevertheless this is not so when reinforcements are stopped off part way along the length of a member, and α then necessitates tying in with stirrups, while γ is generally taken care of by the concrete mass. An end cover of 2 in. is normally regarded as sufficient to take care of β.

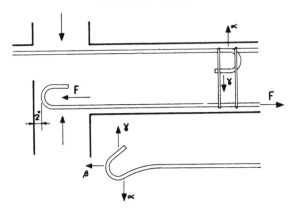

FIG. 3.42. DEFORMATION OF HOOKED END ON TENSION STEEL

(ii) *Bars in compression*

Bars in compression require shorter bond lengths than bars in tension. This is partly because the bar in compression swells laterally and achieves a better grip (like a wedge); it is also due to the end bearing where the bar stops and butts against the concrete. Thus

$$l = \frac{D \times f_{sc}}{5 \times \text{permissible average bond stress}}.$$

$$(3.95)$$

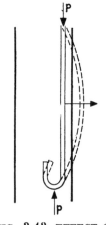

FIG. 3.43. EFFECT OF HOOKED END ON COMPRESSION STEEL

C.P. 114 (1957) further makes the provision that a bar should extend at least $12D$ beyond the point at which it is no longer required to resist stress. From formula 3.95 the minimum embedment lengths can be calculated for different steels in different concretes. Such results are given in Table 3.3.

Hooked ends for compression bars serve but little purpose and may indeed be detrimental. They deprive the bar of its proper axial end bearing, and tend to cause outward buckling of the bar with consequent spalling as shown in Fig. 3.43.

LOCAL BOND STRESS

In a member in bending, such as a beam, it is clear that because the bending moment varies along the length of the member, the stress in the tension reinforcements must also vary. This is achieved by a transfer of force to or from the concrete by development of bond stress. This is associated with the axial shear complementary to the transverse shear. By considering the rate of change of bending moment in unit length of the member (i.e. the shear), it can be shown that for a member of constant depth the

$$\text{local bond stress} = \frac{Q}{l_a o} \tag{3.96}$$

where Q is the total shear across the section, l_a is the lever arm, o is the sum of the perimeters of the bars in the tensile reinforcements. Where the local bond stress is precisely calculated in this manner, the higher local values given in Table 3.2 are permissible.

STOPPING-OFF BARS

Consider the arrangement of tension reinforcements in a member subject to bending as shown in Fig. 3.44. Suppose the bending

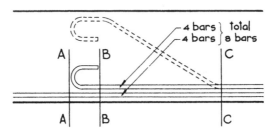

FIG. 3.44. HIGH BOND STRESS WHEN BARS STOPPED OFF

moment decreases from right to left, and that at CC the moment has reduced so that the inner layer of bars can apparently be terminated as shown. Clearly at BB, due to the strain in the concrete, both layers of reinforcements will be stressed approximately equally. And at AA, supposing BB and AA are close to one another, the stress in the outer layer bars will be suddenly doubled, while clearly the stress in the inner layer bars has had to tail off abruptly. Under these circumstances very high bond stresses may arise unless the inner bars are terminated approximately at a point of contraflexure. Alternatively the surplus bars can be terminated by bending them at an easy slope out of the tension zone and anchoring them in the compression zone as shown dotted. Bars bent in this manner can of course be utilised to resist shear forces as described in Article 3.4.

LAPS IN BARS

For convenience it frequently becomes necessary to transfer stress from one bar to another and this is done by lapping the bars side by side over a bond length as given by formulae 3.94 or 3.95. C.P. 114 (1957) further requires that the minimum lap lengths are $30D$ for bars in tension and $24D$ for bars in compression. Now formulae 3.94 and 3.95 depend on the bars being properly embedded in the concrete, i.e. intimate and continuous contact must be achieved between the bars and the concrete for the whole of their lengths. In this way the stress passes from one bar into the concrete and then from the concrete into the second bar. This clearly is not achieved if the lapped bars are spliced closely together over the bond length; and the bars should either be kept apart by about one bar diameter or alternatively the bond length should strictly be increased by some 15 per cent.

STIRRUPS, LINKS, ETC.

In normal binding it is impracticable to provide the bond lengths indicated in Table 3.2. Instead it is generally adequate to bend the binding bar through an angle of 180° round a bar of at least its own diameter, and continue the bar beyond the end of the curve for a length not less than $8D$. Hooked ends of this type should be situated in the compression zone of the member, where the concrete gets an enhanced mechanical grip. For members in torsion, more positive anchorage is necessary than described above: *see* Article 3.5.

REVERSED MOMENTS IN FRAMES

Bond stress considerations can prove critical in the rigid joints of frames. Consider a multi-storey building subject to horizontal

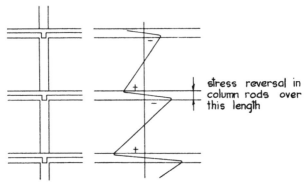

stress reversal in column rods over this length

FIG. 3.45. HIGH BOND STRESS AT MOMENT REVERSAL

sway from wind effects. The sway forces have to be resisted by moments in the columns, the moments reaching maximum positive and negative values immediately above and below each floor level, as shown in Fig. 3.45. Thus within the depth of the incoming floor beam the stress in the column reinforcements has to change from full tension to full compression. In order to avoid excessive bond stresses it may be necessary to use many small bars for the column reinforcement, or provide deeper floor beams than would be required by other considerations.

ART. 3.7. STIFFNESS OF MEMBERS

The greater part of the analysis of reinforced concrete structures concerns frames with rigid joints. At each joint the apportionment of bending moments between the various incoming members

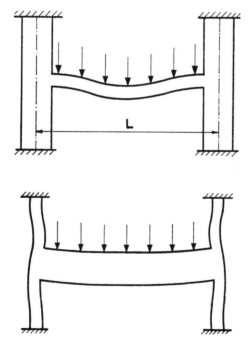

FIG. 3.46. RELATIVE STIFFNESSES OF BEAMS AND COLUMNS

depends on their relative stiffnesses. Clearly, by reference to Fig. 3.46, if the columns are very much stiffer than the beam,

the moments at the ends of the beam will approach $\dfrac{WL}{12}$, whereas
if the beam is much stiffer than the columns, the moment at the
ends of the beams will be much less, in the limit approaching zero.

Strictly the relative stiffnesses of members depends on the ratio
of their flexural rigidity and length, and the conditions of fixity
at the remote ends of the members. Flexural rigidity is given by
the product EI; and where the proportions of reinforcements in
the members do not vary excessively, E can be regarded as near
enough constant so that the relative stiffnesses becomes a function
of the quotients $\dfrac{I}{L}$ for the various members, provided of course
the members do not vary in cross-section along their lengths. I,
the second moment of area, may conveniently be calculated in all
cases on the entire concrete section, ignoring the reinforcement.

At any point where beams frame into a column, the effects of
fixity at the remote ends of the various members is catered for in
C.P. 114 (1957) by means of a simplified table. This is reproduced
as Table 6.1. Alternatively, more exact estimates can be made of
the distribution of moments. These matters are considered further
in Chapter 6, where worked examples are given.

REFERENCES

1. British Standard Code of Practice C.P. 114 (1957) Amended 1965,
   The Structural Use of Reinforced Concrete in Buildings. British
   Standards Institution (1965).
2. HOGNESTAD, HANSON and MCHENRY. Concrete Stress Distribution in
   Ultimate Strength Design. *Journal of American Concrete Institute.*
   Vol. 4. (December, 1955).
3. FABER, OSCAR. *Reinforced Concrete Beams in Bending and Shear.* Concrete Publications Ltd. (1924).
4. TURNER, LESLIE and DAVIES, V. C. *Plain and Reinforced Concrete in
   Torsion, with particular reference to Reinforced Concrete Beams.*
   Selected Engineering Paper, No. 165. Institution of Civil Engineers
   (1934).
5. ERNST, G. C. Ultimate Torsional Properties of Rectangular Reinforced
   Concrete Beams. *Journal of the American Concrete Institute,* No. 4.
   Vol. 20. (October, 1957).
6. COWAN, H. J. An Elastic Theory for Torsional Resistance of Rectangular Reinforced Concrete Beams. *Magazine of Concrete Research.*
   Cement and Concrete Association, London. No. 4. (July, 1950).
7. COWAN, H. J. Torsion in Reinforced and Prestressed Concrete Beams.
   Journal of the Institution of Engineers, Australia. No. 9. Vol. 28. (September, 1956).
8. HAWKES and EVANS. Bond Stresses in Reinforced Concrete Columns
   and Beams. *The Structural Engineer.* No. 12. Vol. 29. (December
   1951).

CHAPTER 4

BEAM DESIGN

ART. 4.1. NON-CONTINUOUS BEAMS

THE greatest benefits of reinforced concrete appear in monolithic structures cast-in-situ, where the beams and columns develop continuity, and where ribs and slabs function in unison to give T-beam action. Nevertheless non-continuous beams occur in such cases as single-span bridges or floors supported on masonry walls; and rectangular beams occur as lintols, or members of space-frames. A variety of worked examples of such cases now follow, all based on the theory given in Chapter 3. The calculations are made both by the Elastic method and by the Plastic method, and in order that the results may be directly compared, the following design stresses are used throughout:

$$p_{cb} = 1,000 \text{ lb/sq. in.},$$
$$p_{st} = 20,000 \text{ lb/sq. in.},$$
$$p_{sc} = 18,000 \text{ lb/sq. in.}$$

The effect of working to different design stresses for members in flexure is considered later in Article 5.2 for slabs; but the same principles shown there apply equally to beams.

RECTANGULAR BEAMS: SINGLY REINFORCED

Consider the case of a lintol, effective span 14 ft. to carry a uniformly distributed superimposed load of 12,000 lb. (*see* Fig. 4.1).

Total load carried by beam is

Superimposed load	=	12,000 lb.
Self-weight, say 14 ft. × 135 lb/ft.	=	1,900 lb.
		13,900 lb.

Centre moment is $M = \dfrac{Wl}{8} = 13,900 \times \dfrac{14}{8} \times 12 = 292,000$ in. lb.

The shear at the ends is $Q = \dfrac{13,900}{2} = 6950$ lb.

(i) *Elastic design*

Try a lintol size 16 in. deep × 8 in. wide overall. From equation 3.10

$$A_{\mathrm{st}} = \frac{M}{p_{\mathrm{st}}l_{\mathrm{a}}} = \frac{292{,}000}{20{,}000 \times (0\cdot86 \times 14\frac{1}{4})} = 1\cdot19 \text{ sq. in.}$$

This is given by two $\frac{7}{8}$ in. dia. rods (area $1\cdot20$ sq. in.).

$$\frac{M}{bd_1{}^2} = \frac{292{,}000}{8 \times 14\frac{1}{4}{}^2} = 180.$$

This is less than 184, the economic value for the design stresses to which we are working, as seen by reference to Fig. 3.8. Therefore the compressive stress in the concrete is something slightly

Design Method	Section	Reinforcement
Elastic	16" x 8"	2- 7/8" dia
Plastic	16" x 8"	2- 7/8" dia
Plastic	14" x 8"	2- 1" dia

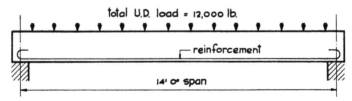

FIG. 4.1. SINGLY-REINFORCED RECTANGULAR BEAM

less than 1000 lb/sq. in. But the slightest reduction in size of the beam would clearly cause overstressing of the concrete. Thus the size chosen is the best for a singly reinforced beam.

The shear stress $q = \dfrac{Q}{bl_{\mathrm{a}}}$, equation 3.86,

giving $q = \dfrac{6950}{8 \times (0\cdot86 \times 14\frac{1}{4})} = 71 \text{ lb/sq. in.}$

This is quite low: but some stirrup reinforcement should be provided.

(ii) *Plastic design*

(a) As a first exercise, let us now design by the plastic theory a beam of the same size as has been shown above to be the most economical according to the elastic theory.

From equations 3.53 and 3.54 we have

$$a_1 = \frac{1 \cdot 5 \times \alpha}{1 - \sqrt{(1 - 3\alpha)}}, \text{ where } \alpha = \frac{M}{bd_1^2 p_{cb}}. \quad (4.1)$$

In our case we have

$$\alpha = \frac{292,000}{8 \times 14\frac{1}{4}^2 \times 1000} = 0 \cdot 18,$$

so that $\quad a_1 = \dfrac{1 \cdot 5 \times 0 \cdot 18}{1 - \sqrt{(1 - 0 \cdot 54)}} = \dfrac{0 \cdot 27}{1 - 0 \cdot 678} = 0 \cdot 838.$

As an alternative method, since the relationship between a_1 and $\dfrac{M_r}{bd_1^2}$ approximates to a straight line, we have roughly

$$a_1 = 1 \cdot 03 - \frac{1 \cdot 12}{p_{cb}} \times \frac{M_r}{bd_1^2} \quad (4.2)$$

which is accurate enough for the complete range of singly reinforced rectangular beams.

From this equation we have $a_1 = 1 \cdot 03 - (1 \cdot 12 \times 0 \cdot 18) = 0 \cdot 83$ showing good enough agreement with the more accurate figure obtained above.

Since $l_a = a_1 d_1$, we have $l_a = 0 \cdot 83 \times 14\frac{1}{4} = 11 \cdot 9$ in. Thus from equation 3.53

$$A_{st} = \frac{M_r}{p_{st} l_a} = \frac{292,000}{20,000 \times 11 \cdot 9} = 1 \cdot 23 \text{ sq. in.}$$

This compares with the area of steel required by the elastic design of $1 \cdot 19$ sq. in., and is again adequately provided by two $\frac{7}{8}$ in. dia. rods.

The value $0 \cdot 83$ for a_1 being greater than $0 \cdot 75$ indicates that no compression reinforcement is required; indeed it is clear that according to the plastic theory the 16 in. $\times$ 8 in. beam is not working to full capacity.

(b) Let us now determine the maximum total load the beam 16 in. $\times$ 8 in. would carry singly reinforced, according to the plastic theory. From equation 3.57

$$M_r = bd_1^2 \frac{p_{cb}}{4}, \text{ giving } M_r = 8 \times 14\frac{1}{4}^2 \times \frac{1000}{4} = 406,000 \text{ in. lb.}$$

Thus $406,000$ in. lb. $= W \times \dfrac{14}{8} \times 12$, giving $W = 19,300$ lb.

less self-weight of beam $= \quad 1,900$ lb.

$\overline{\qquad\qquad}$

$17,400$ lb.

Thus by plastic design the singly reinforced beam will carry $17,400$ lb. (as against $12,000$ lb. by the elastic method), i.e. an increase of 45 per cent. At this increased load the end shear-stress q is 99 lb/sq. in.

Where singly reinforced rectangular beams are worked up to the limit in the plastic design method, n_1 is limited by C.P. 114 (1957) to 0.5, so that l_a is limited to $0.75 d_1$. Thus

$$A_{st} = \frac{M_r}{p_{st} \times 0.75 d_1} = \frac{406,000}{20,000 \times 0.75 \times 14\frac{1}{4}} = 1.91 \text{ sq. in.}$$

This is given by two $1\frac{1}{8}$ in. dia. rods (area 2.00 sq. in.). As compared with elastic design, this is 65 per cent more steel for only 45 per cent more load, but it must be remembered that according to the elastic theory it would be necessary for the same enhanced load either to increase the size of the beam, or to introduce some compression reinforcement as well as increase the tension reinforcement.

(c) Let us now consider the minimum size of a singly reinforced lintol required to carry the original total load of $13,900$ lb. according to the plastic design method. From equation 3.57

$$\frac{M_r}{bd_1{}^2} = \frac{p_{cb}}{4} = \frac{1000}{4} = 250.$$

Therefore $\qquad d_1 = \sqrt{\dfrac{292,000}{8 \times 250}} = 12.1$ in.

Using a beam 14 in. $\times$ 8 in., where $d_1 = 12\frac{1}{4}$ in., from equation 4.2

$$a_1 = 1.03 - \left(\frac{1.12}{1000} \times \frac{292,000}{8 \times 12\frac{1}{4}{}^2} \right) = 0.76,$$

giving $\qquad A_{st} = \dfrac{292,000}{20,000 \times (0.76 \times 12\frac{1}{4})} = 1.57$ sq. in.

This is given by two 1 in. dia. rods (area 1.57 sq. in.).

This design of beam 14 in. $\times$ 8 in. with two 1 in. rods (by plastic design) compares with the beam 16 in. $\times$ 8 in. with two

$\frac{7}{8}$ in. rods (by elastic design), showing a reduction of concrete and shuttering by 13 per cent and 10 per cent respectively, but an increase in steel of 28 per cent.

RECTANGULAR BEAMS: DOUBLY REINFORCED

Consider a roof beam, effective span 26 ft., carrying a uniformly distributed load of 1500 lb/ft. run. Other conditions limit the beam depth to 24 in. overall. Total load carried by beam is

> Superimposed load 26 × 1500 = 39,000 lb.
> Self-weight, say 26 × 250   = 6,500 lb.
> _____
> 45,500 lb.

Centre moment is

$$M = \frac{Wl}{8} = 45,500 \times \frac{26}{8} \times 12 = 1,780,000 \text{ in. lb.}$$

The shear at the ends is $Q = \dfrac{45,500}{2} = 22,750 \text{ lb.}$

(i) *Elastic design (using design curve)*

$$A_{st} = \frac{M}{p_{st} \times l_a} = \frac{1,780,000}{20,000 \times (0 \cdot 86 \times 21\frac{1}{4})} = 4 \cdot 86 \text{ sq. in.}$$

This is given by eight $\frac{7}{8}$ in. dia. rods (area 4·8 sq. in.).

For a beam 24 in. × 10 in. $\dfrac{M}{bd_1{}^2} = \dfrac{1,780,000}{10 \times 21\frac{1}{4}{}^2} = 394.$ As this exceeds 184, clearly compression reinforcement is required with the design stresses we are using.

Now $r_{p1} = \dfrac{A_{st}}{bd_1} \times 100 = \dfrac{4 \cdot 86}{10 \times 21\frac{1}{4}} \times 100 = 2 \cdot 29 \text{ per cent.}$

From Fig. 3.12 when r_{p1} equals 2·29, and $\dfrac{M}{bd_1{}^2}$ equals 394, we see we require r_s equal to something a little under 1·0.

Since $A_{sc} = r_s A_{st}$, we require an area of compression steel in this case just less than the area of tension steel and it is convenient here to use eight $\frac{7}{8}$ in. dia. rods. Compression reinforcement should be effectively anchored over the length it is required to act in compression at centres not exceeding twelve times the diameter of the compression reinforcements. In our case this is $12 \times \frac{7}{8} = 10\frac{1}{2}$ in.

II

The shear stress at the ends of the beam is

$$q = \frac{Q}{bl_a} = \frac{22{,}750}{10 \times (0{\cdot}86 \times 21\frac{1}{4})} = 124 \text{ lb/sq. in.}$$

This beam would not be regarded by the Code as of secondary importance, and shear reinforcement is clearly required. A

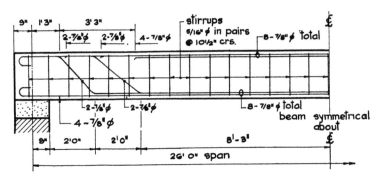

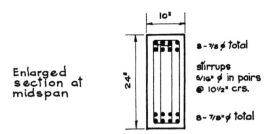

FIG. 4.2. DOUBLY-REINFORCED RECTANGULAR BEAM
(ELASTIC DESIGN)

suitable arrangement is indicated in Fig. 4.2. An example of the design of shear reinforcement is given in Article 4.2.

(ii) *Elastic design (without use of design curve)*

Figure 3.12 was prepared from equations 3.17, 3.18 and 3.22 for the design stresses to which we are working in these examples. However, where a design curve is not available, the compression reinforcement is calculated in the following manner.

From equation 3.17, for an economic design $n_1 = 0{\cdot}428$, so that $d_n = 0{\cdot}428 \times 21\frac{1}{4} = 9{\cdot}1$ in. If the centre of compression steel is

$1\frac{3}{4}$ in. down from the top of the beam, the stress in the compression steel (equation 3.13) is

$$15 \times 1000 \times \left(\frac{9 \cdot 1 - 1 \cdot 75}{9 \cdot 1}\right) = 12,100 \text{ lb/sq. in.}$$

and the lever arm for the compression steel is $21\frac{1}{4} - 1\frac{3}{4} = 19\frac{1}{2}$ in.

Now the excess moment to be carried by the compression reinforcement is

$$(394 - 184)bd_1{}^2 = 210 \times 10 \times 21\frac{1}{4}{}^2 = 950,000 \text{ in. lb.}$$

Whence $\qquad A_{sc} = \dfrac{950,000}{12,100 \times 19 \cdot 5}\left(\dfrac{15}{14}\right) = 4 \cdot 35 \text{ sq. in.}$

The term $\left(\dfrac{15}{14}\right)$ allows for the reduced area of concrete available due to the space occupied by the compression reinforcement. The area $4 \cdot 35$ sq. in. agrees with the value determined in the previous example from the design curve in Fig. 3.12.

(iii) *Plastic design (using design curve)*

Taking the same size of beam as in the previous examples, $\dfrac{M}{bd_1{}^2} = 394$. Setting this amount across from the vertical axis of the curve in Fig. 3.26, we see immediately that $r_{p1} = 2 \cdot 46$ per cent, and $r_s = 0 \cdot 36$. Thus we have

$$A_{st} = \frac{2 \cdot 46}{100} \times 10 \times 21\frac{1}{4} = 5 \cdot 21 \text{ sq. in.}$$

and $\qquad A_{sc} = 0 \cdot 36 \times 5 \cdot 21 = 1 \cdot 86 \text{ sq. in.}$

Comparing these with the elastic design values, it is seen that the area of tensile reinforcement requires to be slightly increased whereas the area of compression reinforcement can be more than halved.

(iv) *Plastic design (without use of design curve)*

Taking again the same size of beam, the maximum moment a singly reinforced beam can resist is

$$bd_1{}^2 \times \frac{p_{cb}}{4} = 10 \times 21\frac{1}{4}{}^2 \times \frac{1000}{4} = 1,130,000 \text{ in. lb.}$$

for which the tension reinforcement required, from equation 3.58 is

$$\frac{1}{0 \cdot 03t_1} = \frac{1}{0 \cdot 6} = 1 \cdot 67 \text{ per cent}$$

which in our case is

$$\frac{1\cdot67}{100} \times 10 \times 21\tfrac{1}{4} = 3\cdot54 \text{ sq. in.}$$

Now the additional moment to be accommodated as a result of reinforcing in compression is

$$1,780,000 - 1,130,000 = 650,000 \text{ in. lb.}$$

and this requires additional tension steel of

$$\frac{M}{p_{st}l_a} = \frac{650,000}{20,000 \times 19\tfrac{1}{2}} = 1\cdot67 \text{ sq. in.,}$$

making a total tension steel of $3\cdot54 + 1\cdot67 = 5\cdot21$ sq. in. Similarly the total compression reinforcement is

$$\frac{M}{p_{sc}l_a} = \frac{650,000}{18,000 \times 19\tfrac{1}{2}} = 1\cdot86 \text{ sq. in.}$$

These of course are the same answers as derived graphically in the previous example.

(v) *Steel beam theory*

Considering again our roof beam of the same dimensions and carrying the same load, we have by the steel beam theory,

Lever arm = 24 in. $-$ 2$\tfrac{3}{4}$ in. $-$ 2$\tfrac{3}{4}$ in. = 18$\tfrac{1}{2}$ in.

Therefore area of steel reinforcement, top and bottom,

$$= \frac{1,780,000}{18,000 \times 18\tfrac{1}{2}} = 5\cdot35 \text{ sq. in.}$$

This is seen to be a little more than required by the elastic theory, and is explained by the facts that the compressive stress of the concrete is ignored, and the method assumes the neutral axis to be a little lower down the section.

RECTANGULAR BEAM SUBJECT TO TORSION

Consider a beam, effective span 20 ft. between thin reinforced concrete walls, subject to a uniform superimposed load of 450 lb/ft. run acting at an eccentricity of 10·5 in. from the longitudinal axis of the beam.

Total load carried by beam is

Superimposed load 20 ft. $\times$ 450 lb/ft. =	9,000 lb.
Self-weight, say 20 ft. $\times$ 250 lb/ft. =	5,000 lb.
	14,000 lb.

End torque $T = 450$ lb/ft. × 10 ft. × 10·5 in. = 47,200 lb. in.

End shear is $\dfrac{14,000}{2} = 7000$ lb.

Let us try a beam 20 in. deep × 12 in. wide overall (Fig. 4.3).

Centre moment is $14,000 \times \dfrac{20}{8} \times 12 = 420,000$ in. lb.

whence $A_{st} = \dfrac{420,000}{20,000 \times (0·86 \times 18)} = 1·36$ sq. in.

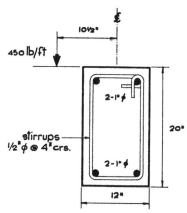

FIG. 4.3. RECTANGULAR BEAM SUBJECT TO TORSION

This is given by two 1 in. dia. rods (area 1·57 sq. in.). At mid-span where the reinforcement is fully engaged resisting the maximum bending moment, the vertical and torsional shear stresses are zero. At the supports, the vertical shear stress is

$$q = \frac{7000}{12 \times (0·86 \times 18)} = 38 \text{ lb/sq. in.}$$

This leaves $100 - 38 = 62$ lb/sq. in. shear stress available for resisting torsion.

From equation 3.91 the unreinforced concrete section could therefore take

$$T = \frac{b^2}{2}\left(a - \frac{b}{3}\right)q = \frac{9^2}{2}\left(17 - \frac{9}{3}\right) \times 62 = 35,200 \text{ lb. in.,}$$

a little less than the applied torque.

If the two 1 in. dia. rods from midspan are carried through to the supports, and two 1 in. dia. top rods are provided for the full length of the beam, the total percentage by volume of longitudinal steel is

$$\tfrac{1}{2}p = \frac{4 \times 0.785}{12 \times 20} \times 100 = 1.3 \text{ per cent.}$$

This is only effective in enhancing the torsional resistance of the member if it is able to act in conjunction with transverse steel of an equal percentage by volume. If stirrups $\tfrac{1}{2}$ in. diameter are provided at 4 in. centres, the percentage by volume is

$$\tfrac{1}{2}p = \frac{56 \text{ in.} \times 0.196 \text{ sq. in.} \times \dfrac{12 \text{ in.}}{4 \text{ in.}} \times 100}{20 \text{ in.} \times 12 \text{ in.} \times 12 \text{ in.}} = 1.15 \text{ per cent.}$$

This is just less than the percentage of longitudinal steel, so that the total percentage by volume of reinforcement effective in torsional resistance is taken as $p = 2 \times 1.15 = 2.30$ per cent.

The total safe effective torsional resistance of the concrete plus steel is therefore, from equation 3.92

$$T_{c+s} = T_c(1 + 0.25p)$$
$$= 35,200[1 + (0.25 \times 2.30)] = 55,500 \text{ lb. in.}$$

This is in excess of the applied torque of 47,200 lb. in. and therefore satisfactory.

ART. 4.2. CONTINUOUS BEAMS

To a large extent, the problem of reinforced concrete design lies in evaluating the bending-moments in continuous structures, and this is probably truer of beams than any other type of member because the combinations of loading in the different spans can be varied without limit. Usually beams are continuous through several supports and monolithic with the supporting columns, so that the loading at any span affects, and is affected by, the adjacent spans and columns. Strictly speaking, in a continuous frame, the loading at any one span will affect every beam and column in the frame, but it is usually accurate enough for practical design to consider the effects of adjacent spans and columns only. If the columns are flexible in relation to the beams they carry, they do not greatly interfere with the rotation of the beams; and the beams may be considered as continuous at the supports and free from rotational restraint. This is more nearly true at internal columns (as opposed to external columns), particularly where the structural anatomy and loading is roughly symmetrical and tendency for rotation is therefore only small.

At external columns the conditions are necessarily asymmetrical and some attention is required to moment distribution: in other words, the beam cannot be regarded as freely-supported (due to the stiffness of the column), nor on the other hand can it be regarded as fully fixed or *encastré*. This is considered further in Article 4.3.

In the present article consideration is given to freely-supported continuous beams. This clearly might be a line of beams in a building supported at intervals on masonry walls, or a bridge system supported on suitable rocker-bearings. But as shown above, the same approach may be made in solving beam-runs in a continuous frame of reasonable symmetry, with the exception of the end-spans.

The worst bending-moments which need to be considered are usually given by the following arrangements of superimposed loadings:

(i) *Alternate spans loaded and all other spans unloaded.* (This generally gives the worst positive moments at or near midspan.)

(ii) *Any two adjacent spans loaded and all other spans unloaded.* (This gives severe negative moments at the supports.)

These conditions can apply only to the *superimposed loads*. The *dead load* due to the self-weight of the beams, slabs and finishes is of course fixed in amount and in position, and as the moment-factors arising from the more even distribution of the dead load are less than the moment-factors due to the worst conditions for superimposed load as given above at (i) and (ii), it is correct to consider these separately.

The moments may be calculated by any of the methods of analysis of redundant systems given in books on structural analysis. It will be found that the negative moments at the supports work out to be greater than the positive moments at midspan. This is sometimes inconvenient, particularly with T-beams which are strong at mid-span (having part of the slab available to act as a compression flange) but weaker at the supports where only the narrow rib is available. Furthermore, at columns where reinforcements from beams on perhaps four sides have to be accommodated between the column reinforcements, the congestion may be such that it becomes difficult to place and consolidate the concrete effectively. To assist in these matters C.P. 114 (1957) permits the negative moments at the supports to be decreased by up to 15 per cent provided that the numerical value of such reduction is added to the positive mid-span moment for the same loading system.

The justification of such a ruse is exemplified by Fig. 4.4 where loads are shown to be in equilibrium, either with complete fixity at the supports and zero moment at mid-span, or with all the bending taken at midspan and complete freedom at the supports, the numerical values of the moments at support and midspan separately being equal. It is clear therefore that a little judicious

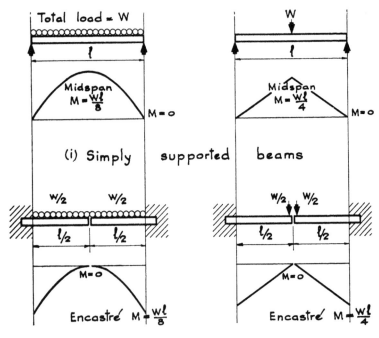

FIG. 4.4. ADJUSTMENT OF MOMENT POSITIONS

adjustment of moments – as between supports and mid-span – will not be amiss, particularly with a material as plastic as concrete, so well endowed with the quality of creep.

To avoid the labour of full mathematical analysis where the loading intensity and adjacent span lengths do not vary by more than 15 per cent it is satisfactory to use the approximate moment factors given in Figs. 4.5 and 4.6. Although these have been rounded off a little for convenience, they are nevertheless very well within the limits of accuracy set by the many other uncertainties in design, such as actual loading intensity, and strengths and behaviour of materials. Furthermore, the premise that interior

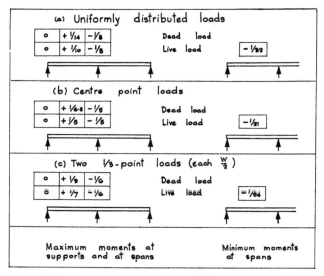

FIG. 4.5. MOMENT FACTORS × Wl: 2 BAYS ONLY

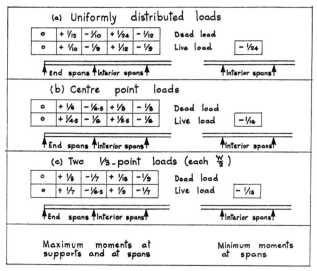

FIG. 4.6. MOMENT FACTORS × Wl: 3 OR MORE BAYS

supports are free from rotational restraint is never strictly true: columns of course do have some stiffness, and where beams rotate on masonry-bearings the support becomes eccentric so that the actual spans are varied.

The values in Fig. 4.5 apply to beams of two spans only. The values in Fig. 4.6 apply to beams of three or more spans. The values for point-loading cater for main-beams supporting secondary beams; and the whole of the dead load is then taken as acting at the positions of the point-loads. In all cases:

W_d is the *total* dead load per span,
W_s is the *total* superimposed load per span.

The tables on the right-hand side of Figs. 4.5 and 4.6 give maximum *negative* moments at mid-span, when the span in question is unloaded but the spans to each side are loaded. Negative mid-span moments can only arise when W_s is well in excess of W_d.

An example of the design of a continuous beam now follows, working to the design stresses of:

$$p_{cb} = 1,000 \text{ lb/sq. in.,}$$
$$p_{st} = 20,000 \text{ lb/sq. in.,}$$
$$p_{sc} = 18,000 \text{ lb/sq. in.}$$

DESIGN OF CONTINUOUS T-BEAM

Consider the case of a run of continuous floor beams at 10 ft. centres one from another, each spanning 22 ft. (Fig. 4.7). The superimposed floor load including movable partitions is 115 lb/sq. ft.; and the floor finish weighs 12 lb/sq. ft. A suitable slab thickness for spanning the 10 ft. between beams is 4 in. The design of one of the internal beams is as follows:

Total dead load of floor is:

4 in. slab	50 lb/sq. ft.
Finishes	12 lb/sq. ft.
Beams, say	15 lb/sq. ft.

77 lb/sq. ft.

Maximum total load per beam is:

Dead load 22 ft. × 10 ft. × 77 lb/sq. ft.   = 17,000 lb.
Superimposed load 22 ft. × 10 ft. × 115 lb/sq. ft. = 25,300 lb.

42,300 lb.

Shear at the supports is

$$Q = \frac{42,300}{2} = 21,200 \text{ lb.}$$

From Fig. 4.6 the maximum mid-span moment to be designed for is

$$M = \frac{W_d l}{24} + \frac{W_s l}{12}$$

$$= \left(17,000 \times \frac{22}{24} \times 12\right) + \left(25,300 \times \frac{22}{12} \times 12\right)$$

$$= 187,000 + 557,000 = 744,000 \text{ in. lb.}:$$

and the maximum support moment to be designed for is

$$M = \frac{W_d l}{12} + \frac{W_s l}{9}$$

$$= \left(17,000 \times \frac{22}{12} \times 12\right) + \left(25,300 \times \frac{22}{9} \times 12\right)$$

$$= 374,000 + 742,000 = 1,116,000 \text{ in. lb.}$$

which, allowing for the width of the support, may be reduced to 928,000 in. lb. (*see* Fig. 4.7).

In the following different methods of calculation, a T-beam of overall rib dimensions 18 in. deep × 10 in. wide is used. In arriving at the width of the rib it is important to see that there is sufficient width to accommodate the rods so that the spaces between the rods are such as to allow the concrete to be placed and properly consolidated. Sufficient attention is not always paid to this consideration.

(i) *Elastic design*

For those who like a full calculation, the area of tension steel at mid-span is calculated by processes as follows. From equation 3.23, n_1 is determined, and equals 0·427. Then from equation 3.27, a_1 is obtained, and equals 0·88. Thus we have

$$A_{st} = \frac{744,000}{20,000 \times (0·88 \times 15·75)} = 2·68 \text{ sq. in.}$$

This is given by six $\frac{3}{4}$ in. dia. rods (area 2·64 sq. in.). It will be found that the labour in working out n_1 and a_1 is considerable.

Unless one can judge by experience, it is necessary to check that the breadth of slab available to act as the compression flange of the T-beam is such that the stress in. the concrete is kept within the permissible value. From equation 3.23a, we have

$$t_1 = \frac{bd_s}{2A_{st}} \frac{(2d_n - d_s)}{d_n}.$$

This gives $b = 19 \cdot 2$ in. for the breadth of slab required in our case.

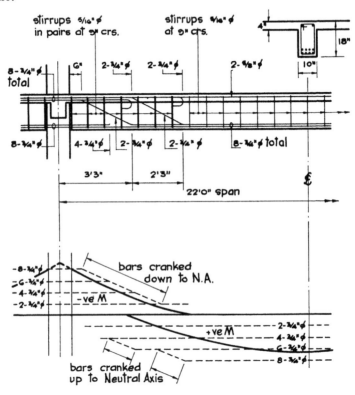

Bending Moment

FIG. 4.7. TEE-BEAM WITH CONTINUITY

C.P. 114 (1957) recommends that the breadth of flange considered available for taking compression should not exceed the least of the following:

(i) one-third of the effective span of the T-beam;
(ii) the distance between the centres of the ribs of the T-beams;

(iii) the breadth of the rib plus twelve times the thickness of the slab.

In our case the least of these is (iii) being 10 in. + (12 × 4) = 58 in. The 19·2 in. breadth required is well within this value.

Alternative method using design curve. The labour in arriving at the area of tension steel and breadth of compression flange required may be reduced by the use of the design curves given in Fig. 3.15. Knowing $n_1 = 0.427$, and having $\dfrac{d_s}{d_1}$ equal in our case to $\dfrac{4}{15.75} = 0.254$, we have (from the lower curve) r_{p1} equals 0·89 per cent and (from the upper curve) a_1 equals 0·88.

Thus
$$A_{st} = \frac{744,000}{20,000 \times (0.88 \times 15.75)} = 2.68 \text{ sq. in.}$$

as before, and
$$b = \frac{A_{st}}{r_{p1}d_1} \times 100 = \frac{2.68 \times 100}{0.89 \times 15.75} = 19.2 \text{ in.}$$

as before.

Alternative approximate method. The labour in determining the mid-span tension steel by direct calculation is greatly reduced by using the approximate formula 3.31, so that

$$l_a = d_1 - \frac{d_s}{2} = 15.75 - 2.0 = 13.75 \text{ in.}$$

Then from equation 3.32

$$A_{st} = \frac{M}{\left(d_1 - \dfrac{d_s}{2}\right)p_{st}} = \frac{744,000}{13.75 \times 20,000} = 2.70 \text{ sq. in.}$$

It is seen that this is very little different from the result obtained by the previous methods.

Considering now the support moment, we have

$$A_{st} = \frac{928,000}{20,000 \times (0.86 \times 15\frac{1}{2})} = 3.48 \text{ sq. in.}$$

This is given by eight $\frac{3}{4}$ in. dia. rods (area 3·52 sq. in.). But $\dfrac{M_r}{bd_1{}^2} = \dfrac{928,000}{10 \times 15\frac{1}{2}{}^2} = 387$. This difficulty may be met either:

(i) by deepening the beam at the support by the provision of a *haunch*, so that the enhanced depth of beam will take the large negative moment on the rectangular section without compression reinforcement, or

(ii) by designing at the support as for a rectangular beam with compression reinforcement.

Aesthetic and other practical considerations frequently rule out the use of haunches, and in our example we will provide compression steel.

$$r_{p1} = \frac{3 \cdot 48 \times 100}{10 \times 15\frac{1}{2}} = 2 \cdot 25 \text{ per cent.}$$

Then from Fig. 3.12 we see that we require $r_s = 0 \cdot 9$. Eight $\frac{3}{4}$ in. dia. rods are conveniently used.

Let us consider now the shear stress in the beam rib.

$$q = \frac{Q}{b_r l_a} = \frac{21,200}{10 \times (0 \cdot 86 \times 15\frac{1}{2})} = 159 \text{ lb/sq. in.}$$

We will arrange bent-up reinforcements at the supports, and stirrups equally spaced throughout, as shown in Fig. 4.7. The shear resistance of the reinforced beam at the support is therefore made up as follows:

From diagonal tension in two $\frac{3}{4}$ in. bars bent-up at
30° (equation 3·88) 0·88 × 20,000 × sin 30°
$\qquad\qquad\qquad\qquad\qquad = F_T = \quad$ 8,800 lb.
From associated diagonal compression ($F_C = F_T$)= $\quad$ 8,800 lb.
From $\frac{5}{16}$ in. stirrups in pairs at 9 in. centres
$\qquad\qquad\qquad\qquad\qquad$ (equation 3.87)

$$\frac{20,000 \times (4 \text{ strands} \times 0 \cdot 077) \times (0 \cdot 86 \times 15\frac{1}{2})}{9} = \quad 9,200 \text{ lb.}$$

$\qquad\qquad\qquad\qquad\qquad\qquad\qquad\qquad$ 26,800 lb.

Note the stirrups alone will resist a shear force of 9,200 lb. and are therefore adequate by themselves to a distance from the centre-line of the beam equal to 11 ft. $\times \dfrac{9,200}{21,200} = 4 \cdot 75$ ft.
Allowing for a little direct inclined compression the arrangement proposed is clearly satisfactory. From Fig. 4.7 it is clear that the bars bent diagonally to assist in taking shear leave sufficient area of steel in the flanges to take the bending moments at all points along the beam.

It is perhaps worth noting in passing, that had the load on the beam been a central point load, perhaps from an incoming secondary beam as against a uniform load, the shear would have been almost constant for the full distance between the support and mid-span. Under these circumstances there would be very little reduction in shear stress at sections nearer mid-span, and bent-up bars would be required over the full length of the beam being at a constant slope to suit the constant shear.

(ii) *Plastic design*

As with the elastic method we will first outline the more complete method of calculation for the tension steel at mid-span. This will be found tough going. By rearranging equation 3.66 the necessary breadth of flange can be expressed

$$b = 3\left(\frac{M_r}{p_{cb}d_1{}^2} - \frac{b_r}{4}\right)\left(\frac{1}{\frac{d_s}{d_1}\left(2 - \frac{d_s}{d_1}\right)}\right) + b_r$$

and in our case

$$b = 3\left(\frac{744}{1000 \times 15{\cdot}75^2} - \frac{10}{4}\right)\left(\frac{1}{\frac{4}{15{\cdot}75}\left(2 - \frac{4}{15{\cdot}75}\right)}\right) + 10 = 13{\cdot}4\,\text{in.}$$

With a flange breadth of 13·4 in., the lever arm ratio a_1 is given by equation 3.70 as under

$$a_1 = \left(1 - \frac{d_s}{2d_1}\right) + \left\{\frac{b_r}{4b}\frac{\left[3 - 2\left(2 - \frac{d_s}{d_1}\right)\right]}{\left[\frac{b_r}{b} + \frac{2d_s}{d_1}\left(1 - \frac{b_r}{b}\right)\right]}\right\},$$

giving

$$a_1 = \left(1 - \frac{4}{2 \times 15{\cdot}75}\right) +$$
$$+ \left\{\frac{10}{4 \times 13{\cdot}4} \times \frac{\left[3 - 2\left(2 - \frac{4}{15{\cdot}75}\right)\right]}{\left[\frac{10}{13{\cdot}4} + 2 \times \frac{4}{15{\cdot}75}\left(1 - \frac{10}{13{\cdot}4}\right)\right]}\right\} = 0{\cdot}768.$$

Thus the lever arm $l_a = a_1 d_1 = 0{\cdot}768 \times 15{\cdot}75 = 12{\cdot}1$ in., and

$$A_{st} = \frac{744,000}{20,000 \times 12{\cdot}1} = 3{\cdot}07\,\text{sq. in.}$$

Alternative approximate method. The lever arm ratio may be obtained simply from equation 3.71 as

$$a_1 = 1 - \frac{d_s}{2d_1},$$

giving $\qquad\qquad a_1 = 1 - \frac{4}{2 \times 15{\cdot}75} = 0{\cdot}873.$

Thus the lever arm $l_a = a_1 d_1 = 0 \cdot 873 \times 15 \cdot 75 = 13 \cdot 7$ in., and

$$A_{st} = \frac{744,000}{20,000 \times 13 \cdot 7} = 2 \cdot 71 \text{ sq. in.}$$

The lever arm ratio taken in equation 3.71 is based on the flange taking the total compression. The breadth of flange required is therefore

$$b = \frac{M}{l_a \times d_s \times (\tfrac{2}{3} p_{cb})}$$

$$= \frac{744,000}{13 \cdot 7 \times 4 \times 0 \cdot 67 \times 1000} = 20 \cdot 3 \text{ in.}$$

From the above it will be seen that the flange breadth and area of tension steel given by the plastic approximate method agree closely with the findings of the elastic method: whereas the outcome of the more laborious plastic design shows more tension steel required on the basis of a reduced flange breadth. This is because the fuller plastic calculation (as opposed to the approximate plastic calculation) takes into account the compression in the upper part of the beam rib at full stress, and this reduces the amount of the lever arm. The assumption of full compression in the rib is of course quite absurd in our case where only $13 \cdot 4$ in. flange breadth is in operation out of a total available shown earlier to be 58 in.: in other words other things would be happening long before plastic behaviour developed to the fullest extent with inclusion of full permissible stresses in the rib. And it should be appreciated that equations 3.67 and 3.72 are incompatible as regards length of lever arm; nevertheless, so long as this is understood no harm will result. This may not be clear from the manner the equations are given in C.P. 114 (1957).

The design for the support moments by the plastic method is carried out exactly as for rectangular beams with compression reinforcement; and the shear reinforcement is designed as given before for the elastic design.

ART. 4.3. BEAMS IN CONTINUOUS FRAMES

FRAMES NOT SUBJECT TO HORIZONTAL SHEAR

In a continuous rigid-jointed frame where all loading is vertical, the beams at internal columns may be regarded as continuous and capable of free rotation unless the columns are stiff in relation to the beams they support. This is discussed in Article 4.2 and an example is worked out there in accordance with the theories given in Chapter 3. However, at *external* columns the conditions are normally asymmetrical, and bending is induced into the columns as shown in Article 6.4. The sum of the bending moments in the columns immediately above and below any incoming beam then provides the negative (or end-fixity) moment to the beam at its point of support; and this in turn affects the amount of positive midspan moment in the beam. This is best illustrated in an example as follows.

Consider the arrangement of an office-block building as shown in Fig. 6.2. The total inclusive floor loading is 170 lb/sq. ft. made up of 110 lb/sq. ft. dead load (including partitions) and 60 lb/sq. ft. superimposed load. We will design the main beam at first-floor level.

Maximum load per beam is:

Dead load 20 ft. $\times$ 14 ft. $\times$ 110 lb/sq. ft. = 30,800 lb.
Superimposed load 20 ft. $\times$ 14 ft. $\times$ 60 lb/sq. ft. = 16,800 lb.

$$\overline{\underline{\underline{47,600 \text{ lb.}}}}$$

Thus the end shear (ignoring continuity) is 23,800 lb. Now from Fig. 4.5 for a two-span system without end-fixity, we have

Maximum mid-span moment (positive)

$$= \left(30,800 \times \frac{20}{14} \times 12\right) + \left(16,800 \times \frac{20}{10} \times 12\right) = 930,000 \text{ in. lb.}$$

and maximum internal-support moment (negative)

$$= \left(30,800 \times \frac{20}{8} \times 12\right) + \left(16,800 \times \frac{20}{8} \times 12\right) = 1,428,000 \text{ in. lb.}$$

But in Article 6.4 it is shown that moments of 176,000 in. lb. occur in the external column both above and below first-floor level. These two moments together provide end-fixity to the beam at the external support of $(-M) = 176,000 + 176,000 = 352,000$ in. lb. Now the least effect this end-fixity could have on the mid-span moment would occur if the internal column were completely held against rotation. The carry-over moment to the internal column would then be $\left(+\dfrac{M}{2}\right)$; and the nett effect of $(-M)$ at the external end and $\left(+\dfrac{M}{2}\right)$ at the internal end would be a reduction to the mid-span moment of $\dfrac{M}{4}$. And the least effect

12

the external end-fixity $(-M)$ could have on the negative internal-support moment would occur when the internal column is completely free to rotate. Then the carry-over moment of $\left(+\dfrac{M}{2}\right)$ would be shared by the two beams on either side of the column, giving a nett reduction of $\dfrac{M}{4}$.

In other words, the end-fixity of 352,000 in. lb. at the external column reduces both the midspan moment and the internal-support moment by not less than $\dfrac{352,000}{4} = 88,000$ in. lb. If a fuller mathematical analysis of the problem is pursued, it will be

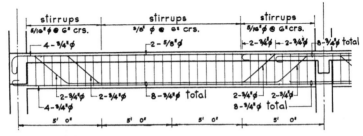

FIG. 4.8. BEAM IN END-BAY OF CONTINUOUS FRAME

found that the error in the above approximation does not exceed 5 per cent.

Thus we have beam moments as follows:

External support moment $- 352,000$ in. lb.
Mid-span moment
         $930,000 - 88,000 = \quad + 842,000$ in. lb.
Internal support moment
       $- 1,428,000 + 88,000 = - 1,340,000$ in. lb.

Allowing for the width of the supporting columns, the moments for which the beam is to be designed are:

External support moment $- 300,000$ in. lb.
Mid-span moment (no reduction) $+ 842,000$ in. lb.
Internal support moment $- 1,050,000$ in. lb.

Let us take a beam 20 in. $\times$ 8 in. overall (*see* Fig. 4.8):

A_{st} required at mid-span

$$= \frac{842,000}{20,000 \times (0 \cdot 86 \times 17 \cdot 5)} = 2 \cdot 80 \text{ sq. in.}$$

This is provided by eight $\frac{3}{4}$ in. dia. bottom rods (area 3·52 sq. in.)
A_{st} required at the external end

$$= \frac{300,000}{20,000 \times (0·86 \times 17·5)} = 1·00 \text{ sq. in.}$$

This is provided by four $\frac{3}{4}$ in. dia. top rods (area = 1·76 sq. in.)
A_{st} required at the internal end

$$= \frac{1,050,000}{20,000 \times (0·86 \times 17·5)} = 3·48 \text{ sq. in.}$$

This is provided by eight $\frac{3}{4}$ in. dia. top rods (area 3·52 sq. in.)

At the inner end $\qquad \dfrac{M}{bd_1^2} = \dfrac{1,050,000}{8 \times 17\frac{1}{2}^2} = 429.$

With $\qquad r_{p1} = \dfrac{3·52}{8 \times 17\frac{1}{2}} \times 100 = 2·52 \text{ per cent,}$

we have from the design curve in Fig. 3.12 $r_s = 1·0$ approx., i.e.
eight $\frac{3}{4}$ in. dia. rods are required in the bottom of the beam for
compression at the internal column.

Allowing for continuity, the shear at the internal column is

$$23,800 \text{ lb.} + \frac{(1,340,000 - 352,000) \text{ in. lb.}}{(20 \times 12) \text{ in.}}$$

$$= 23,800 + 4100 = 27,900 \text{ lb.}$$

This would give a shear stress on the concrete alone of

$$\frac{27,900}{8 \times (0·86 \times 17\frac{1}{2})} = 232 \text{ lb/sq. in.}$$

With the arrangement of shear reinforcement shown in Fig. 4.8
we have a resistance at the support as follows.

Two $\frac{3}{4}$ in. bent-up bars
$F_T = (2 \times 0·44) \times 20,000 \times 0·6$ $\qquad = 10,500 \text{ lb.}$
Diagonal compression
$F_C = F_T$ $\qquad = 10,500 \text{ lb.}$
$\frac{5}{16}$ in. stirrups at 6 in. crs.

$$(2 \times 0·077) \times 20,000 \times \frac{0·86 \times 17\frac{1}{2}}{6} = \frac{7,600 \text{ lb.}}{28,600 \text{ lb.}}$$

This is satisfactory.

For the centre part of the beam the $\frac{3}{8}$ in. stirrups at 6 in. centres are good for

$$(2 \times 0\cdot11) \times 20,000 \times \frac{0\cdot86 \times 17\frac{1}{2}}{6} = 11,000 \text{ lb.}$$

and are therefore adequate on their own to a distance from the centre-line of the beam equal to 10 ft. $\times \dfrac{11,000}{27,900} = 4$ ft. Allowing for a little direct inclined compression, the arrangement shown in Fig. 4.8 is therefore clearly satisfactory.

FRAMES SUBJECT TO HORIZONTAL SHEAR

Where frames are subject to horizontal shear, as, for instance, from wind, earthquake or other forces, and there are no solid-panels or cross-bracings to bring the shear forces down to foundation level, the shears have to be taken up in bending in the frame members. The effect this has on the columns is discussed in Article 6.5. From Figs. 6.6 and 6.7 it is clear that moments are also induced into the beams, being a maximum at the ends of the beams and reducing to zero at contraflexure points along the length. The moments are of course reversed if the wind direction changes. These wind-moments are additional to those due to vertical loading; but the permissible stresses in the materials may be increased by 25 per cent where the increase is solely the result of the wind-loading.

CHAPTER 5

SLAB DESIGN

ART. 5.1. INTRODUCTION

IN reinforced-concrete work, the design of slabs is normally more straightforward than the design of beams, and for the following reasons. There is only the depth of the slab which the designer can vary by choice whereas with beams he has within his control the depth and the breadth of the beam rib, as well as parts of the slab where T-beam action is developed. The question of shear stress in slabs is frequently not significant; and where it is (as for example at columns in flat slab construction), it is unusual to provide shear reinforcements, and the slab thickness is settled so that $\frac{Q}{bl_a}$ is kept within the permissible value. It is unusual also to make use of compression reinforcement in slabs, except perhaps at the supports where continuity is developed. Thus slab analysis normally consists simply of the calculation of tension reinforcement in a singly reinforced member, and the depth of the slab is generally determined either by consideration of acceptable deflections or the compressive stress in the concrete.

The main problem in designing reinforced concrete slabs lies in determining the applied bending moments. Where the slabs are spanning in one direction and continuous over the supports, the bending moments may be taken from Figs. 4.5 and 4.6 for cases where the loading intensity and adjacent spans do not vary by more than 15 per cent. This is true even where the slabs are supported on beam ribs with which they are cast monolithically, unless the beams are particularly stiff in torsion in relation to the slabs in bending. This latter is unusual; and indeed it is the direct tension or compression in the slab which is normally relied upon to steady the beam rib against torsional displacement.

However, where the slabs are spanning in two directions, the determination of bending moments becomes more problematical, and calculations may be based on empirical values as given in Tables 5.1 and 5.2 later in this chapter.

For flat slabs, where the columns support the slabs directly without the use of beams, the true analysis becomes yet more problematical, and unless the designer wishes to go in for a detailed analysis of the structure as a continuous frame using Hardy Cross or other suitable methods, empirical design factors may be used as given in Table 5.3. There is normally little to be gained in economy by pursuing the fuller analytical method of design, and the empirical basis is satisfactory for most cases.

It is interesting to note that Figs. 4.5 and 4.6, for slabs spanning in one direction only, very properly take cognizance of the fact that for superimposed loads greater moment coefficients apply than for dead loads. Yet for cases of slabs spanning in two directions and for flat slabs, where the structural actions become more obscure, this differentiation between the effects of superimposed and dead loads is glossed over in C.P. 114 (1957), and only the total unit load considered.

Direct comparisons between elastic and plastic designs for members in flexure are given in Chapter 4, and accordingly, to save space, this will not be repeated here. Indeed the whole of the examples in the present chapter are limited to the elastic method of design. The comparisons in Article 5.2 of the effects of working to different steel and concrete stresses are ther more readily appreciated.

ART. 5.2. SOLID SLABS SPANNING IN ONE DIRECTION

NORMAL DESIGN PROCEDURE

Consider the design of a floor-slab to carry 75 lb/sq. ft. on a 15 ft. span, freely supported at both ends (Fig. 5.1). The concrete

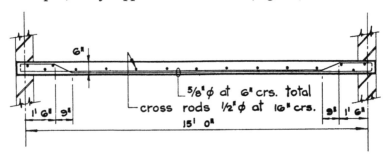

FIG. 5.1. SLAB SPANNING IN ONE DIRECTION

is to be 1:2:4 nominal (p_{cb} = 1000 lb/sq. in.) and the reinforcement is to be mild steel (p_{st} = 20,000 lb/sq. in.).

A suitable thickness for a slab with mild steel reinforcement is 6 in., i.e. one-thirtieth of the span.

Total load intensity is therefore:

| Superimposed load | 75 lb/sq. ft. |
| Self-weight of slab | 75 lb/sq. ft. |

$$150 \text{ lb/sq. ft.}$$

Per foot width, total load carried is

$$150 \text{ lb/sq. ft.} \times 15 \text{ ft.} = 2250 \text{ lb.}$$

Centre moment is

$$M = 2250 \times \frac{15}{8} \times 12 = 50{,}600 \text{ in. lb.}$$

Taking 0·86 for the lever arm ratio, and allowing one inch for bottom cover plus half the rod diameter, we have the area of steel required

$$A_{st} = \frac{M}{p_{st}l_a} = \frac{50{,}600}{20{,}000 \times (0{\cdot}86 \times 5)} = 0{\cdot}59 \text{ sq. in.}$$

This is given by $\frac{5}{8}$ in. dia. rods at 6 in. centres (area 0·61 sq. in.). A check on the concrete stress is given by

$$\frac{M}{bd_1^2} = \frac{50{,}600}{12 \times 5^2} = 168.$$

This is less than 184, and indicates that the compressive stress in the concrete is less than the permissible value of 1000 lb/sq. in. The shear stress in a slab of this duty would be known by any experienced designer to be insignificant, and normally no check calculations would be made. In fact the shear stress is

$$q = \frac{Q}{bl_a} = \frac{1125}{12 \times (0{\cdot}86 \times 5)} = 22 \text{ lb/sq. in.}$$

(as against 100 lb/sq. in. permissible, or 80 lb/sq. in. preferable).

DETERMINATION OF STRESSES IN A GIVEN SLAB

Frequently an engineer has to decide whether some given reinforced slab is adequate for a particular duty. For example he might be checking drawings prepared by some other designer, or he might be considering the suitability of an existing structure where the original drawings are available. In such cases he will have to work backwards to determine what stresses will be developed in the slab when the required loads are applied, and then check whether these are permissible.

Consider a 5 in. slab reinforced with $\frac{1}{2}$ in. rods at $4\frac{1}{2}$ in. centres ($\frac{1}{2}$ in. cover to steel). This slab is required to resist a total moment of 35,000 in. lb/ft. width, including the effect of its self-weight. Taking the cross-sectional area of a $\frac{1}{2}$ in. rod as 0·2 sq. in., the total area of steel per foot width of slab is

$$0\cdot2 \times \frac{12}{4\frac{1}{2}} = 0\cdot53 \text{ sq. in.}$$

Then from equation 3.10

$$f_{st} = \frac{M}{A_{st}l_a} = \frac{35,000}{0\cdot53 \times (0\cdot86 \times 4\cdot25)} = 18,100 \text{ lb/sq. in.}$$

and from equation 3.4

$$f_{cb} = f_{st}\frac{2A_{st}}{bd_n} = 18,100 \times \frac{2 \times 0\cdot53}{12 \times 1\cdot8} = 900 \text{ lb/sq. in.}$$

The value of 1·8 for d_n is obtained from Fig. 3.5 for the known percentage of steel.

DETERMINATION OF SAFE LOAD A GIVEN SLAB CAN CARRY

Suppose a 6 in. slab reinforced with $\frac{5}{8}$ in. dia. rods at 6 in. centres is freely supported at both ends on a 12 ft. 6 in. span. The concrete is 1:2:4 nominal ($p_{cb} = 1000$ lb/sq. in.) and the reinforcement is mild steel ($p_{st} = 20,000$ lb/sq. in.). What superimposed load can the slab carry?

The steel percentage $r_{p1} = \dfrac{0\cdot61}{12 \times 5} \times 100 = 1\cdot02$ per cent.

Therefore from Fig. 3.7 it is seen that the limiting capacity of the slab in bending will be determined by the tension reinforcement; and we have

$$M_r = A_{st}p_{st}\,l_a = 0\cdot61 \times 20,000 \times (0\cdot86 \times 5) = 52,500 \text{ in. lb.}$$

Now if w is the inclusive load per sq. ft. of slab causing bending, the maximum applied moment at mid-span will be

$$M = \frac{wl^2}{8} = \frac{w \times (12\frac{1}{2})^2}{8} \times 12 = 235w \text{ in. lb.}$$

Equating M and M_r, we have

$$235w = 52,500,$$

so that $\qquad\qquad\qquad w = 225 \text{ lb/sq. ft.}$

Deducting from this the self-weight of the slab, we have the maximum useful load it is permissible to apply $= (225 - 75) = 150$ lb/sq. ft. In cases of doubt a check should be made on the shear stress in the concrete: but here the shear stress is clearly quite low.

EFFECTS OF VARYING DESIGN STRESSES

A floor slab carrying 200 lb/sq. ft., inclusive of its own weight, spans 20 ft. between walls, the ends being non-continuous. Designs for this are now given in a somewhat academic fashion, using different permissible stresses for the concrete, and reinforcing steel; and comparisons are made later to illustrate the effects of working to these different design stresses. In order that the comparisons should be real and fair, the following calculations are a little more meticulous than one would do for normal practical design work. And any question of limiting the deflections by adopting a suitable minimum thickness for the slab is not considered at this stage.

The design stresses to be taken are as follows:

(a) 1:2:4 concrete $\qquad$ $p_{cb} = 1,000$ lb/sq. in.
$\quad$ Mild steel reinforcement $\quad$ $p_{st} = 20,000$ lb/sq. in.
(b) 1:1:2 concrete $\qquad$ $p_{cb} = 1,500$ lb/sq. in.
$\quad$ Mild steel reinforcement $\quad$ $p_{st} = 20,000$ lb/sq. in.
(c) 1:2:4 concrete $\qquad$ $p_{cb} = 1,000$ lb/sq. in.
$\quad$ Cold-worked reinforcement $\quad$ $p_{st} = 30,000$ lb/sq. in.
(d) 1:1:2 concrete $\qquad$ $p_{cb} = 1,500$ lb/sq. in.
$\quad$ Cold-worked reinforcement $\quad$ $p_{st} = 30,000$ lb/sq. in.

Considering a foot width of slab 20 ft. long, the load W is $20 \times 1 \times 200$ lb. $= 4000$ lb.

The centre moment is

$$M = \frac{Wl}{8} = \frac{4000 \times (20 \times 12)}{8} = 120,000 \text{ in. lb.}$$

Case (a). $p_{cb} = 1000$ lb/sq. in. $\quad p_{st} = 20,000$ lb/sq. in. From Fig. 3.8 (or equation 3.12), for economic design

$$M_r = 184 \, bd_1{}^2$$

whence

$$\text{minimum } d_1 = \sqrt{\frac{M}{184b}} = \sqrt{\frac{120,000}{184 \times 12}} = 7 \cdot 38 \text{ in.}$$

To allow for $\frac{3}{4}$ in. round rods and $\frac{3}{4}$ in. cover, the total thickness of slab required is then $d = 7 \cdot 38 + 0 \cdot 375 + 0 \cdot 75 = 8 \cdot 51$ in. Now from Fig. 3.8 (or equations 3.5 and 3.6) $r_{p1} = 1 \cdot 07$ per cent, whence

$$A_{st} = \frac{r_{p1}}{100} \times bd_1 = \frac{1 \cdot 07}{100} \times 12 \times 7 \cdot 38 = 0 \cdot 946 \text{ sq. in.,}$$

requiring a spacing of $\frac{3}{4}$ in. round rods (area 0·44 sq. in.) of

$$s = \frac{12 \text{ in.} \times 0\cdot 44}{0\cdot 946} = 5\cdot 57 \text{ in.}$$

Case (b). $p_{cb} = 1500$ lb/sq. in. $p_{st} = 20,000$ lb/sq. in.
From Fig. 3.8, $M_r = 328\, bd_1^2$

whence $d_1 = \sqrt{\dfrac{120,000}{328 \times 12}} = 5\cdot 52 \text{ in.,}$

giving a total slab thickness $d = 6\cdot 65$ in.
From Fig. 3.8, $r_{p1} = 1\cdot 99$ per cent

whence $A_{st} = \dfrac{1\cdot 99}{100} \times 12 \times 5\cdot 52 = 1\cdot 32 \text{ sq. in.,}$

requiring a spacing of $\frac{3}{4}$ in. round rods of $s = 4$ in.
   Case (c). $p_{cb} = 1000$ lb/sq. in. $p_{st} = 30,000$ lb/sq. in.
We have $M_r = 148\, bd_1^2$ whence $d_1 = 8\cdot 21$ in., and for $\frac{3}{4}$ in. square
rods twisted (which have to be accommodated on the diagonal)
$d = 9\cdot 49$ in. Now $r_{p1} = 0\cdot 56$ per cent, whence $A_{st} = 0\cdot 55$ sq. in.
requiring a spacing of $\frac{3}{4}$ in. square rods (area 0·563 sq. in.) of
$s = 12\cdot 3$ in.
   Case (d). $p_{cb} = 1500$ lb/sq. in. $p_{st} = 30,000$ lb/sq. in.
We have $M_r = 276\, bd_1^2$, whence $d_1 = 6\cdot 01$ in. and $d = 7\cdot 29$ in.
Now $r_{p1} = 1\cdot 07$ per cent, whence $A_{st} = 0\cdot 77$ sq. in. and $s = 8\cdot 78$ in.
   These results can now be collected for comparison. In practice
the dimensions would of course be rounded off.

	Concrete		Depth to reinft. d_1	Slab thickness d	Reinforcement		Area of reinft. per ft. A_{st}
	Nominal mix	Stress p_{cb}			Quality	Stress p_{st}	
		lb/sq. in.	in.	in.		lb/sq. in.	sq. in.
(a)	1:2:4	1000	7·38	8·51	Mild steel	20,000	0·95
(b)	1:1:2	1500	5·52	6·65	Mild steel	20,000	1·32
(c)	1:2:4	1000	8·21	9·49	Cold worked	30,000	0·55
(d)	1:1:2	1500	6·01	7·29	Cold worked	30,000	0·77

From the above table we may note the following:
   (1) Comparing (a) with (c), or (b) with (d), it is seen that the
effect of using cold-worked steel is naturally enough to reduce the
quantity of steel, but only at the expense of an increased thickness
of slab with consequent greater dead weight. Thus of the 200 lb/
sq. ft. inclusive design load taken, more of this is accounted for

by the slab itself leaving a reduced balance available to meet useful superimposed loading. Or considering this another way, to make our comparisons strictly true we should have designed the slab with high-strength steels for a slightly greater inclusive load, and this would have resulted in a slab thickness requirement greater than shown above.

(2) Comparing (a) with (b), or (c) with (d), it is seen that the effect of using greater strength concrete is to reduce the slab thickness, as one would of course expect, but only at the expense of an increased weight of steel.

(3) A slab simply supported on a 20 ft. span will deflect, and this deflection will increase with time. Whether this deflection will be objectionable depends on circumstances. If the slab is supporting block partitions with plaster or similar finishes, deflection more than a certain amount would be liable to cause diagonal cracking in the partitions. Of the four designs, cases (b) and (d) will deflect the most due to the reduced thickness and the increased stresses.

For a simply supported slab spanning 20 ft. a reasonable minimum thickness would be one-thirtieth of 20 ft., equals 8 in. Thus the slab given by designs (b) and (d) would not be suitable, and a perfectly reasonable result comes from the use of the popular 1:2:4 concrete and ordinary mild steel reinforcement.

(4) Which choice of materials and stresses will give the cheapest design depends on many factors. Clearly cold-twisted steel must cost more than ordinary mild steel, so that the steel cost at (c) will not be in proportion to the reduction of steel area required as compared with (a), whereas at (a) the amount and weight of concrete required is nearly 10 per cent less than at (c).

And comparing (a) with (b), the cost of 8·51 in. of 1:2:4 concrete placed may well be about the same as 6·65 in. of 1:1:2 concrete, whereas 0·95 sq. in. of steel will clearly cost much less than 1·32 sq. in. Thus the slab of 1:2:4 concrete may well be cheaper than the slab of 1:1:2 concrete.

ART. 5.3. HOLLOW TILE SLABS SPANNING ONE DIRECTION

The principal advantage of hollow tile floor slabs lies in their reduced weight (see Fig. 5.2). The clay blocks replace a proportion of the slab concrete below the neutral axis where it contributes nothing to the bending strength. As against the saving of weight and concrete, there is the greater labour of setting the clay blocks and reinforcements in position and of placing the concrete in the narrow ribs between the blocks. For short spans with thin slabs, the saving in weight and cost of concrete is generally less than the cost of the clay blocks and the additional labour. But for spans of 16 ft. or more, where the slab thickness is 6 in. or more, the hollow tile slab generally shows an overall saving in cost; particularly when account is taken of the effect of weight-reduction on the remainder of the structure, including all framing and foundations.

Unless special arrangements are made for stiffening hollow tile slabs at right angles to the lines of the ribs, concentrated

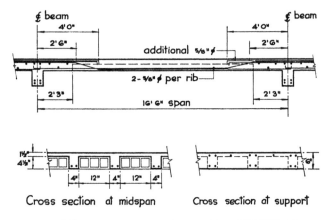

Cross section at midspan　　　Cross section at support

FIG. 5.2. HOLLOW-TILE SLAB, CONTINUOUS AT
BOTH ENDS

loads – as from partitions – may affect locally the stresses in the reinforcements very much more severely than in the case of solid slabs. This point should not be overlooked in preparing design calculations.

Consider the following design example. A slab continuous over several spans each of 16 ft. 6 in. is required to carry movable partitions equivalent to 40 lb/sq. ft., floor and ceiling finishes totalling 20 lb/sq. ft., and an applied floor-loading of 60 lb/sq. ft. The concrete is to be 1:2:4 nominal mix, and the reinforcements mild steel.

The minimum practical thickness for the slab would be 6 in. Clay tiles 12 in. wide and $4\frac{1}{2}$ in. deep would be suitable; and would need to be spaced 4 in. apart to give enough concrete between the tiles to accommodate the reinforcements (Fig. 5.2). Thus the ribs are at 16 in. centres, and it is convenient therefore to calculate throughout for a 16 in. width of slab. Then per foot run of 16 in. width we have:

Floor construction: $1\frac{1}{2}$ in. topping: $16 \times 1\frac{1}{2} \times \dfrac{150}{144} = 25$ lb.

Concrete rib: $4 \times 4\frac{1}{2} \times \dfrac{150}{144} = 19$ lb.

Clay tile: 16 lb.

_____

60 lb.

Floor and ceiling finishes: 20 lb/sq. ft. $\times \dfrac{16}{12} = 27$ lb.

Total dead load: 87 lb.

Superimposed loads:

Applied floor-load   60 lb/sq. ft.
Movable partitions   40 lb/sq. ft.

_____

Total superimposed load: 100 lb/sq. ft. $\times \dfrac{16}{12} = 133$ lb.

Midspan moment is

$$\left(87 \times \frac{16\frac{1}{2}^{2}}{24} \times 12\right) + \left(133 \times \frac{16\frac{1}{2}^{2}}{12} \times 12\right) = 48,300 \text{ in. lb.}$$

A_{st} required is $\dfrac{48,300}{20,000 \times (0.86 \times 5)} = 0.56$ sq. in., which is given by two $\frac{5}{8}$ in. dia. rods (area 0.61 sq. in.).

The Support-moment is

$$\left(87 \times \frac{16\frac{1}{2}^{2}}{12} \times 12\right) + \left(133 \times \frac{16\frac{1}{2}^{2}}{9} \times 12\right) = 72,200 \text{ in. lb.}$$

A_{st} required is $\dfrac{72,200}{20,000 \times (0.86 \times 5)} = 0.84$ sq. in. This needs three $\frac{5}{8}$ in. dia. rods (area 0.91 sq. in.) and is provided by one bent-up rod from ribs on opposite sides of the support, and one additional rod arranged midway between ribs.

There are two critical sections near the support where the concrete compressive stress due to bending needs to be checked. At the support itself we have $\dfrac{M}{bd_1^{2}} = \dfrac{72,000}{16 \times 5^{2}} = 180$ which is satisfactory. At the point where the clay tiles terminate (in our case about 2 ft. 3 in. from the centre-line of the support), only the concrete in the 4 in. rib is available to act in compression. Here the negative moment has reduced to about 30,000 in. lb., so that

$$\frac{M}{bd_1^2} = \frac{30,000}{4 \times 5^2} = 300.$$

This requires compression reinforcement and the arrangement shown in Fig. 5.2 is satisfactory.

The critical section for shear is normally at the ends of the ribs, and in our case we have per rib

$$Q = (87 + 133) \times \left(\frac{16\frac{1}{2}}{2} - 2\frac{1}{4}\right) = 220 \times 6 = 1320 \text{ lb.}$$

whence
$$q = \frac{1320}{4 \times (0.86 \times 5)} = 77 \text{ lb/sq. in.}$$

which is satisfactory.

Note that the partition allowance of 40 lb/sq. ft. was based on the weight of any single partition parallel to the ribs being carried by two ribs only.

ART. 5.4. SOLID SLABS SPANNING IN TWO DIRECTIONS AT RIGHT ANGLES

Let w be the total load per unit area,
l_y be the length of the longer side,
l_x be the length of the shorter side,

and let α and β be empirical bending-moment coefficients as given in Tables 5.1 and 5.2 respectively (based on C.P. 114 (1957)).

TABLE 5.1. *Bending moment coefficients for two-way slabs simply supported on four sides*

l_y/l_x	1·0	1·1	1·2	1·3	1·4	1·5	1·75	2·0
α_x	0·062	0·074	0·084	0·093	0·099	0·104	0·113	0·118
α_y	0·062	0·061	0·059	0·055	0·051	0·046	0·037	0·029

Table 5.1 is for cases of slabs simply supported at all sides, where no provision is made to resist torsion or to prevent the corners of the slab from lifting. The bending moments at mid-span on strips of unit width may then be taken as:

$$M_x = \alpha_x w l_x^2 \text{ on span } l_x, \tag{5.1}$$

and
$$M_y = \alpha_y w l_x^2 \text{ on span } l_y. \tag{5.2}$$

TABLE 5.2. *Bending moment coefficients for two-way slabs supported on four sides with provision for torsion at corners*

Type of panel and moments considered	Short span coefficients β_x								Long span coefficients β_y for all values of $\frac{l_y}{l_x}$
	Values of $\frac{l_y}{l_x}$								
	1·0	1·1	1·2	1·3	1·4	1·5	1·75	2·0	
Case 1. Interior panels. Negative moment at continuous edge	0·033	0·040	0·045	0·050	0·054	0·059	0·071	0·083	0·033
Positive moment at mid-span	0·025	0·030	0·034	0·038	0·041	0·045	0·053	0·062	0·025
Case 2. One short or long edge discontinuous. Negative moment at continuous edge	0·041	0·047	0·053	0·057	0·061	0·065	0·075	0·085	0·041
Positive moment at mid-span	0·031	0·035	0·040	0·043	0·046	0·049	0·056	0·064	0·031
Case 3. Two adjacent edges discontinuous. Negative moment at continuous edge	0·049	0·056	0·062	0·066	0·070	0·073	0·082	0·090	0·049
Positive moment at mid-span	0·037	0·042	0·047	0·050	0·053	0·055	0·062	0·068	0·037
Case 4. Two short edges discontinuous. Negative moment at continuous edge	0·056	0·061	0·065	0·069	0·071	0·073	0·077	0·080	
Positive moment at mid-span	0·044	0·046	0·049	0·051	0·053	0·055	0·058	0·060	0·044
Case 5. Two long edges discontinuous. Negative moment at continuous edge									0·056
Positive moment at mid-span	0·044	0·053	0·060	0·065	0·068	0·071	0·077	0·080	0·044
Case 6. Three edges discontinuous (one short or long edge continuous) Negative moment at continuous edge	0·058	0·065	0·071	0·077	0·081	0·085	0·092	0·098	0·058
Positive moment at mid-span	0·044	0·049	0·054	0·058	0·061	0·064	0·069	0·074	0·044
Case 7. Four edges discontinuous Positive moment at mid-span	0·050	0·057	0·062	0·067	0·071	0·075	0·081	0·083	0·050

Table 5.2 is for cases of any conditions of end-fixity, where provision is made by the addition of suitable reinforcements to resist torsion at the corners of the slab and to prevent the corners of the slab from lifting. Then, with reference to Fig. 5.3, the

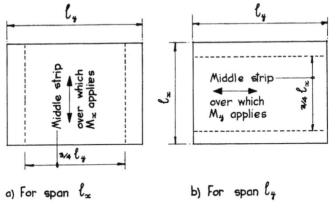

a) For span l_x b) For span l_y

FIG. 5.3. WIDTHS OF STRIPS REQUIRING FULL REINFORCEMENT IN SLABS SPANNING IN TWO DIRECTIONS

maximum bending moments on strips of unit width may be taken as:

$$M_x = \beta_x w l_x^2 \text{ in direction of span } l_x, \qquad (5.3)$$

$$M_y = \beta_y w l_x^2 \text{ in direction of span } l_y. \qquad (5.4)$$

An example follows showing the use of Table 5.2.

An office floor-slab is divided by beams into panels 16 ft. × 11 ft. 6 in. Partitions are equivalent to 50 lb/sq. ft., floor and ceiling finishes amount to 25 lb/sq. ft., and the superimposed load is 60 lb/sq. ft. The proportions of this slab are well suited to two-way spanning; and 5 in. thickness would be appropriate. Consider the design of a floor panel at the corner of the building:

Self-weight of slab	65 lb/sq. ft.
Partitions	50 lb/sq. ft.
Floor and ceiling finishes	25 lb/sq. ft.
Superimposed load	60 lb/sq. ft.
	200 lb/sq. ft.

The ratio $\dfrac{l_y}{l_x} = \dfrac{16}{11 \cdot 5} = 1 \cdot 4$ approx.

Consider first the l_x span of 11 ft. 6 in. $M_x = \beta_x w l_x^2$.
Moment at internal support

$$= 0 \cdot 070 \times 200 \times 11\tfrac{1}{2}^2 \times 12 = 22,300 \text{ in. lb.}$$

which requires

$$A_{st} = \frac{22,300}{20,000 \times (0 \cdot 86 \times 4\tfrac{1}{4})} = 0 \cdot 30 \text{ sq. in. of mild steel.}$$

Moment at mid-span

$$= 0 \cdot 053 \times 200 \times 11\tfrac{1}{2}^2 \times 12 = 16,800 \text{ in. lb.}$$

which requires $A_{st} = \dfrac{16,800}{20,000 \times (0 \cdot 86 \times 4\tfrac{1}{4})} = 0 \cdot 23 \text{ sq. in.}$

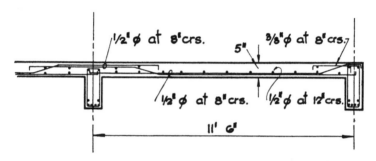

FIG. 5.4. SLAB SPANNING IN TWO DIRECTIONS

These areas of steel are provided by $\frac{1}{2}$ in. dia. rods at 8 in. centres (area $0 \cdot 30$ sq. in.). Alternate bars from mid-span are bent up at the internal support from the slabs at either side of the beam as shown in Fig. 5.4.

$$\frac{M}{bd_1^2} = \frac{22,300}{12 \times 4\tfrac{1}{4}^2} = 103 \text{ which is satisfactory for } 1:2:4 \text{ concrete.}$$

At the discontinuous edge-support it is reasonable to provide two-thirds of the mid-span steel area, i.e. $\frac{2}{3}$ of $0 \cdot 23$, equals $0 \cdot 15$ sq. in., which is provided by $\frac{3}{8}$ in. dia. stirrups from the beam at 8 in. centres (area $0 \cdot 16$ sq. in.) bent over into the slab.

Consider now the l_y span of 16 ft. $M_y = \beta_y w l_x^2$.

Moment at internal support

$$= 0 \cdot 049 \times 200 \times 11\tfrac{1}{2}^2 \times 12 = 15,500 \text{ in. lb.}$$

which requires $A_{st} = \dfrac{15,500}{20,000 \times (0 \cdot 86 \times 4\tfrac{1}{4})} = 0 \cdot 21 \text{ sq. in.}$

Moment at mid-span
$$= 0.037 \times 200 \times 11\tfrac{1}{2}^2 \times 12 = 11{,}700 \text{ in. lb.}$$

which requires $A_{st} = \dfrac{11{,}700}{20{,}000 \times (0.86 \times 3\tfrac{3}{4})} = 0.18$ sq. in.

These are provided by $\frac{1}{2}$ in. dia. rods at 12 in. centres (area 0.20 sq. in.). The support steel required is actually less than 0.21 sq. in. due to the falling off of the bending moment diagram at the edge of the beam.

At the discontinuous edge-support, $\frac{3}{8}$ in. dia. stirrups at 8 in. centres are provided from the beam as before.

At the discontinuous corner suitable torsion reinforcement would be provided by mats of $\frac{1}{2}$ in. dia. bars in the top and bottom of the slab, the spacing for convenience being 8 in. and 12 in. centres to marry in with the main reinforcements already selected. These torsion reinforcements would extend 2 ft. 6 in. into the slab.

ART. 5.5. FLAT-SLAB CONSTRUCTION

Greater use is now being made of flat-slab construction. The beamless soffite which this gives has many advantages. From the architect's point of view the appearance is in keeping with modern trends and avoids the need for false ceilings; and great flexibility in planning is given. From the engineer's point of view, for industrial structures such as the roofs of reservoirs and chemical stores the avoidance of the beam-rib arrises reduces the risk of spalling from exposure and rust just at the points where the bulk of the reinforcement is situated: and the absence of beam-ribs allows free circulation of air and consequently facilitates ventilation of foul or humid atmosphere. From the contractor's point of view the fixing of shuttering and reinforcement is very much simpler as also is the placing of the concrete. And there are frequently economic advantages in that floor and roof heights can be reduced, thus showing a saving in the height of walls, with consequent saving of weight and cost.

The simplest form of flat-slab is of one thickness throughout, supported on simple prism columns. Generally the simplicity of this form leads to extravagance, as the slab thickness required to take the shear and negative moments at the columns is in excess of requirements to take the smaller positive moments in the spans. For this reason the slab is often thickened out to about the quarter-span points at the columns, and these thickenings on the soffite are known as *drops*. Even with the drops, the shear stresses at the columns may be considerable, and in such cases the heads of the columns are then splayed out in the manner shown in the examples given later.

Design of flat slabs by detailed analysis of the structure as a continuous rigid frame is seldom worth the labour it involves. The empirical method of design given in C.P. 114 (1957) is known from experience to be satisfactory but the whole of the detailed

TABLE 5.3. *Apportionment of M_0 (equation 5.5) in flat-slab panels*

	Apportionment of moments between the column and middle strips expressed as percentages of M_0	
	Column strip	Middle strip
Interior panels:		
with drops		
Negative moments	50	15
Positive moments	20	15
without drops		
Negative moments	46	16
Positive moments	22	16
Exterior panels:		
with drops		
Exterior negative moments	45	10
Positive moments	25	19
Interior negative moments	50	15
without drops		
Exterior negative moments	41	10
Positive moments	28	20
Interior negative moments	46	16

requirements of this empirical method cannot be included here, and the reader is referred to the Code itself. Nevertheless the following broad principles are of interest. Where there are three or more rectangular bays in two directions at right-angles, and the bay lengths do not exceed the bay widths by more than $33\frac{1}{3}$ per cent, and the dimensions of adjacent bays do not vary by more than 10 per cent, then the total of all bending moments, positive and negative, per panel width may be taken as

$$M_0 = \frac{wL_2}{10} \left(L_1 - \frac{2D}{3} \right)^2 \tag{5.5}$$

where w is the total load per unit area,

L_1 is the length of the panel in the direction of the span,
L_2 is the width of the panel at right-angles,
D is the diameter of the column, or column head.

Each panel is then assumed to be divided into *column strips* and *middle strips*, each generally being equal to $\frac{L_2}{2}$. The apportionment of M_0 as between column strip and middle strip, and as between positive moments (along the centre-lines of the panel) and negative moments (along the edges of the panel on lines joining the centres of the columns and around the perimeter of the column heads), is then taken from Table 5.3.

HEAVY INDUSTRIAL SLAB

Let us consider the design of the floor slab of a bulk store for crushed clinker, as might be required as part of a cement-works.

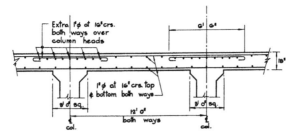

FIG. 5.5. HEAVY FLAT-SLAB WITHOUT DROPS

To suit the arrangements of outlets in the slab serving the under-floor extract conveyors, the columns are arranged at 12 ft. centres in both directions. The superimposed load from the clinker is 2500 lb/sq. ft., and taking a slab thickness of 18 in. without drops, we have:

Self-weight of slab   225 lb/sq. ft.
Superimposed load 2500 lb/sq. ft.

_____

2725 lb/sq. ft.

Suppose the columns are splayed out to form heads 3 ft. square (Fig. 5.5). The critical section for shear will be at a distance from the edge of the column-head equal to half the slab thickness, so that the shear perimeter will be $4 \times (36 + 18) = 216$ in. The total load per column will be

$$12 \text{ ft.} \times 12 \text{ ft.} \times 2725 \text{ lb/sq. ft.} = 393,000 \text{ lb.}$$

and deducting the load within the shear perimeter of

$$4\tfrac{1}{2} \text{ ft.} \times 4\tfrac{1}{2} \text{ ft.} \times 2725 \text{ lb/sq. ft.} = \underline{55,000 \text{ lb.}}$$

we have the total shear $Q = \underline{\underline{338,000 \text{ lb.}}}$

Thus the shear stress

$$q = \frac{338,000}{216 \times 0 \cdot 86 \times 16} = 114 \text{ lb/sq. in.}$$

With the substantial support-moment in this situation, and the bottom steel running through, a $1:1\tfrac{1}{2}:3$ nominal mix concrete will carry this shear satisfactorily.

Considering now the bending in an internal bay, and using the empirical method of design (equation 5.5),

$$M_0 \text{ both ways} = \frac{wL_2}{10}\left(L_1 - \frac{2D}{3}\right)^2 = \frac{2725 \times 12}{10}\left(12 - \frac{2 \times 3}{3}\right)^2$$
$$= 327,000 \text{ ft. lb.} = 3,900,000 \text{ in. lb.}$$

Then from Table 5.3 we have by apportionment,

Column strip:
    Support moment per strip = 46 per cent = 1,790,000 in. lb.
    Centre moment per strip   = 22 per cent =   860,000 in. lb.

Middle strip:
    Support and centre
      moments per strip   = 16 per cent =   625,000 in. lb.

Dividing these by the strip widths of 6 ft., we have
Column strip:
    Support moment   = 300,000 in. lb/ft.
    Centre moment   = 143,000 in. lb/ft.

Middle strip:
    Support and centre moments = 104,000 in. lb/ft.

The compressive stress in the concrete is checked by

$$\frac{M}{bd_1{}^2} = \frac{300,000}{12 \times 16^2} = 98$$

which is satisfactory.

Using mild steel reinforcement, the area required per foot over the column is

$$\frac{300,000}{20,000 \times (0 \cdot 86 \times 16)} = 1 \cdot 09 \text{ sq. in.,}$$

and for the remainder of the slab is well provided by

$$\frac{143,000}{20,000 \times (0 \cdot 86 \times 16)} = 0 \cdot 52 \text{ sq. in.}$$

Thus an overall mesh, top and bottom, of 1 in. dia. bars at 16 in. centres, both ways, would give 0·58 sq. in.; and this would be doubled up in the top face only over the columns to give 1 in. dia. bars at 8 in. centres (area 1·16 sq. in.). Note from Table 3.2 that the permissible average bond stress with $1:1\frac{1}{2}:3$ concrete is 135 lb/sq. in., and from Table 3.3 a bond length is required of 37 dia. beyond the line of the edge of the column head. A U-type hook is equivalent to 16 dia., so the top rods have to extend $(37 - 16) = 21$ in. minimum beyond the edge of the column head.

In a slab of this thickness the bars generally should not be closer than about 16 in. centres so as to allow the men to get their feet down in between. The simplicity afforded by providing uniform meshes of reinforcement in this way often more than pays for the small extra weight of steel involved; and a great part of the labour of bending and placing the steel is avoided.

It should be noted that the maximum value for $\dfrac{M}{bd_1{}^2}$ was only 98, yet the concrete mix is $1:1\frac{1}{2}:3$ and capable of a value up to 254, even ignoring the compression reinforcement which is available in this case. This arises from the slab thickness and concrete mix being determined on considerations of shear stress at the column head, *there being no drop provided*. The great merit of providing drops in appropriate cases is thus evident; though in the present case of a heavy industrial slab, where considerations of simplicity of shuttering and ease of placing large quantities of concrete quickly are of first importance, the design shown is practical and satisfactory. Indeed for a store of this nature, where heavy materials are deposited in bulk from a considerable height, the thickness of 18 in. would be the minimum desirable from the point of view of absorbing the shock of impact.

WAREHOUSE FLOOR-SLAB

A warehouse floor is to carry 150 lb/sq. ft. The floor-finish weighs 18 lb/sq. ft. Columns spaced at 16 ft. centres in both directions are convenient. The depth of beams under the floor would be objectionable, so that a flat-slab design is indicated: indeed for the loading and spans indicated, the flat-slab will present the cheapest form of construction. To avoid damage from trucking, the columns will with advantage be made of octagonal section; and 12 in. across flats is suitable. Column heads, also octagonal, will be made 24 in. across flats (see Fig. 5.6).

Consider the design of an internal bay of this floor. The slab is taken as 5 in. thick, with drops 8 ft. × 8 ft. of $2\frac{1}{2}$ in. projection below the general level of the slab soffite. We have

Self-weight of slab	65 lb/sq. ft.
Self-weight of drops	7 lb/sq. ft.
Floor finish	18 lb/sq. ft.
Superimposed load	150 lb/sq. ft.
	240 lb/sq. ft.

M_0 both ways

$$= \frac{wL_2}{10}\left(L_1 - \frac{2D}{3}\right)^2 = \frac{240 \times 16}{10}\left(16 - \frac{2 \times 2}{3}\right)^2 \times 12$$

$$= 990,000 \text{ in. lb/8 ft. strip} = 124,000 \text{ in. lb/ft.}$$

From Table 5.3 we have by apportionment

Column strip:

Support moment = 50 per cent = 62,000 in. lb/ft.
Centre moment = 20 per cent = 24,800 in. lb/ft.

Middle strip:

Support and centre moments
= 15 per cent = 18,600 in. lb/ft.

$\dfrac{M}{bd_1{}^2}$ at the support $= \dfrac{62,000}{12 \times 6\frac{1}{4}{}^2}$ (second layer) $= 132;$

at centre of col. strip $= \dfrac{24,800}{12 \times 4\frac{1}{4}{}^2} = 114;$

at the middle strip $= \dfrac{18,600}{12 \times 3\frac{3}{4}{}^2}$ (second layer) $= 110.$

The area of mild steel required to resist the column-strip centre-moment is

$$A_{st} = \frac{24,800}{20,000 \times (0.86 \times 4\frac{1}{4})} = 0.34 \text{ sq. in.}$$

This is provided by $\frac{1}{2}$ in. dia. rods at 6 in. centres (area 0·39 sq. in.), alternate rods being bent up over the supports as shown in Fig.

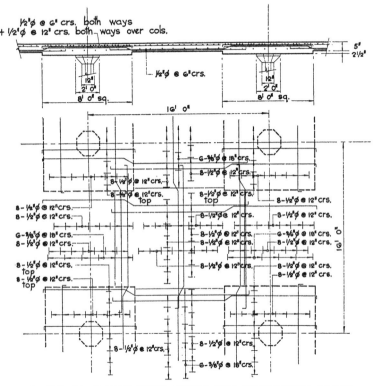

FIG. 5.6. FLAT-SLAB WITH DROPS FOR WAREHOUSE FLOOR

5.6. rods The middle-strip moments are also catered for by $\frac{1}{2}$ in. dia. at 6 in. centres.

At the column head $A_{st} = \dfrac{62,000}{20,000 \times (0.86 \times 6\frac{1}{4})} = 0.58 \text{ sq. in.}$

This is provided by the bent-up centre-span bars = 0·39 sq. in. plus additional $\frac{1}{2}$ in. dia. rods at 12 in. centres = 0·19 sq. in.

$$= 0.58 \text{ sq. in.}$$

The critical section for shear in the 5 in. slab is at a distance of half the slab thickness from the edge of the drop. In our case this is 4 ft. plus $2\frac{1}{2}$ in. from the column centre, i.e. 4 ft. $2\frac{1}{2}$ in.; and the shear perimeter is therefore $4 \times$ (8 ft. 5 in.) = 404 in. The average shear force along this line is

$$Q = (16^2 - 8{\cdot}4^2)240 = 44{,}600 \text{ lb.}$$

whence the shear stress is

$$q = \frac{44{,}600}{404 \times (0{\cdot}86 \times 4\frac{1}{4})} = 30 \text{ lb/sq. in.}$$

The critical section for shear in the $7\frac{1}{2}$ in. drop is at a distance of half the drop thickness from the edge of the column head. This is 12 in. $+ 3\frac{3}{4}$ in. $= 15\frac{3}{4}$ in. from the column centre; and the shear perimeter is $(4 \times 31\frac{1}{2}) = 126$ in. The average shear force along this line is

$$Q = (16^2 - 2{\cdot}62^2)240 = 60{,}000 \text{ lb.}$$

whence the shear stress is

$$q = \frac{60{,}000}{126 \times (0{\cdot}86 \times 6\frac{1}{4})} = 89 \text{ lb/sq. in.}$$

This shear stress, and a maximum $\dfrac{M}{bd_1^2}$ value of 132, are both accommodated by a 1 : 2 : 4 nominal concrete mix.

ART. 5.6. LIFT-SLAB CONSTRUCTION

A development of the flat-slab is found in the technique of lift-slab construction. Here all slabs for floors and roof (without drops) are cast one upon the other on the ground, and later lifted one at a time to their final required levels. The Authors have successfully used this form of construction on a large industrial project in Great Britain; though otherwise, up to the present time, the technique has more generally been limited to works in America. The procedure is as follows.

Structural steel columns of standard rolled sections are first erected with rigid base connections so as to stand free for the full height of the building as individual vertical cantilevers. The ground-floor slab is then concreted, trowelled to a true and even finish, and left to set and harden. The quality of this surface is important, because the ground-floor slab subsequently serves as the soffite form for the first-floor slab, and, apart from aesthetic considerations, any roughness or pitting will increase the key between the slabs, and later prevent them lifting apart. The ground-floor slab is then covered with a paraffin solution or similar medium to prevent adhesion between the ground and first-floor slabs. Next, a special steel collar is threaded over each column and lowered on to the ground-floor slab. The first-floor slab is then poured at ground level incorporating the steel collars (which act

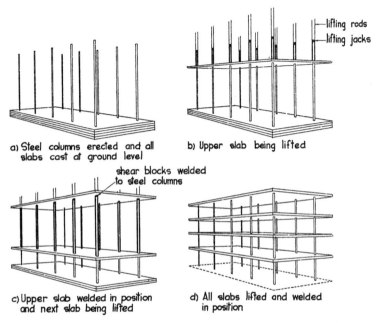

a) Steel columns erected and all slabs cast at ground level

b) Upper slab being lifted

shear blocks welded to steel columns

c) Upper slab welded in position and next slab being lifted

d) All slabs lifted and welded in position

FIG. 5.7. STAGES IN THE OPERATION OF "LIFT-SLAB" CONSTRUCTION

now as formwork keeping the new concrete clear of the steel columns, and later as shear connections from which the whole of the slab load will be supported). And so the procedure repeats itself as many times as necessary until all the slabs have been concreted at ground level, one on top of the other, separated only by the film of paraffin solution.

Hydraulic jacks are now placed one on top of each column; and two threaded lifting-rods are connected from the jacks to the lifting-collars in the roof-slab still lying on the stack at ground level. The stroke of the jacks is normally 3 in., and when the roof-slab has been lifted by this amount, nuts on the rods at the jack take a temporary hold of the load, and the cylinder of the jack is dropped back ready to make another 3 in. lift. In this way the roof slab is lifted, stage by stage, to its final position at the top of the steel columns, and steel seating plates are welded to the sides of the steel columns immediately beneath the lifting-collars, so that the slab is firmly held in its final elevated position. The lifting rods are then released and connected to the collars on the upper-most floor-slab, which in turn is jacked up to its final level; and so the procedure continues. The various stages of construction by the lift-slab method are indicated diagrammatically in Fig. 5.7.

The slab at each floor or roof level is divided in plan into separate sections so as to limit the area of work lifted in any one operation. Normally the divisions are arranged so that there are not more than twelve columns in any section; thus there are only twelve jacks to be controlled at a time. In this way it is easier to ensure a uniform rate of lifting, and so avoid unequal loading of the jacks or excessive distortion and stressing of the slab. A check can be put on the relative travel of the jacks after each 3 in. lift, and corrections made where necessary. Between the sections lifted in separate operations, a gap is left about 3 ft. wide with continuity reinforcements projecting from the slabs on both sides; and after adjacent sections have been lifted to their final level this gap is filled in with concrete placed in-situ on formwork in the normal way.

The advantages of lift-slab construction include the following. All slabs are concreted at ground level with economy of normal hoisting or pumping equipment, and with reduced risk of damage from exposure. Soffite shuttering and all attendant propping is avoided, the only formwork required being initially at the slab edge, and later for the make-up strips between the lifted slab sections. These features can result in a considerable saving of time, and force further saving in that electricians and other trades who have equipment to be concreted into the slabs are compelled to complete their work expeditiously in order not to hold up the concreting operations. The quicker erection of the slabs then provides earlier cover from weather. Following trades who are acquainted with this method of construction have learnt the ad-vantage of placing their equipment and other materials on the slabs before they are lifted. Thus the roof-finisher, by placing his equipment and materials on the roof slab while it is still lying at ground level, can have everything hoisted for him to the full height of the building in the normal course of construction procedure.

With lift-slab constructions it is desirable to set the external columns in some distance from the edge of the building. In this way an edge cantilever-moment is developed in the slab and continuity achieved. Indeed to effect maximum economy, the edge columns should be set in by such an amount that the cantilever-moments roughly balance the full support-moments from the adjoining span. In any event the external columns have to be kept in some 12 in. so that the lifting-collar can be accommodated in the slab and properly held without risk of its shearing off.

In addition to the calculations required for the building in its completed state, separate calculations are necessary to check the conditions that arise during the lifting operations. For example when lifting starts on the uppermost slab, the columns have a slenderness ratio and freedom considerably greater than in the final state of the building: although on the other hand the loading conditions will be less. At this stage, also, the slabs require fuller analysis than given by the empirical method: and as the lifting rods clearly can offer no stiffness or restraint at the points of support it is inadmissible to provide only for the reduced negative moments at the critical sections for shear immediately adjacent to the columns – and the full values have to be taken. Later, when the collars are welded to the columns in their final positions, it becomes possible to develop some stiffness in the joint.

In the case of tall buildings in America where it would be inconvenient in the initial stages of construction to have the steel columns standing free for the full height of the building, the following arrangement has been adopted. Columns of something more than (say) three storey heights of the building are first erected free-standing in the normal way, and then all suspended slabs cast one upon the other at ground level. The upper slabs are then lifted together to the top of the columns and temporarily secured. The remaining three lower slabs are then raised to their final positions and secured; and later the columns are extended in stages as suitable, and the upper slabs subsequently jacked to their final levels and secured.

ART. 5.7. STAIR SLABS

Stairs of normal arrangement are designed in the same manner as described in Article 5.2 for slabs spanning in one direction. In the cases of stairs with open wells, where the ends of spans cross one another at right angles to form quarter-landings, the reinforcement from both spans has of course to carry through, so that the quarter-landings become reinforced in two directions. The effect of this, combined with the enhanced torsional stiffness arising from the continuous support given on two sides of the quarter-landings, enables the effective spans to be regarded as extending

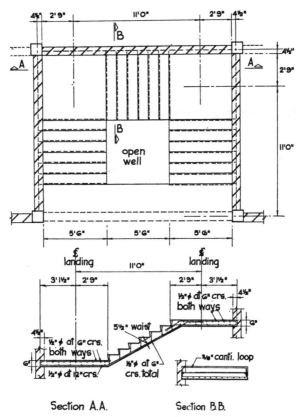

Section A.A.   Section B.B.

FIG. 5.8. STAIR WITH QUARTER-LANDINGS

only to the centres of the quarter-landings, and at these points complete end-fixity may be assumed.

Consider for example a staircase arranged as shown in Fig. 5.8. The slabs are supported continuously at the outer edges of the stair-well by $4\frac{1}{2}$ in. bearing in chases left in the 9 in. brick walls. In this way the beams in the building-frame can all be arranged at floor levels and horizontally, greatly simplifying the work of the bricklayer. Allowing now for the increased dead-load from the concrete forming the treads above the $5\frac{1}{2}$ in. thick slab-waist, an inclusive design load of 250 lb/sq. ft. is appropriate. Each flight spans effectively 11 ft. 0 in.; and with full continuity we have

$$M = 250 \times \frac{11^2}{12} \times 12 = 30{,}200 \text{ in. lb/ft.}$$

requiring an area of mild steel reinforcement of

$$A_{st} = \frac{30{,}200}{20{,}000 \times (0\cdot86 \times 4\frac{3}{4})} = 0\cdot37 \text{ sq. in.}$$

which is provided by $\frac{1}{2}$ in. dia. rods at 6 in. centres (area $0\cdot39$ sq. in.). The effective depth of $4\frac{3}{4}$ in. allows for certain of the reinforcements at the 6 in. slabs at the quarter-landings being in a second layer.

Cross-rods in the sloping slabs are provided by canti-loops $\frac{3}{8}$ in. dia. which are arranged just behind each riser. In this position the loops help knit together the concrete in the treads when the riser-forms are removed: at the same time they give some indeterminable amount of cantilever action from the brick walls, taking advantage of the maximum depth of the member.

CHAPTER 6

COLUMN DESIGN

ART. 6.1. INTRODUCTION

CONCRETE is strong in compression: and since the basic function of columns in most normal frameworks is to carry vertical loads down to some lower level (generally to the foundations) it might seem at first that here at least was a component it should be possible to design without a great deal of intricate calculation. And yet in fact the design of columns frequently proves to be as laborious as any other part of a reinforced-concrete frame, if not more so. This is because in continuous rigid-jointed construction, part of the effect of loads on beams or slabs is the transmission of bending into the columns. Even internal columns in frameworks of structural symmetry are subject to out-of-balance bending moments, as for example when superimposed loads are applied on the floor on one side of the columns but not on the other. And the empirical method of design for flat-slabs (Article 5.5) is very reasonably dependent on internal columns being designed to resist bending moments equal to 50 per cent of the negative moment in the column strips of the slabs they support.

Clearly at external columns (unless the columns are set in some distance) the matter of bending becomes more severe. Here the dead load of beams and slabs are themselves also out-of-balance; and the abrupt termination of the beam or slab system places the whole burden of any end-fixity on to the columns alone. Thus the deflection of the incoming beams or slabs makes for eccentricity or rotation of the columns, the amount of this depending on the relative stiffnesses of the beams and the columns (as shown in Fig. 3.46) so that the larger a column, the stiffer it becomes and the greater the bending-moment it picks up. (This is considered further in using Table 6.1 in Article 6.4.) Thus the size of a column influences the amount of bending for which it has to be designed; so that a certain amount of skill and experience are necessary if the process is not to become too wearisome, particularly as the analysis of members when subject jointly to bending and direct compression, is itself somewhat arduous (*see* Bending and Direct Compression Combined in Articles 3.2 and 3.3).

Even for the limited number of cases of columns supposedly

subject to a truly concentric load, it is not prudent to rely on
plain concrete unreinforced, and C.P. 114 (1957) stipulates that a
minimum percentage of reinforcement should be provided, the
longitudinal steel being not less than 0·8 per cent of the gross
cross-sectional area of the column required to transmit all loading.
Thus long columns are prevented from buckling (a phenomenon
arising from lack of uniformity of the member, and resulting –
curiously enough – in tension on one side of the compression
member) and columns in redundant frameworks are protected
against failure in tension in such circumstances as when the

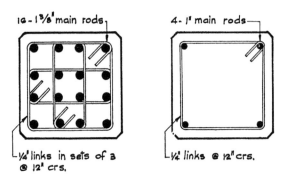

FIG. 6.1. MAXIMUM AND MINIMUM REINFORCEMENTS IN
COLUMN 18 IN. SQUARE

surrounding floors near to the column are unloaded above but
heavily loaded below, or when the frames are subject to unequal
settlements or the like. C.P. 114 (1957) also sets an upper limit
of 8 per cent for the longitudinal steel in columns. The object of
this is to preclude such a concentration of steel as would make it
difficult to place the concrete and consolidate it properly without
requiring an undesirably high water/cement ratio.

Thus a column of 18 in. square cross-section (area 324 sq. in.)
should have not less than four 1 in. dia. rods (area 3·14 sq. in.,
being 0·97 per cent), nor more than sixteen 1⅜ in. dia. rods (area
23·8 sq. in., being 7·4 per cent). These two arrangements are
indicated in Fig. 6.1 where the difficulty of placing concrete
between the rods of the heavily reinforced section is apparent.
Normally designers avoid reinforcing columns beyond about 4 per
cent in order to ensure good sound concrete. The arrangement of
transverse links in Fig. 6.1 provides restraint against outward
buckling of each of the longitudinal reinforcements. Links are
frequently made one-quarter the diameter of the main rods they

tie, though for practical reasons they should generally be no less than $\frac{5}{16}$ in. diameter, otherwise they tend to allow the main rods to wander in the formwork, and may themselves be pushed out of shape in the process of concreting.

At the prevailing relative costs of concrete, steel and formwork, concentrically loaded columns with a low percentage of steel are more economical per ton of load supported than columns with a higher percentage of steel. It is true that a high steel content makes for smaller columns, which is often a convenience; but the use of a richer concrete mix tends towards the same end, and more economically. Sometimes, however, it becomes necessary to have both: high steel content, as well as rich concrete mix. For example, in multi-storey buildings it is a great advantage to retain one column size through two or more storey-heights: this simplifies the formwork to the columns, as well as to the beams and slabs, and allows also for greater repetition in partitioning and the like.

ART. 6.2. CONCENTRICALLY LOADED COLUMNS

In Article 3.1 the Code formula for the safe concentric load on a reinforced concrete column is given as

$$P_0 = p_{cc}A_c + p_{sc}A_{sc}.$$

The justification for the use of this formula, based on the plastic behaviour of the concrete, is also given. The true behaviour of a column under load within the range of normal working stresses has been discussed in Article 2.7 on Creep.

Consider now the design of an internal column for an office-block building as shown in Fig. 6.2. The arrangement of floor framing is symmetrical, and the superimposed loading is not a large part of the total which therefore can be regarded as reasonably balanced about the column. Such a column would normally be designed as though it were concentrically loaded. The working stress in the concrete would then be limited to p_{cc} (the permissible stress in direct compression), whereas if bending has been taken into account, the working stress could have been taken as p_{cb} (the permissible compressive stress in bending) which is about 1·3 times greater than p_{cc}. Thus there is a hidden margin here of 30 per cent to cater for some inevitable bending effects ignored by the simplifying assumption.

Each internal column supports 280 sq. ft. of roof or floor area. The total inclusive loadings for roof and floors are in this case 110 lb/sq. ft. and 170 lb/sq. ft. respectively. Floor heights are 11 ft. The column loads are therefore as follows:

			lb.	lb.
Load from roof:	280 sq. ft. × 110 lb/sq. ft.	=	30,800	
	Self wt. column (say)		1,200	
				32,000
	Load above 6th floor			
Load from 6th floor:	280 sq. ft. × 170 lb/sq. ft.	=	47,600	
	Self wt. column (say)		1,200	
				48,800
	Load above 5th floor			80,800
Load from 5th floor:				48,800
	Load above 4th floor			129,600
Load from 4th floor:				48,800
	Load above 3rd floor			178,400
Load from 3rd floor:				48,800
	Load above 2nd floor			227,200
Load from 2nd floor:				48,800
	Load above 1st floor			276,000
Load from 1st floor:				48,800
	Load to foundation			324,800.

If a $1:2:4$ nominal concrete mix is used, we have $p_{cc} = 760$ lb/ sq. in. The reinforcements will be mild steel so that $p_{sc} = 18,000$ lb/sq. in. Using equation 3.2 we have $P_0 = p_{cc}A_c + p_{sc}A_{sc}$. For a 10-in. square column with four $\frac{5}{8}$ in. dia. rods (1·2 per cent)

$$P_0 = (760 \times 98·8) + (18,000 \times 1·2) = 96,800 \text{ lb.}$$

This will suit well from roof level down to 5th floor level. Actually a smaller column, say 8 in. square, would suffice to carry the roof load, but the difficulty of accommodating the rods from the incoming beams, together with the trouble of making special shuttering for the columns of one storey-height only, would more than outweigh any savings in cost of concrete. For convenience we will adhere to the 10-in. square column under the 5th floor also, but additional reinforcement will be required. With four 1 in. dia. rods (3·1 per cent) we have

$$P_0 = (760 \times 96·86) + (18,000 \times 3·14) = 130,000 \text{ lb.}$$

which is satisfactory.

Below 4th floor level a change in column section is necessary. Using a 12-in. square section with four $1\frac{1}{8}$ in. dia. rods (2·8 per cent) we have

$$P_0 = (760 \times 140) + (18,000 \times 4) = 178,500 \text{ lb.}$$

And below 3rd floor the 12-in. square column with eight $1\frac{1}{8}$ in. dia. rods (5·6 per cent) will carry 249,000 lb.

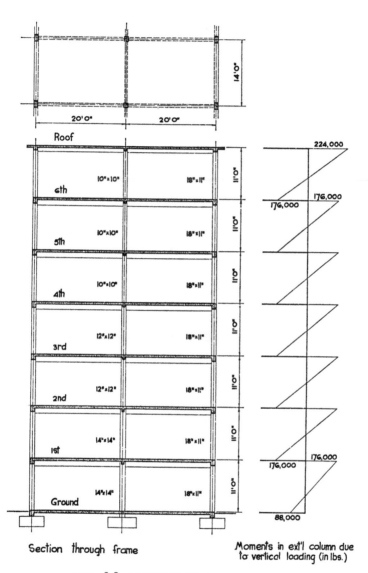

FIG. 6.2. COLUMNS IN OFFICE-BLOCK

Another change in section, to 14 in. square, becomes necessary below 2nd floor level, and with eight $1\frac{1}{8}$ in. dia. rods (4·1 per cent) this will carry 287,000 lb. Below 1st floor it is convenient to adhere to the same size of column and reinforcements, but to increase the concrete strength. Thus by using a $1:1\frac{1}{2}:3$ nominal concrete mix ($p_{cc} = 950$ lb/sq. in.) we have

$$P_0 = (950 \times 188) + (18,000 \times 8) = 323,000 \text{ lb.}$$

These results may be summarised as follows:

Position	Size	Concrete mix	M.S. reinft.	Applied load	Safe P_0
				(lb.)	(lb.)
Under roof	10 in. × 10 in.	1:2:4	$4 - \frac{5}{8}$ in.	32,000	
Under 6th floor	10 in. × 10 in.	1:2:4	$4 - \frac{5}{8}$ in.	80,800	96,800
Under 5th floor	10 in. × 10 in.	1:2:4	$4 - 1$ in.	129,600	130,000
Under 4th floor	12 in. × 12 in.	1:2:4	$4 - 1\frac{1}{8}$ in.	178,400	178,500
Under 3rd floor	12 in. × 12 in.	1:2:4	$8 - 1\frac{1}{8}$ in.	227,200	249,000
Under 2nd floor	14 in. × 14 in.	1:2:4	$8 - 1\frac{1}{8}$ in.	276,000	287,000
Under 1st floor	14 in. × 14 in.	$1:1\frac{1}{2}:3$	$8 - 1\frac{1}{8}$ in.	324,800	323,000

In the above no allowance has been made for the fact that the concrete in the lower columns will necessarily be some 6 months old before the upper part of the building can be completed and loaded. Thus, using the age-factors as given in Table 2.2, the column under 1st floor will be good for

$$P_0 = (1·20 \times 950 \times 188) + (18,000 + 8) = 358,000 \text{ lb.}$$

This shows a surplus of strength of 11 per cent over requirements; and clearly it would have sufficed to use eight 1 in. rods, giving

$$P_0 = (1·20 \times 950 \times 189·7) + (18,000 \times 6·3) = 329,000 \text{ lb.}$$

In the same way some small savings could be made on the other lower storey-height columns, though the effect becomes less as we go higher up the building because the age of the concrete before it receives its design load reduces progressively.

Nor in the previous calculations was any allowance made for the reduction of superimposed floor-loads as permitted for multi-storey buildings. Such an allowance is granted to cater for the improbability of all floors at any column being loaded simultaneously to the full design amount. This of course is a matter the designer is likely to have views on; and clearly if he knows it is the intention that all floors will be loaded simultaneously to the full amount of the design-loading, he will make no reduction when collecting the loads for column design. In the present case the maximum permissible saving on this score could amount to about 6 per cent on the lower storey-height columns, but progressively less higher up the building.

ART. 6.3. ECCENTRICALLY LOADED COLUMNS

It has already been discussed that in most practical cases of column design, direct load is associated with bending. This bending may be regarded as an eccentricity of the direct load, so that if

P is the direct load on the column,

M is the applied bending moment, then

e is the eccentricity, equals $\dfrac{M}{P}$.

It is shown in Chapter 3 that the method of column calculation varies depending on whether the eccentricity is sufficient to develop tension on one side of the member. The examples which now follow are for both cases, i.e. where tension does develop, and where it does not. In both cases, analyses are made by the Elastic Theory (Art. 3.2) and by the Plastic Theory (Art. 3.3). Stresses throughout are limited to

$$p_{cb} = \quad 1000 \text{ lb/sq. in.,}$$
$$p_{st} = 20{,}000 \text{ lb/sq. in.,}$$
$$p_{sc} = 18{,}000 \text{ lb/sq. in.}$$

CASE WHERE TENSION DOES NOT DEVELOP

Consider the case of a column subject to a direct load P of 200,000 lb. and a bending moment M of 240,000 in. lb. Let us try a column 16 in. deep by 12 in. wide with six $1\frac{1}{8}$ in. dia. rods as shown in Fig. 6.3.

(i) *Elastic design (without use of design curve)*

From equation 3.34 we have the equivalent area of concrete

$$A_E = bd + (m - 1)A_s = (16 \times 12) + (14 \times 6) = 276 \text{ sq. in.}$$

From equation 3.35 the second moment of area is

$$I_E = \frac{bd^3}{12} + (m - 1)A_s\left(\frac{d}{2} - d_2\right)^2 = \frac{12 \times 16^3}{12} + (14 \times 6 \times 6^2)$$
$$= 7110 \text{ in.}^4$$

Then from equation 3.36 the stress in the concrete is

$$f = \frac{P}{A_E} \pm \frac{My}{I_E} = \frac{200{,}000}{276} \pm \frac{240{,}000 \times 8}{7110} = 725 \pm 270$$

$= 995$ lb/sq. in. max. (or 455 lb/sq. in. min., showing that no tension develops).

The maximum compressive stress in the reinforcement is

$$\left[725 + \left(270 \times \frac{6 \text{ in.}}{8 \text{ in.}} \right) \right] \times 15 = 13{,}900 \text{ lb/sq. in.}$$

This is given here for interest only, since clearly in the limit, with

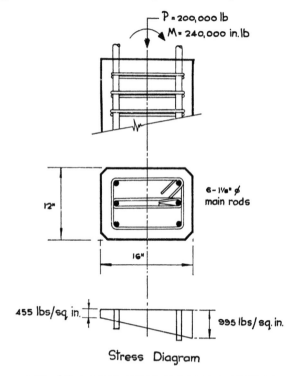

FIG. 6.3. ECCENTRICALLY LOADED COLUMN
(NO TENSION DEVELOPS)

p_{cb} restricted to 1000 lb/sq. in., it cannot exceed $1000 \times 15 = 15{,}000$ lb/sq. in. which is always within the permissible range.

(ii) *Elastic design (using design curve)*

It would not be normal to use a design curve for such a simple problem. But for those who prefer, the method would be as follows.

Eccentricity $\quad e = \dfrac{M}{P} = \dfrac{240{,}000}{200{,}000} = 1{\cdot}20$ in.

whence we have $\qquad \dfrac{e}{d} = \dfrac{1{\cdot}2}{16} = 0{\cdot}075.$

Now $\quad r_p = \dfrac{A_s}{bd} \times 100 = \dfrac{6}{16 \times 12} \times 100 = 3{\cdot}13$ per cent.

Then reading from the design curve (Fig. 3.17), we have

$$\frac{P}{bdf_{cb}} = 1{\cdot}05$$

whence $f_{cb} = 995$ lb/sq. in. as before.

(iii) *Plastic design*

The simplified theory described in Article 3.3 assuming a rectangular stress block does not strictly apply to the present case, since this theory is based on considerations of an average uniform compressive stress extending from the outer fibre to the neutral axis. In the present example the neutral axis lies well outside the boundaries of the section. However, the adequacy of the section can be checked using the semi-empirical formula given in equation 3.82. First, however, it is necessary to determine P_0, P_b, and e_b.

From equation 3.2 we have

$$P_0 = p_{cc}A_c + p_{sc}A_{sc} = (760 \times 186) + (18{,}000 \times 6) = 2\,49{,}000 \text{ lb}$$

Then from equation 3.77 (and equation 3.76 for X) we have

$$P_b = bd_1Xp_{cc} - A_{sc}(p_{st} - p_{sc})$$
$$= (12 \times 14 \times 0{\cdot}625 \times 760) - 3(20{,}000 - 18{,}000)$$
$$= 73{,}900 \text{ lb.}$$

Since $P = 200{,}000$ lb. is greater than $P_b = 73{,}900$ lb. the design is controlled by compression (as in fact was obvious).

From equation 3.78 we have

$$P_b e_b = bd_1Xp_{cc}\left(\frac{d - Xd_1}{2}\right) + A_{sc}(p_{sc} + p_{st})\left(\frac{d}{2} - d_2\right)$$

$$= 12 \times 14 \times 0{\cdot}625 \times 760\left(\frac{16 - 0{\cdot}625 \times 14}{2}\right) +$$
$$+ 3(18{,}000 + 20{,}000)6$$

$$= 973{,}000 \text{ in. lb. Whence } e_b = \frac{973{,}000}{73{,}900} = 13{\cdot}17 \text{ in.}$$

Thus with a known eccentricity of $\dfrac{240{,}000}{200{,}000}$ equals $1{\cdot}20$ in., we have the permissible load from equation 3.82 as

$$P = \frac{P_0}{1 + \left(\dfrac{P_0}{P_b} - 1\right)\dfrac{e}{e_b}} = \frac{249{,}000}{1 + \left(\dfrac{249{,}000}{73{,}900} - 1\right)\dfrac{1{\cdot}20}{13{\cdot}17}} = 205{,}000 \text{ lb.}$$

This is just in excess of the actual load to be carried and is therefore satisfactory.

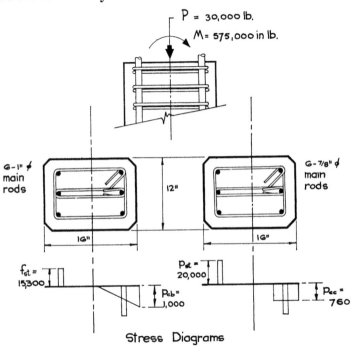

P = 30,000 lb.

M = 575,000 in lb.

6-1" ⌀ main rods

12"

6-7/8" ⌀ main rods

16"

16"

$f_{st} =$ 15,300

$P_{st} =$ 20,000

$P_{cb} =$ 1,000

$P_{cc} =$ 760

Stress Diagrams

(i) Elastic Design   (ii) Plastic Design

FIG. 6.4. ECCENTRICALLY LOADED COLUMN
(TENSION DEVELOPS)

CASE WHERE TENSION DOES DEVELOP

Consider the case of a column subject to a direct load P of 30,000 lb. and a bending moment M of 575,000 in. lb. Here the eccentricity is $\dfrac{M}{P} = \dfrac{575{,}000}{30{,}000} = 19{\cdot}2$ in., bringing the load well outside the section of any appropriately sized column. Let us decide upon a column section 16 in. deep by 12 in. wide, and determine the amounts of reinforcement required, first according to the Elastic Theory and later according to the Plastic Theory. In a problem with considerable bending such as this, it is convenient to make use of design curves, whichever theory is followed.

(i) *Elastic design*

The assumed stress distribution is shown in Fig. 6.4(i). The appropriate design curve is derived in Article 3.2, and given in Fig. 3.18.

We have $\dfrac{e}{d} = \dfrac{19 \cdot 2}{16} = 1 \cdot 20$: therefore tension will develop. If the concrete is to be stressed up to the full permissible value we have

$$\frac{M}{bd^2 p_{cb}} = \frac{575{,}000}{12 \times 16^2 \times 1000} = 0 \cdot 187.$$

Then from the design curve we read off $r_p = 2 \cdot 4$ per cent, so that

$$A_s = 2 \cdot 4 \times \frac{16 \times 12}{100} = 4 \cdot 6 \text{ sq. in.}$$

This is provided by six 1 in. dia. rods (area $4 \cdot 7$ sq. in.). From the design curve we can also read $n_1 = 0 \cdot 495$, and from equation 3.43

$$f_{st} = \left(\frac{1 - n_1}{n_1} \right) m f_{cb}$$

whence $f_{st} = \left(\dfrac{1 - 0 \cdot 495}{0 \cdot 495} \right) 15 \times 1000 = 15{,}300$ lb/sq. in.

(ii) *Plastic design*

The assumed stress distribution is shown in Fig. 6.4(ii). As before, $\dfrac{e}{d} = 1 \cdot 20$: and stressing the concrete to the full permissible value we have

$$\frac{M}{bd^2 p_{cc}} = \frac{575{,}000}{12 \times 16^2 \times 760} = 0 \cdot 246.$$

From the design curve in Fig. 3.29, we have $\dfrac{r_p}{p_{cc}} = 0 \cdot 0023$

whence $r_p = 0 \cdot 0023 \times 760 = 1 \cdot 75$ per cent

so that $A_s = 1 \cdot 75 \times \dfrac{16 \times 12}{100} = 3 \cdot 36$ sq. in.

This is provided by six $\frac{7}{8}$ in. dia. rods (area $3 \cdot 6$ sq. in.).

In this case, where design is controlled by tension in the steel, the reinforcement will be stressed fully to 20,000 lb/sq. in.; whereas in the previous design (by Elastic Theory) the steel stress was only 15,300 lb/sq. in., accounting for the extra area there required.

ART. 6.4. COLUMNS IN CONTINUOUS FRAMES

Beams and slabs bend and deflect under the action of their own weight and the loads they support, and by reason of the stiffness of the joints with the columns, the bending of the beams or slabs induces the columns to bend also. In Article 6.3 it has been shown how to analyse columns subject to such bending: the problem now is to determine the amount of the bending. In other words, how are the bending moments in the columns to be evaluated?

In the case of internal columns which support an arrangement of beams and loading which are reasonably symmetrical, it is normal to design as though the columns were loaded concentrically, as shown in Article 6.2. In the case of flat-slab construction designed in accordance with the empirical method described in Article 5.5, it is a requirement of C.P. 114 (1957) that internal and truly external columns should be designed above and below to resist total bending moments equal to 50 and 90 per cent respectively of the negative moment in the column strips of the slabs they support: this being intended to cater for the worst arrangement of out-of-balance of the superimposed load, i.e. when alternate panels only are loaded.

In the case of external columns which support an unsymmetrical arrangement of beams, the bending moments in the columns may be estimated using the expressions given in Table 6.1 where

M_e is the bending moment at the end of the beam framing into the column, assuming fixity at both ends of the beam,

K_b is the stiffness of the beam,

K_l is the stiffness of the lower column,

K_u is the stiffness of the upper column.

The stiffness of members has already been discussed in Article 3.7. M_e is sometimes known as the *encastré*, or end-fixity moment. It has the following values:

for a uniformly distributed load $\dfrac{Wl}{12}$,

for a point load at mid-span $\dfrac{Wl}{8}$,

for point loads at the two third-points $\dfrac{Wl}{9}$,

where W is the total load on the beam and l is the span.

The expressions given in Table 6.1 assume the columns above and below to be fixed at their remote ends. This may sometimes be near the truth, though certainly is not always so. Nevertheless the expressions are generally accurate enough to give safe results when taken in conjunction with the other provisions of the Code. The difficulty of using the expressions in Table 6.1 is immediately apparent. The moment in the column cannot be determined until K_b, K_l and K_u are known. But the values of K_l and K_u

TABLE 6.1. *Moments in external (and similarly loaded) columns*

	Moments for frames of one bay	Moments for frames of two or more bays
Moment at foot of upper column	$M_e \dfrac{K_u}{K_l + K_u + 0\cdot5\,K_b}$	$M_e \dfrac{K_u}{K_l + K_u + K_b}$
Moment at head of lower column	$M_e \dfrac{K_l}{K_l + K_u + 0\cdot5\,K_b}$	$M_e \dfrac{K_l}{K_l + K_u + K_b}$

are dependent on our knowing first the section required for the column; and we cannot determine this without knowing the value of the moment to be resisted. Thus we have to proceed by a process of trial and error: and the amount of error may be considerable until a certain amount of experience has been obtained. The details of the process are best indicated by following through a worked example.

In Article 6.2 the design was given for an internal column of a seven-storey office-block. Let us now consider the design of one of the external columns of the same building (*see* Fig. 6.2). The inclusive loadings for roof and floors are 110 and 170 lb/sq. ft. respectively. In addition to these, the external beams at roof level carry 200 lb/ft. run being overhang and self, and the external beams at floor levels carry 650 lb/ft. run being brickwork, windows and self. Floor heights are 11 ft.

To suit architectural requirements, the external columns are to be 11 in. one way so as to be concealed flush within an 11 in. cavity wall: and to suit a uniform width of glazing the full height of the building will be made 18 in. the other way. Any small saving which could be made structurally in the 18 in. direction would be more than offset by the additional cost and fuss of making up in brickwork and render to the clear opening widths required to suit the glazing. Thus the size of the external columns is dictated by the architecture to be constant for the full height of the building.

This is quite a common experience, and very suitable structurally as the calculations will show. Furthermore the repetition in form-work, not only to the columns but also to the beams and slabs, makes for economy and saving of time.

The external-column loads are as follows:

			lb.	lb.
Load from roof:				
Main beam	140 sq. ft. × 110 lb/sq. ft.	=	15,400	
Ext. beams	14 ft. × 200 lb/ft.	=	2,800	
Self wt. column	(say)	=	2,000	
	Load above 6th floor			20,200
Load from 6th floor:				
Main beam	140 sq. ft. × 170 lb/sq. ft.	=	23,800	
Ext. beams	14 ft. × 650 lb/ft.	=	9,100	
Self wt. column	(say)		2,000	
				34,900
	Load above 5th floor			55,100
Load from 5th floor:				34,900
	Load above 4th floor			90,000
Load from 4th floor:				34,900
	Load above 3rd floor			124,900
Load from 3rd floor:				34,900
	Load above 2nd floor			159,800
Load from 2nd floor:				34,900
	Load above 1st floor			194,700
Load from 1st floor:				34,900
	Load to foundation			229,600.

It will be convenient to work to the same design stresses as used for the internal column, i.e. $p_{cc} = 760$ lb/sq. in. and $p_{sc} = 18,000$ lb/sq. in. Also $p_{cb} = 1000$ lb/sq. in. and $p_{st} = 20,000$ lb/sq. in.

Roof level:

At roof level we have for the main beam,

total load $W = 20$ ft. × 14 ft. × 110 lb/sq. ft. = 30,800 lb.

giving an end-fixity $M_e = 30,800 \times \dfrac{20}{12} \times 12 = 616,000$ in. lb.

A suitable size for this beam would be 16 in. × 8 in. overall. If this acts with the 5 in. slab as a T-beam, the breadth of slab to be taken into account (*see* Art. 4.2) is the least of

(i) one-third of beam span $= \dfrac{20 \text{ ft.} \times 12}{3} = 80$ in.;

(ii) distance between rib centres = 14 ft. × 12 ft. = 168 in.;

(iii) rib breadth + (slab thickness × 12) = 8 + (5 × 12) = 68 in.

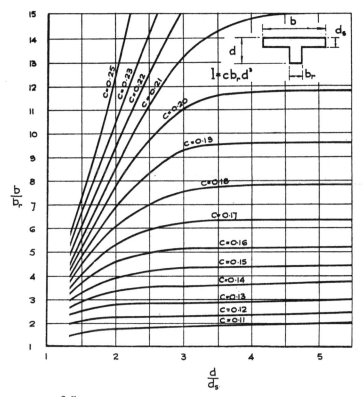

FIG. 6.5. SECOND MOMENT OF AREA OF TEE-BEAMS

The controlling breadth is therefore 68 in.

Thus we have $\dfrac{b}{b_r} = \dfrac{68}{8} = 8 \cdot 5$, and $\dfrac{d}{d_s} = \dfrac{16}{5} = 3 \cdot 2$, whence, from Fig. 6.5, $I_b = 0 \cdot 185 \times 8 \times 16^3 = 6380$ in.[4], so that

$$K_b = \frac{I_b}{L_b} = \frac{6380}{20 \times 12} = 26 \cdot 6 \text{ in.}^3$$

Considering now the upper length of column, we have

$$I = \frac{18 \times 11^3}{12} = 2000 \text{ in.}^4$$

so that $\qquad\qquad K = \dfrac{I}{L} = \dfrac{2000}{11 \times 12} = 15 \cdot 2 \text{ in.}^3$

Then from Table 6.1 the moment at the head of the column is

$$M = M_e \frac{K_1}{K_1 + K_u + K_b}$$

$$= 616,000 \frac{15 \cdot 2}{15 \cdot 2 + 0 + 26 \cdot 6} = 224,000 \text{ in. lb.}$$

We have $P = 20,200 \text{ lb.}$

Therefore $e = \dfrac{M}{P} = \dfrac{224,000}{20,200} = 11 \cdot 1 \text{ in.}$

and $\dfrac{e}{d} = \dfrac{11 \cdot 1}{11} = 1 \cdot 01.$

Now a practical minimum for the reinforcement in a column 18 in. × 11 in. would be six $\frac{3}{4}$ in. dia. rods, so that

$$r_p = \frac{6 \times 0 \cdot 44}{11 \times 18} \times 100 = 1 \cdot 33 \text{ per cent.}$$

Reading then from the design curve in Fig. 3.18, we have

$$\frac{M}{bd^2 f_{cb}} = 0 \cdot 14$$

whence $f_{cb} = \dfrac{224,000}{18 \times 11^2 \times 0 \cdot 14} = 735 \text{ lb/sq. in.}$

Also from Fig. 3.18 we have $n_1 = 0 \cdot 45$,

so that $f_{st} = 735 \times 15 \left(\dfrac{0 \cdot 55}{0 \cdot 45} \right) = 13,500 \text{ lb/sq. in.}$

Thus the column section at roof level is satisfactory, but by no extravagant margin.

6th floor level:

For the main floor beam we have

total load $W = 20 \text{ ft.} \times 14 \text{ ft.} \times 170 \text{ lb/sq. ft.} = 47,600 \text{ lb.}$

giving an end-fixity

$$M_e = 47,600 \times \frac{20}{12} \times 12 = 953,000 \text{ in. lb.}$$

A suitable size for this would be 20 in. × 8 in. overall. The breadth of slab acting with this is 68 in. as at roof level, giving $I_b = 12,500 \text{ in.}^4$, whence $K_b = 52 \cdot 1 \text{ in.}^3$

Considering now the moment at the head of the column under 6th floor, we have from Table 6.1

$$M = M_e \frac{K_1}{K_1 + K_u + K_b} = 953{,}000 \frac{15\cdot2}{15\cdot2 + 15\cdot2 + 52\cdot1}$$

$$= 176{,}000 \text{ in. lb.}$$

We have $\qquad\qquad P = 55{,}100 \text{ lb.}$

Therefore $\qquad e = \dfrac{M}{P} = \dfrac{176{,}000}{55{,}100} = 3.2 \text{ in.}$

and $\qquad\qquad\qquad \dfrac{e}{d} = \dfrac{3\cdot2}{11} = 0\cdot29.$

With six $\frac{3}{4}$ in. dia. rods as before, giving $r_p = 1\cdot33$ per cent, we have from Fig. 3.18

$$\frac{M}{bd^2 f_{cb}} = 0\cdot13,$$

so that $f_{cb} = 620$ lb/sq. in. and $f_{st} = 2000$ lb/sq. in. Thus both the concrete and steel stresses are less than at roof level.

Here it will be reasonable to check the effect of bending at the foot of the column above 6th floor level. From Table 6.1

$$M = M_e \frac{K_u}{K_1 + K_u + K_b} = 176{,}000 \text{ in. lb., as before.}$$

$$P = 20{,}200 \text{ lb.}$$

Therefore $\quad e = \dfrac{176{,}000}{20{,}200} = 8\cdot7 \text{ in. and } \dfrac{e}{d} = \dfrac{8\cdot7}{11} = 0\cdot80.$

From Fig. 3.18, with $r_p = 1\cdot33$ per cent, we have $\dfrac{M}{bd^2 f_{cb}} = 0\cdot14$, giving $f_{cb} = 580$ lb/sq. in. and $f_{st} = 9500$ lb/sq. in. Note that the stress in the concrete due to bending is less immediately above 6th floor than immediately below. And the tensile stress in the steel is very reasonable. Lower down the building where the direct loads in the column increase (whereas the bending-moments do not) the tensile stress in the steel will clearly be less. In fact the eccentricities become so small that no tension develops at all. Checks on the effect of bending in the columns immediately above the lower floors is therefore unnecessary.

5th floor level:

The end-fixity moment in the beam is the same as for 6th floor. So also is the distribution of this into the column. Thus $M = 176{,}000$ in. lb. and $P = 90{,}000$ lb., so that $e = 1\cdot9$ in. and

$\dfrac{e}{d} = 0.17$ in. At this point in the building the direct stress in the column exceeds the bending stress, so that no tension develops. Therefore the appropriate design curve is represented by Fig. 3.17. With six $\frac{3}{4}$ in. dia. rods as before ($r_p = 1.33$ per cent) we have $\dfrac{P}{bdf_{cb}} = 0.63$, so that $f_{cb} = 720$ lb/sq. in.; f_{st} is of course nil.

4th floor level:

As before $M = 176,000$ in. lb. but $P = 124,900$ lb. Thus $e = 1.4$ in. and $\dfrac{e}{d} = 0.13$. With six $\frac{3}{4}$ in. rods as before, we have, using Fig. 3.17, $f_{cb} = 880$ lb/sq. in.

3rd floor level:

$M = 176,000$ in. lb. and $P = 159,800$ lb. $e = 1.1$ in. and $\dfrac{e}{d} = 0.10$. Here it becomes necessary to increase the reinforcement to six 1 in. dia. rods so that $r_p = 2.37$ per cent. Using Fig. 3.17, we have $f_{cb} = 890$ lb/sq. in.

2nd floor level:

$M = 176,000$ in. lb. and $P = 194,700$ lb. In the same way, with six 1 in. dia. rods ($r_p = 2.37$ per cent), and using Fig. 3.17, we have $f_{cb} = 1000$ lb/sq. in.

1st floor level:

$M = 176,000$ in. lb. and $P = 229,600$ lb. Here we have the choice of either increasing further the steel percentage, or increasing the concrete strength. For the internal column we used a $1:1\frac{1}{2}:3$ nominal concrete mix below 1st floor level, and it will avoid confusion on the works to do the same for the external column. Thus with $e = 0.77$ in. and $\dfrac{e}{d} = 0.07$ and r_p still 2.37 per cent, we have $\dfrac{P}{bdf_{cb}} = 1.02$, so that f_{cb} is 1130 lb/sq. in., which is satisfactory.

Note in all the above we have been working to p_{cb}, the permissible compressive stress of the concrete in bending. In the lower storey-heights, the proportion of stress attributable to direct load increases, and it is necessary to check that the permissible stress in direct compression is not exceeded. At 1st floor level we have (equation 3.2)

$$P_0 = p_{cc}A_c + p_{sc}A_{sc}$$
$$= (950 \times 193.3) + (18,000 \times 4.7) = 269,000 \text{ lb.}$$

which is in excess of the 229,600 lb. required to be carried, and therefore satisfactory.

At 2nd floor level we have

$$P_0 = (760 \times 193 \cdot 3) + (18,000 \times 4 \cdot 7) = 232,000 \text{ lb.}$$

which again is satisfactory.

These results may be summarised as follows:

Position	Size	Concrete mix	M.S. reinft.	Applied load	Moment above	Moment below
				(lb.)	(in. lb.)	(in. lb.)
Under roof	18 in. × 11 in.	1:2:4	6 – $\frac{3}{4}$ in.	20,200	Nil	224,000
Under 6th floor	18 in. × 11 in.	1:2:4	6 – $\frac{3}{4}$ in.	55,100	176,000	176,000
Under 5th floor	18 in. × 11 in.	1:2:4	6 – $\frac{3}{4}$ in.	90,000	176,000	176,000
Under 4th floor	18 in. × 11 in.	1:2:4	6 – $\frac{3}{4}$ in.	124,900	176,000	176,000
Under 3rd floor	18 in. × 11 in.	1:2:4	6 – 1 in.	159,800	176,000	176,000
Under 2nd floor	18 in. × 11 in.	1:2:4	6 – 1 in.	194,700	176,000	176,000
Under 1st floor	18 in. × 11 in.	1:1½:3	6 – 1 in.	229,600	176,000	176,000

Comparing these with the results obtained in Article 6.2 for the internal column, the following general observations may be made. Whereas the internal column increases in size lower down the building, the external column is conveniently and economically maintained one constant size all the way down. This is because in the external column bending is predominant near the top where the direct load is least; whereas near the bottom, where the load is a maximum, the bending is of less significance. Note also that although the total external-column load is only about two-thirds the total internal-column load, both have roughly the same cross-sectional area. In some cases therefore it may be convenient to make the internal columns near the bottom of a building to the same dimensions as the external columns. Corner columns usually have yet less load than the other external columns, but more bending, since the incoming beams are out-of-balance in two directions.

The same comments regarding age-factors and load reductions, as given in the two concluding paragraphs of Article 6.2 for concentrically loaded columns, apply equally to eccentrically loaded columns.

ART. 6.5. FRAMES SUBJECT TO LATERAL FORCES

In Article 6·4 the effect was considered of bending in columns resulting from the deflection of beams subject to vertical loading. The present article considers the effect on rigid-jointed frames of horizontal shear forces as can result from wind or earthquakes or other forces. Suppose at the two ends of the office-block indicated in Fig. 6.2 there were rigid flank-walls of reinforced concrete or brickwork; these would act as webs of vertical cantilevers

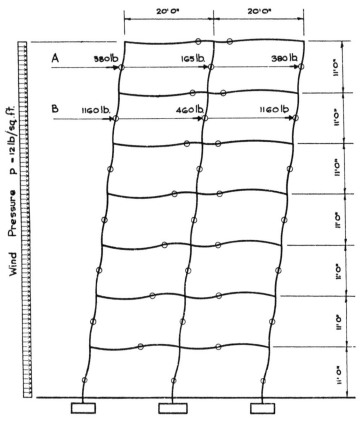

FIG. 6.6. SWAY OF FRAME SUBJECT TO LATERAL
FORCES

springing from ground level and would so stiffen the building against wind forces that no lateral flexing of the structure could occur. If, on the other hand, the flanks of the building were copiously glazed and could afford no bracing effect, the stability of the structure against lateral collapse would depend solely on the flexural rigidity of the beams and columns, and the stiffness of their joints one with another (*see* Fig. 6.6). Let us study this.

If we take the horizontal force of the wind on the face of the building as 12 lb/sq. ft. (*British Standard Code of Practice*, C.P. 3, Chapter V (1952)), the force per 14 ft. wide bay per foot height of building is 14 ft. × 12 lb/sq. ft. = 168 lb. Then with reference to Fig. 6.6 the total horizontal shear per 14 ft. bay down to level A is 5½ ft. × 168 lb/ft. = 925 lb. Level A is the mid-height of the uppermost-storey, and is assumed to be the level of contraflexure in the columns when the roof sways laterally in relation to the 6th floor. The shear force of 925 lb. will be shared between the three columns across the building in proportion to their stiffnesses.

For the external columns $\dfrac{I}{l}$ is 15·2 in.³ (as shown in Article 6.4),

and for the internal column $\dfrac{I}{l}$ under the roof is 6·3 in³. Thus at A each external column takes a shear of

$$\frac{15\cdot2}{2(15\cdot2)+6\cdot3} \times 925 = 380 \text{ lb.}$$

while the internal column takes a shear of 165 lb. The moment at top and bottom of the respective columns is then the shear force times half the column height, which for the external columns is 380 lb. × 5½ ft. × 12 = 25,000 in. lb. and for the internal column is 165 lb. × 5½ ft. × 12 = 10,900 in. lb.

Similarly at B, mid-height between 5th and 6th floors, the shear per bay is 16½ ft. × 168 lb/ft. = 2780 lb., which is shared between external and internal columns as 1160 and 460 lb. respectively, giving moments of 1160 × 5½ ft. × 12 = 77,000 in. lb. and 460 × 5½ ft. × 12 = 30,500 in. lb. respectively. The full results at all levels of the building are tabulated below, taking into account the increased shears lower down the building and the varying stiffnesses of the internal columns.

Position	Total shear per 14 ft. bay	Moments in columns	
		External	Internal
	(lb.)	(in. lb.)	(in. lb.)
Roof to 6th floor	925	25,000	10,900
6th floor to 5th floor	2,780	77,000	30,500
5th floor to 4th floor	4,620	127,000	52,500
4th floor to 3rd floor	6,500	150,000	129,000
3rd floor to 2nd floor	8,350	192,000	166,000
2nd floor to 1st floor	10,200	190,000	292,000
1st floor to foundation	12,000	221,000	352,000

These moments are indicated in Fig. 6.7.

In passing, it is fair to note that in taking the contraflexure points in the columns at their mid-heights, the matter has been oversimplified. This approximation assumes that the relative stiffnesses and rotations at each joint are equal. Clearly this

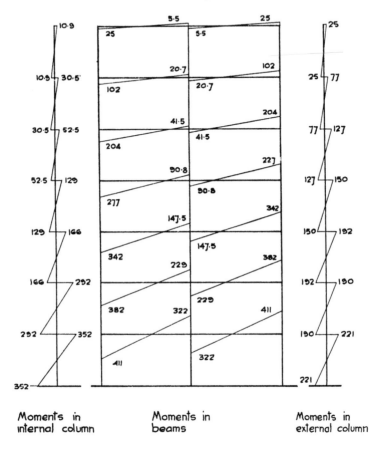

Moments in   Moments in   Moments in
internal column   beams   external column

All moments given in lbs. x 10⁻³

FIG. 6.7. MOMENTS IN FRAME SUBJECT TO LATERAL
SWAY

is not strictly true. For example the uppermost-storey columns are better fixed at their lower ends than at the top where there is no column over: thus the contraflexure point in this length would be nearer the roof than mid-height. Nevertheless the errors involved in this are no greater than other assumptions made, as for example, the force taken to represent the wind!

A check has now to be made to ascertain the effect of these additional moments on the stresses previously calculated in Articles 6.2 and 6.4. Consider first the external columns.

Under 1st floor :

M due to vertical load 176,000 in. lb.
    due to wind   221,000 in. lb.

Maximum $M =$ 397,000 in. lb.

$$e = \frac{397,000}{229,000} = 1\cdot73 \text{ in.} \qquad \frac{e}{d} = \frac{1\cdot73}{11} = 0\cdot157. \qquad r_p = 2\cdot37 \text{ per cent.}$$

Therefore from Fig. 3.17, $\dfrac{P}{bdf_{cb}} = 0\cdot77$, whence $f_{cb} = 1500 \text{ lb/sq. in.}$

Now design-stresses may be increased by 25 per cent, where such increase is caused solely by wind loading, so that with the $1:1\frac{1}{2}:3$ nominal concrete mix, the maximum permissible value for f_{cb} would be $1250 \times 1\cdot25 = 1560 \text{ lb/sq. in.}$ and the design is therefore satisfactory.

Under 2nd floor :

M due to vertical load 176,000 in. lb.
    due to wind   190,000 in. lb.

Maximum $M =$ 366,000 in. lb.

$$e = \frac{366,000}{194,700} = 1\cdot88 \text{ in.} \qquad \frac{e}{d} = \frac{1\cdot88}{11} = 0\cdot171. \qquad r_p = 2\cdot37 \text{ per cent.}$$

$$\frac{P}{bdf_{cc}} = 0\cdot74, \text{ whence } f_{cb} = 1330 \text{ lb/sq. in.}$$

With the $1:2:4$ nominal concrete mix the maximum permissible value for wind is $1000 \times 1\cdot25 = 1250 \text{ lb/sq. in.}$, some 6 per cent less than the actual value. But this is amply taken care of at this level of the building by the age-factor allowance (*see* Table 2.2).

Similarly f_{cb} under 3rd floor works out to be 1300 lb.; and again is taken care of by the age-factor allowance. Under 4th floor, f_{cb} is 1200 lb/sq. in. (less than 1250 lb/sq. in.); and under 5th floor f_{cb} is 1070 lb/sq. in.; and so the values decrease.

Similar values may be calculated to check the wind effect on the internal column as follows.

Under 1st floor:

f_{cb} is 1560 lb/sq. in., as against a maximum permissible value for wind of 1500 lb/sq. in. for $1:1\frac{1}{2}:3$ concrete, to which an age-factor may be applied.

Under 2nd floor:

f_{cb} is 1310 lb/sq. in., as against the permissible 1250 lb/sq. in. for $1:2:4$ concrete to which the age-factor may again be applied.

Similarly under 3rd, 4th and 5th floors, f_{cb} is respectively 1280, 1150 and 1060 lb/sq. in., decreasing at each level.

Note that the moments in the columns above and below any beam make for rotation of the column in the same sense, so that the sum of the column moments has to be resisted by an equal and opposite moment in the incoming beam or beams. This was referred to at the end of Article 4.3.

ART. 6.6. SLENDER COLUMNS

Little reference has been made in the earlier articles of this chapter to the question of buckling. This is the phenomenon of

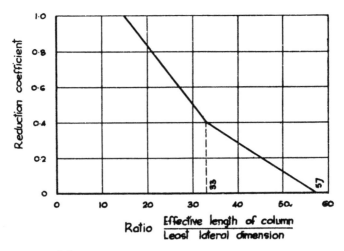

FIG. 6.8. REDUCTION-COEFFICIENTS FOR LOADS ON
LONG COLUMNS

long compression members, when unrestrained along their length, to suffer lateral displacement either as a result of variable elasticity across the width of the section or due to initial curvature of the member as a result of imperfections of workmanship. As the lateral displacement increases, marked bending develops in the column; and a vicious circle is set up, the bending accentuating the displacement and vice versa, until collapse occurs abruptly. With good quality work, and a minimum of 0·8 per cent reinforcement, the likelihood of buckling can normally be discounted in columns where the effective length does not exceed 15 times the width, i.e. up to values of $\dfrac{L}{B} = 15$. Beyond this it is recommended in C.P. 114 (1957) that reduction-coefficients, as given in Fig. 6.8, should be applied to the axial loads as determined by equation 3.2.

Thus a column 12 in. square with four-$\frac{3}{4}$ in. dia. rods which up to 15 ft. effective length would safely carry an axial load of

$$P_0 = (760 \times 142{\cdot}24) + (18{,}000 \times 1{\cdot}76) = 140{,}000 \text{ lb.}$$

would be regarded on an effective length of 33 ft. $\left(\text{where } \dfrac{L}{B} = 33\right)$ capable of carrying safely only $0{\cdot}4 \times 140{,}000 = 56{,}000$ lb.

CHAPTER 7

SPREAD FOUNDATIONS

ART. 7.1. INTRODUCTION

FOUNDATIONS pass the total load from a structure to the ground by direct contact. The load from the superstructure reaches the foundation by means of a number of individual members or units such as columns or walls, and it is the function of the foundation to distribute the load in such a manner that the ground is neither overstressed nor caused to settle more than the super-structure can conveniently accommodate. The problem of foundation design therefore requires a knowledge of the underlying ground, and the nature and requirements of the superstructure; also the effects that both the ground and the foundation have upon one another.

An examination of the ground on which a foundation is to be constructed seldom reveals in simple terms an allowable bearing pressure. This is because in nature the ground occurs in the form of a mixture of many different soil types each with varying properties. Further, even though the ground at any site may be composed of a single soil type to great depth, its index properties may vary considerably throughout its extent, or with the seasons or other outside influences; and all these factors affect the behaviour of the soil when stressed.

It is therefore generally necessary to carry out extensive field and laboratory investigations before foundation design work is started. The results of the investigations then need careful study in order to determine the unit foundation loading which the ground will safely support.

For engineering purposes, ground is considered in two main classes—rocks and soils. Rocks are made up of particles firmly held together so as to form hard and brittle masses. Soils on the other hand are accumulations of separate particles not held rigidly together, the space between the particles being filled generally with water or air.

Although this chapter deals mainly with soils of uniform properties, the behaviour of more complex strata can often be determined by the same test methods and general analyses discussed. Soils may either fail by shearing or may be considered to have failed when there has been undue settlement. If the pressure applied to a soil is just sufficient to cause shear failure, this pressure is known as the *ultimate bearing capacity* of the soil, and if this pressure is then reduced by the application of a suitable load-factor, irrespective of considerations of settlement, the reduced pressure is known as the *safe bearing capacity*. Frequently, however, applied pressures have to be limited by considerations of settlement; otherwise the redistribution of loads and moments within the superstructure result in serious cracking or even structural failure. The *allowable bearing pressure* is the maximum nett loading intensity the soil can support taking into account the safe bearing capacity as well as the amount and type of settlement expected and the ability of the structure to accommodate this. Thus the pressure used for design purposes is controlled not only by soil properties but also by the characteristics of the structure to be supported.

In excavating for a foundation, the weight of the ground removed constitutes an initial relief in pressure at foundation level, and the *nett loading intensity* referred to above is the difference between the actual pressure under the foundation and the original ground pressure at that level due to overburden prior to excavation. This effect is made use of in designing buoyant foundations.

Article 7.2 of this chapter refers to the strength of rocks. Articles 7.3 to 7.9 include a brief introduction to the properties and behaviour of soils, with methods for determining allowable bearing pressures. In Articles 7.10 and 7.11 descriptions are given of various types of foundations suitable for supporting different structures on various types of ground: and calculations are given to determine the sizes of these foundations and the concrete thicknesses and reinforcements.

ART. 7.2. ROCKS

Rocks normally possess great strength, but their safe bearing capacity may depend largely on geological considerations. Safe bearing capacities for rocks evenly bedded and in sound condition are given in Table 7.1, which has been taken from *Civil Engineering Code of Practice*, No. 4 (1954)[1].

Igneous rocks which include granites and basalts have the highest bearing capacities. Metamorphic rocks include gneisses and slates and are also generally of high bearing capacity. Sedimentary rocks are derived from the decomposition of igneous rocks followed by consolidation or cementation and include limestones, sandstones and shales.

Chalks, shales and other softer rocks are liable to vary considerably in strength, and should be tested to ascertain a suitable bearing pressure for foundations. All soft rocks are liable to

TABLE 7.1. *Safe bearing capacities for rocks*

Types of rocks	Safe bearing capacity tons/sq. ft.
Igneous and gneissic rocks in sound condition	100
Massively-bedded limestones and hard sandstones	40
Schists and slates	30
Hard shales, mudstones, and soft sandstones	20
Clay shales	10
Hard solid chalk	6

deteriorate considerably on exposure to air and water and it is important that as soon as the excavations have been bottomed up a protective layer of concrete blinding is laid. It is preferable that the bottoms of excavations in these materials be probed by jack-hammer drilling or similar methods to rule out the possibility of weaknesses immediately underlying the foundation.

All rocks should be inspected in situ for any apparent defects or weaknesses such as heavy shattering, faults, fissures, or steeply

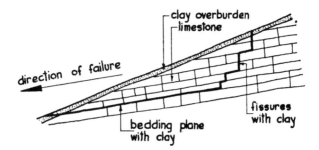

FIG. 7.1. LIMESTONE WEAKENED BY CLAY INCLUSION

dipping bedding-planes. Stratified rocks with bedding-planes dipping towards an exposed face should be treated with suspicion as likely to be unstable: for example a limestone of considerable crushing strength may include bedding planes containing clay washed down from above as shown in Fig. 7.1; the bearing capacity of such a rock formation is clearly influenced by the nature of the clay in terms of slip and squeeze.

ART. 7.3. SOIL TYPES

Soils are formed by the disintegration of rocks due to chemical action, change of temperature, and effects of rain and frost. They are normally classified into types according to their grain size; viz. boulders, gravels, sands, silts and clays. For convenience it is normal to classify soils broadly under two general headings – *cohesionless* and *cohesive*.

Cohesionless soils are made up of individual particles large enough to be seen by the naked eye, and include boulders, gravels and sands. Boulder particles are larger than 60 mm., gravels range

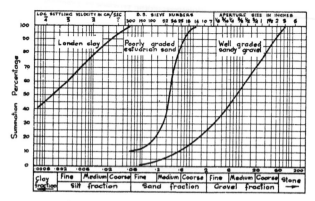

FIG. 7.2. GRAIN SIZE ACCUMULATION CURVES

from 60 mm. to 2 mm., and sand particles range from 2 mm. to 0·06 mm. *Cohesive soils* have finer particles and include silts and clays. Silt size particles range from 0·06 mm. to 0·002 mm., and clay size particles are less than 0·002 mm. Typical grain size accumulation curves for three soil samples are shown on a standard chart in Fig. 7.2.

COHESIONLESS SOILS

Boulders, gravels and sands lack any significant cohesion between the particles; they further lack plasticity – the ability of a soil to undergo appreciable deformation without rupture. Their behaviour under load depends principally on frictional properties which almost alone determine their allowable bearing pressure. The frictional properties depend on the average particle size and shape, the grading of the particles and the degree these are compacted within the soil mass.

Sands have angles of internal friction varying from about

25° in a poorly graded loose state, up to 45° in a well graded state with a high degree of compaction. It is normally sufficiently accurate to assume that the angle of friction of a loose dry sand is equal to its angle of repose. This may be increased by as much as 10° when the same sand is well compacted. Commonly the angle of internal friction of sands is between 30° and 35° in the dry or moist state; and these values are likely to be reduced by a few degrees when the sand is wet.

In the absence of normal pressure, the shearing resistance of cohesionless soils is comparatively small, depending only on the interlocking of the separate particles. Under the action of compression, the internal friction forces increase roughly pro rata to the increase of pressure: and so also, of course, does the shearing resistance. It follows then that the shearing strength of cohesionless soils increases generally with depth.

When submerged, cohesionless soils change in behaviour. The normal compressive pressure between the particles is reduced by hydrostatic buoyancy to about a half, with the consequence that the bearing pressure of the soil is likewise roughly halved. And as a result of the confining pressures being reduced, settlements in cohesionless soils, when submerged, may well be doubled.

COHESIVE SOILS

Silts and clays are susceptible to plastic deformation when loaded. Their strength is derived principally from cohesion between the particles, and this is thought to be attributable to the shearing strength of the boundary zones of water molecules. Since cohesive properties are independent of applied stress, the shearing resistance of cohesive soils may be considerable even in the absence of normal compressive stress.

Silts vary considerably in their properties some approaching those of fine grained cohesionless soils, others exhibiting strong cohesive characteristics. Thus some types of silt may possess angles of internal friction as high as 20°, whereas others may possess practically none. And some silts will exhibit but little plasticity whereas others possess plasticity characteristics of a very high order. Silts generally show an increase of shear strength under applied compressive stress although due to their low permeability this increase may be developed rather slowly. Generally, dense or medium rock flour silts may be considered as similar to very fine grained cohesionless soils, and the more plastic silts as similar to clays. Water flowing through silts can produce an appreciable reduction in stability.

Clays are composed of the finest grained soil particles, and show marked cohesive and plastic properties. Under conditions of no change of moisture content, clay soils behave very much as though they are frictionless, and for many practical purposes this assumption is near enough true. The cohesion of clays may typically vary from less than 100 lb/sq. ft. for extremely soft clays up to more than 4000 lb/sq. ft. for very stiff clays: while medium and stiff clays vary generally between 500 lb/sq. ft. and 2000 lb/sq. ft.

The properties of clays vary considerably with change of moisture content – an excavation in dry clay standing vertically to considerable depths, whereas a clay slurry will subside and flow pretty much as a liquid. Clays also have the property of softening when worked, even though there is no change made in the moisture content. After such change the clay is known as a "remoulded clay" and has a reduced strength; though frequently a large part of the original strength is regained in time. Sensitive clays may be so affected by working that their strength is reduced to between one-quarter to one-eighth of the strength of the original undisturbed clay.

OTHER SOILS

Other soils not referred to above include organic matter such as peat and vegetable matter; and made ground. Peat is a fibrous organic material usually mixed with fine sand or silt, and can occur in various degrees of compaction. Made ground can of course vary between the widest limits, and includes organic materials, industrial wastes, and mixtures of various deposited soils.

Organic soils and made ground vary so widely in composition that foundations constructed on them are liable to undergo considerable settlement. Accordingly such soils are normally considered unsuitable for founding on, and piers or piles are taken through to firmer strata below.

ART. 7.4. SITE INVESTIGATION AND FIELD TESTS ON SOIL

Before any site exploratory work is carried out, the geology of the site has to be fully investigated from geological maps, local knowledge, or previous experience from other engineers working in the same area. Such information reduces the amount of labour and expense in site investigation and will assist in indicating the pattern of exploration to be undertaken.

Unless the scope of investigation can be narrowed by previous knowledge, preliminary exploration should be started at the earliest possible date. The subsequent full-scale investigation is then based on the findings of the preliminary exploration. This fuller investigation is arranged to run simultaneously with the first approximate foundation calculations, since these two guide one another, so that a reasonable opinion of the extent of site investigation can be formed as work proceeds. Inadequate investigation may later cause delay, uncertainty or extravagance in the work: it has also been known to contribute to engineering failures.

On large works the cost of the most comprehensive investigation is usually small compared with the resulting savings on structural design. On smaller jobs it may be more economical to limit the extent of exploration, and increase the factor of safety.

Exploration extends until the nett stresses in the underlying and surrounding strata are proved insignificant. Generally this involves carrying the investigations to a width of about one and a half times the plan dimension of the structure and to a depth of about twice the plan dimension of the structure. The exploration should be sufficiently detailed to reveal all sensible variations within the strata.

TRIAL PITS

The most satisfactory way of examining the ground in its virgin state is from trial pits. Samples may then be conveniently selected and taken for testing. Generally trial pits are not constructed deeper than about 12 ft.; below this it is more usual to make borings.

BORINGS

In unconsolidated soils, shallow borings are made with hand augers working either from ground level or from the bottoms of trial pits. Hand auger bore holes are normally uncased, and are limited in depth to about 20 ft.

For investigations to greater depths and in waterlogged ground, the boreholes are cased by sectional steel tubes. Such boreholes are constructed by one of three methods: *wash boring* in which the soil is brought to the surface by water; *rotary boring* in which a rotating tool cuts the soil; or *percussion boring* in which repeated blows from a chisel tool penetrate the soil or rock.

During boring operations, *disturbed samples* of soil are taken as the boring proceeds by driving a thick-walled sampler tube into the soil at the bottom of the shaft. The sample is recovered from the tube, placed in an airtight jar, and sent to the laboratory for examination. Such samples are useful for general identification and classification purposes only.

To determine the strength properties and consolidation characteristics of a soil, it is necessary to take *undisturbed samples*. In reality it is physically impossible to obtain truly undisturbed samples: nevertheless in cohesive soils the samples are satisfactory for practical testing purposes. The sample is taken by pressing into the soil a thin-walled brass or steel tube, which is then brought to the surface where its ends are sealed with wax, capped and taped. The sample, sealed in its tube, is then sent to the laboratory for testing.

Greater difficulties arise in obtaining undisturbed samples of sand and larger grained sizes because the samples tend to spill out of the tube, and the natural degree of compaction of granular materials is interfered with at the smallest disturbance. These difficulties are accentuated when taking samples below the level of the water table. Many special techniques and tools have been developed in an attempt to overcome these difficulties but these yield no better assessment of the strengths and settlements of sandy soils than do subsurface soundings described later in this article.

Where reliance is placed on the results of tests on samples in determining the allowable bearing pressure on a soil, it cannot be over-emphasised how misleading can be the results of too few isolated samples. The number of samples taken and the number of tests carried out on these should always be generous.

SUBSURFACE SOUNDINGS

Subsurface soundings (or *penetration tests*) are comparative tests only, but very useful in certain circumstances. On sites where the strata are erratic, it may be more important to understand the variations of the strata rather than to determine the precise properties of the soil.

In its simplest form the penetration test consists of gauging the resistance met by a steel rod forced into the ground. Penetrometers are of two main types, *static* and *dynamic*. The static test consists of noting the penetration when the rod with a pointed head is statically loaded by a known dead weight, or by jacking. In the dynamic test the rod is driven into the ground by a hammer of known weight dropping a given height.

The simple procedure described above may be modified by having the driving rod in a loose-fitting steel sleeve so that, all the way down, the rod and sleeve can be forced into the ground independently of one another as desired. In this way the side frictional resistance can be measured separately from the point resistance at all depths. One stage in such a test is shown diagrammatically in Fig. 7.3, which also gives a typical graph of the

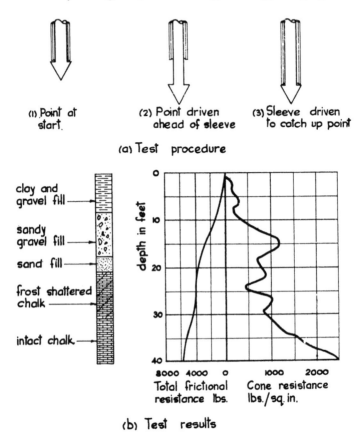

FIG. 7.3. SUBSURFACE SOUNDINGS: PROCEDURE AND RESULTS

point and side friction resistances at all depths as determined from the field results.

Significant changes in ground characteristics, soft pockets or weak layers are thus detected simply and adequately by subsurface soundings. Indeed with cohesionless soils these methods may give as much practical information as any other, though in order to reach a reliable understanding of the findings, it is necessary to sink a small number of borings adjacent to the sub-surface soundings so that the soundings can be interpreted with a knowledge of the soil types through which the tool is passing.

Using a Standard Penetration Test, Terzaghi has related the number of blows per foot of penetration to the relative densities of sands. From this the angle of internal friction, the probable settlement, and the *ultimate* bearing capacity can be deduced. Terzaghi has extended his interpretation to relate the blows per foot to the *allowable* bearing capacity of the sand for footings of different breadths. In Fig. 7.4 Terzaghi's relationship is given

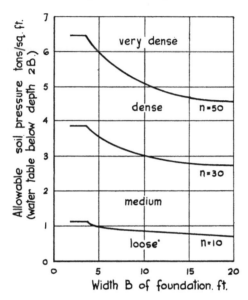

FIG. 7.4. CHART FOR ESTIMATING ALLOWABLE SOIL PRESSURE FOR FOOTINGS ON SAND. (AFTER TERZAGHI.)

for allowable bearing pressures on dry or moist sands, with the maximum settlement limited to 1 in. Allowable pressures on saturated sands should be reduced by 50 per cent for foundations near the surface, and by 33 per cent for foundations at a depth below the surface equal to the foundation breadth. If the saturated sand is very fine or silty further adjustments may be necessary.

LOADING TESTS

Bearing capacities of soils were first determined experimentally from loading tests. This method is still used today, but the test results require skilful interpretation. The simplest arrangement of a loading test is shown in Fig. 7.5a. A stiff bearing plate at the level of the proposed foundation is loaded in stages, and the settlement noted immediately on application of each load increment and subsequently at intervals until settlement at each stage has ceased. Time records are kept. The settlements are measured against a datum outside the influence of the loading test and the

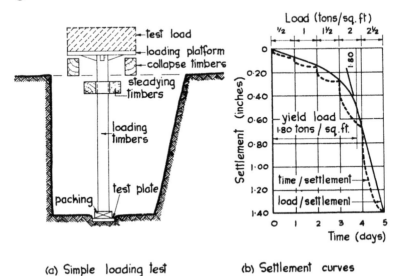

(a) Simple loading test   (b) Settlement curves

FIG. 7.5. LOADING TEST: ARRANGEMENT AND RESULTS

results plotted either in the form of a series of time/settlement curves or as a steady load/settlement curve, as shown in Fig. 7.5b.

The form of the load/settlement graph depends on the nature of the soil. Dense or stiff soils show a sudden increase of settlement once the bearing capacity has been exceeded; whereas loose or soft soils do not display any sudden failure, and it is customary then to take the failure load as the point where the graph dips more steeply or becomes tangential.

The limitations of this method of testing can be appreciated by reference to Fig. 7.6, which compares contours of equal vertical stress for a small loading test and for a full size foundation. It is clear that the test results indicate the nature of the soil to a depth of only about one and a half times the breadth of the test plate; whereas the stress from the actual foundation will extend well into the softer ground below of which the loading test gave no warning.

However, loading tests may well be used to determine the ultimate bearing capacity of plastic clay soils provided the nature of the soil is constant for the stressed depth under the actual foundation: this is because the breadth of foundation has no influence on the bearing capacity of plastic clays. The results will however not give directly the magnitude of probable settlement: but this may be deduced from the fact that the settlement

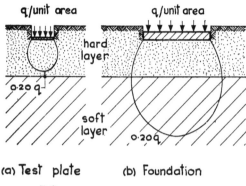

(a) Test plate   (b) Foundation

FIG. 7.6. LIMITATION OF LOADING TEST

of rectangular foundations on a uniform cohesive soil is roughly proportional to the foundation breadth.

For soils where standard sampling and laboratory testing are generally unsuitable, as for example, sands and stoney clays, loading tests may prove most suitable. But for sandy soils the relationship between settlement and foundation breadth is not linear, and as the breadth becomes large, the settlement becomes increasingly independent of the breadth. Terzaghi and Peck[2] give the settlement of a foundation on homogeneous sand as

$$s = s_1 \left(\frac{2b}{b+1} \right)^2 \tag{7.1}$$

where s is the settlement of a foundation of breadth b, and s_1 is the settlement of a test plate 1·0 ft. square under the same intensity of loading.

SHEAR TESTS

As soft and sensitive clays are likely to undergo considerable change when disturbed (as in sampling), it is sometimes convenient to carry out shear tests in situ. For this purpose the cruciform vane shown in Fig. 7.7 is forced into the clay to a depth of about

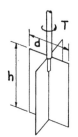

FIG. 7.7.
SHEAR VANE

3 ft. below the bottom of a borehole casing and measurements are made of the torque required to rotate the vane at a specified rate. The shear strength or cohesion of the clay is then given by the equation

$$T = \frac{cd^2}{2}\left(h + \frac{d}{3}\right) \qquad (7.2)$$

where T is the applied torque,

d is the diameter of the vane,
h is the height of the vane,
c is the cohesion of the clay.

ART. 7.5. SOIL TESTING IN THE LABORATORY

Tests on soil samples are split into two main groups, *identification and classification tests* and *engineering property tests*.

IDENTIFICATION AND CLASSIFICATION TESTS

These tests give general information about the soil. The most important include the Index-Property tests which determine the specific gravity of the soil particles, and the bulk density and water content of the soil; and the Mechanical-Analysis tests which determine the sizes of the particles.

With cohesionless soils, it is often useful to know the *relative density* of the soil, since this affects the allowable bearing pressure and settlement. The *voids ratio* of the soil in its natural state is defined as

$$e = \frac{\text{volume of voids}}{\text{volume of soil particles}}, \qquad (7.3)$$

and the *relative density* is defined as

$$\gamma_r = \frac{e_0 - e}{e_0 - e_{min}} \qquad (7.4)$$

where e_0 is the voids ratio of the soil when rearranged into its loosest state,

e_{min} is the voids ratio of the soil when rearranged into its state of maximum density.

It is customary to term soils "loose" when γ_r is less than $\frac{1}{3}$rd, "medium" when γ_r is between $\frac{1}{3}$rd and $\frac{2}{3}$rds, and "dense" if γ_r exceeds $\frac{2}{3}$rds.

With cohesive soils the comparable tests are those to determine the *consistency limits*. These are the *liquid limit* and *plastic limit* tests, which give respectively the moisture content at which the soil passes from the liquid to the plastic state, and from the plastic to the solid state. The moisture content of the actual sample may then be related to these limits, indicating the degree of plasticity of the soil. For this purpose the liquidity index of the soil is given by the expression

$$\text{liquidity index} = \frac{w - w_p}{w_l - w_p}$$

where w is the water content of the sample,

w_p is the water content at the plastic limit,

w_l is the water content at the liquid limit.

If the liquidity index is near to unity, the soil is approaching a liquid state; if the index is zero the soil is then just plastic.

ENGINEERING PROPERTY TESTS

In foundation engineering the more important tests fall into three groups, viz.: *permeability tests, shear strength tests* and *consolidation tests*.

(i) *Permeability tests*

Permeability is the property of a soil in influencing the rate of water flowing through it, and is measured in cms per second. The permeability of a sand may be about 1,000,000 times greater than the permeability of a clay, and this difference affects the relative rates of settlement of the soils. Permeability is determined by standard laboratory or field tests, or more roughly by empirical means following a knowledge of the other index properties of the soil.

(ii) *Shear strength tests*

The shear strength of a soil is made up of the separate resistances due to cohesion and friction.

If c is the cohesion of the soil,

ϕ is the angle of internal friction of the soil,

p_e is the pressure between the soil granules across the shear plane,

s is the shearing resistance,

then $$s = c + p_e \tan \phi. \tag{7.6}$$

This is Coulomb's equation.

The total normal pressure across the shear plane is made up of the sum of the intergranular pressure and the pore water pressure. With some soils under certain drainage conditions, it is necessary to know the pore water pressure in order to interpret the results of shear strength tests; but for many practical purposes the shearing resistance may be taken simply as

$$s = c + p \tan \phi \qquad (7.7)$$

where p is the total normal pressure across the shear plane. The values of c and ϕ given by equation 7.7 are generally satisfactory

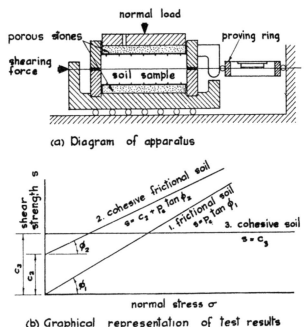

(a) Diagram of apparatus

(b) Graphical representation of test results

FIG. 7.8. SHEAR BOX TEST: APPARATUS AND RESULTS

for use in bearing capacity calculations. The conditions of drainage in shear strength tests are important, and should be kept as near as possible to those likely in the soil under the foundation.

Shearing resistance is measured in the laboratory by one of three tests as follows:

(a) Shear box test

Apparatus for the shear box test is illustrated in Fig. 7.8a. The soil specimen (normally sand) is placed between two stone

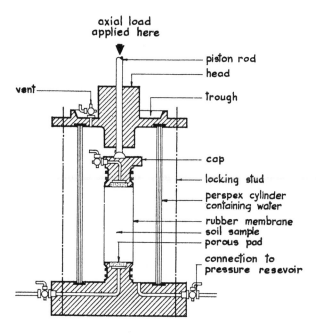

axial load
applied here

piston rod

head

trough

vent

cap

locking stud

perspex cylinder
containing water

rubber membrane
soil sample
porous pad

connection to
pressure resevoir

(a) Apparatus

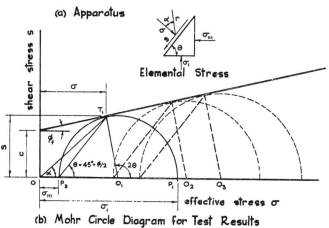

Elemental Stress

shear stress s

effective stress σ

(b) Mohr Circle Diagram for Test Results

FIG. 7.9. TRI-AXIAL COMPRESSION TEST: APPARATUS
AND RESULTS

231

plates within a box split along its horizontal centre-line. One part of the box is moved relative to the other, and the applied shearing force measured. In a series of tests the specimen is sheared under different values of constant normal pressure applied across a plane of rupture. Fig. 7.8b shows the shearing strength results plotted against normal pressure for tests on three different soil types. Each line on the graph represents equation 7.7 – the shear strength value when the normal stress is zero giving the values c, and the inclination of the line giving the values ϕ.

(b) Tri-axial compression tests

In the tri-axial compression test a cylindrical sample contained in a rubber sheath is enclosed in a perspex cylinder filled with water as shown in Fig. 7.9a. By means of a hydraulic pump, the specimen may be subjected to uniform confining pressures; and axial pressure is then applied steadily until the specimen fails by rupture.

The total vertical pressure p_1 and a confining pressure p_3 are the principal stresses within the specimen. Tests are made with different applied pressures, and the results are plotted in the form of Mohr stress circles with diameters along the pressure axis and equal to $p_1 - p_3$ and with centre at $p_3 + \dfrac{p_1 - p_3}{2}$ as indicated in Fig. 7.9b. It can be shown that a line drawn tangential to all the stress circles represents graphically the shear strength equation. In practice not less than three tests should be carried out for each soil sample.

(c) Unconfined compression test

The shear strength of soils with a high clay content may be found more simply from the results of unconfined compression tests. The apparatus used is shown diagrammatically in Fig. 7.10. This test is a special case of the tri-axial test with no applied lateral pressure. The apparatus is fitted with specially calibrated charts so that a stress/strain diagram is recorded automatically as the specimen is loaded axially to failure.

If the material is regarded as frictionless, it is clear that

$$s = c = \frac{q_u}{2} \tag{7.8}$$

where q_u is the unconfined compressive strength of the specimen.

(iii) Consolidation tests

The consolidation test is carried out using an oedometer as

shown in Fig. 7.11a. The soil specimen is contained within a flat
cylindrical ring between two circular porous stones fed with water
from a reservoir. A load is applied to the upper stone, and the compression of the specimen measured on a dial gauge at predetermined intervals of time until no further compression takes place,
usually after twenty-four hours. The load is then increased step by
step and the test repeated for each increment: see Fig. 7.11b. After
the final load has been applied and the maximum settlement
measured, the specimen is unloaded step by step allowing full

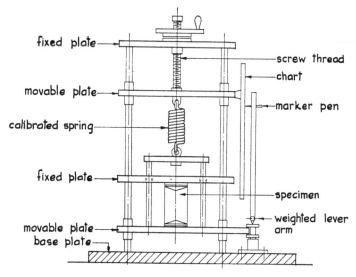

FIG. 7.10. UNCONFINED COMPRESSION TEST
APPARATUS

recovery at each stage, the recovery being measured as before.
When the specimen is fully unloaded its final height is measured.
The application of the results of this test in forecasting the
amounts of settlement of actual structures is given later in
Article 7.7.

ART. 7.6. STRESS DISTRIBUTION IN SOILS

APPROXIMATE METHODS FOR DETERMINING STRESS DISTRIBUTION

In many practical applications it is sufficiently accurate to
assume that the load from a foundation is transferred through the
underlying soil with a constant angle of dispersion of about 30° as

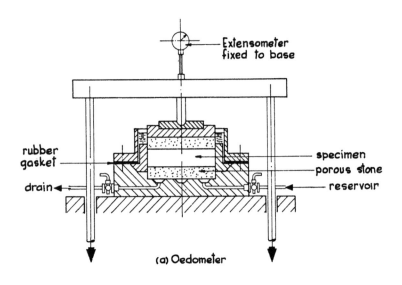

(a) Oedometer

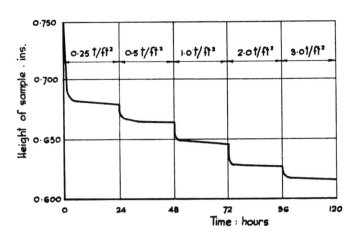

(b) Test Results

FIG. 7.11. CONSOLIDATION TEST: APPARATUS AND
RESULTS

234

shown in Fig. 7.12. The vertical stress p_z on a horizontal plane at depth z below a strip foundation is then given by

$$p_z = \frac{qB}{B + 2z \tan 30} \qquad (7.9)$$

where q is the applied pressure at foundation level, and B is the breadth of the foundation. Equation 7.9 reduces to

$$p_z = q \frac{B}{B + 1.15z}. \qquad (7.10)$$

This method gives no indication of any stress variation across the plane at depth z.

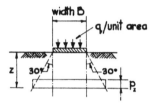

FIG. 7.12. STRESS DISTRIBUTION BY SIMPLE
DISPERSION

A closer approximation to the stress distribution may be obtained quickly by the authors' adaptation of Housel's graphical method as shown in Fig. 7.13. From the two edges a and a' of the strip foundation (width B), lines ab and ac, etc., are drawn at slopes 2 to 1 and 1 to 1 so as to cut any horizontal plane at depth z at points b and c, etc. It is assumed that the maximum stress on the plane occurs over the width aa' and is of such intensity as would occur if the distribution were uniform over width bb'. Thus points d and d' are located, and the line of stress distribution is drawn by joining points c, d, d' and c'. The distance cc' is $2z + B$, and the maximum vertical stress p_z which acts over width aa' is given by

$$p_z = q \frac{B}{B + z}. \qquad (7.11)$$

The stress distribution based on Boussinesq's theory has been shown dotted in Fig. 7.13 for comparison.

BOUSSINESQ'S ELASTIC THEORY FOR STRESS DISTRIBUTION

Boussinesq's theory is based on the soil below the foundation acting as a homogeneous elastic solid. The theory agrees well with the

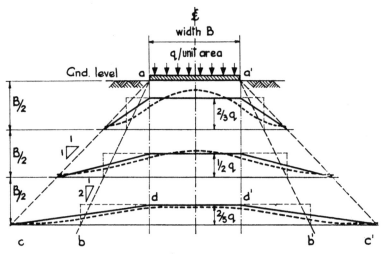

FIG. 7.13. HOUSEL'S GRAPHICAL METHOD OF
STRESS DISPERSION

actual stress distribution at all depths except in the immediate
vicinity of the underside of the foundation.

Adopting the notation indicated in Fig. 7.14a, where P is a
point load at the surface, and p_z is the vertical stress in the soil at
depth z and radius r, Boussinesq's equation is

$$p_z = \frac{-3P}{2\pi} z^3 (r^2 + z^2)^{-\frac{5}{2}} \qquad (7.12)$$

or

$$p_z = k \frac{P}{z^2} \qquad (7.13)$$

where

$$k = \frac{-3}{2\pi} \left[\left(\frac{r}{z} \right)^2 + 1 \right]^{-\frac{5}{2}}.$$

The curve in Fig. 7.14b represents equation 7.13 and gives the
vertical stress in the soil at any depth and radius for a known
applied point load. The shape of the curve in Fig. 7.14b indicates
the stress distribution at any level; and the form of equation 7.13
shows that for a given $\dfrac{r}{z}$ ratio, the stress decreases as the square of
the depth.

It should be noted that Boussinesq's theory relates strictly to a
load applied at a point having no plan dimension. When considering soil stresses at a depth equal to three times the breadth of a

foundation, the error of this assumption becomes negligible. But at lesser depths it is necessary to adapt Boussinesq's theory so as to

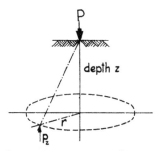

(a) Boussinesq coordinate diagram.

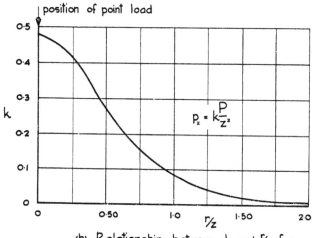

(b) Relationship between k and r/z for vertical soil pressure

FIG. 7.14. BOUSSINESQ'S ELASTIC THEORY FOR STRESS DISTRIBUTION

take account of the physical plan dimensions of the foundation. This follows.

APPLICATION OF BOUSSINESQ'S THEORY TO REAL FOUNDATIONS

Fadum has expressed the stress in a soil under the corner of a rectangular foundation in graphical form as shown in Fig. 7.15a. For any depth z, m and n are known, so that the influence factor k can be read off directly. The vertical stress p_z under the corner of

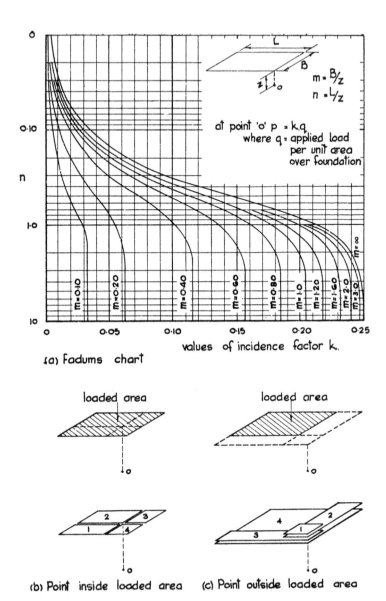

(a) Fadums chart

loaded area

loaded area

(b) Point inside loaded area   (c) Point outside loaded area

FIG. 7.15. FADUM'S METHOD OF STRESS DISTRIBUTION
UNDER REAL FOUNDATIONS

the foundation is then obtained from the equation

$$p_z = kq \qquad (7.14)$$

where q is the unit intensity of pressure under the foundation.
Fadum's method can be extended without limit to determine
stresses at any position (not only under the corners of the foundations) by devices as indicated in Fig. 7.15. For example in Fig.

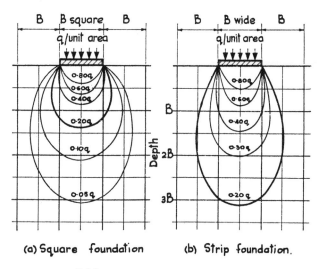

(a) Square foundation   (b) Strip foundation.

FIG. 7.16. BULBS OF CONSTANT STRESS

7.15b, the stress at any point o under the foundation can be
obtained by splitting the foundation area up into four smaller
rectangles and adding the corner stresses from each rectangle to
give the total stress at o. And similarly if o comes outside the area
of the foundation the arrangement indicated in Fig. 7.15c can be
adopted giving the stress at o as the summation of stresses from
areas 1 and 4, less the summation of stresses from areas 2 and 3.
In Fig. 7.16 bulb surfaces of constant vertical soil stress have
been indicated. Anywhere on any bulb of constant stress, the vertical stress in the soil is constant (much as on a map the height
above datum at any contour line is constant). The stress bulbs in
Fig. 7.16 have been derived by Fadum's method and are useful for
determining the extent of necessary site exploration and theoretical analysis. It is often sufficient to limit one's study of a
foundation problem to within the bulb where p_z equals $0.20q$.

Where the shape of a foundation is irregular, or the applied pressure distribution is complicated, an alternative graphical method of working has been devised by Newark. This method makes use of calibrated diagrams over which the outline of the foundation is superimposed, and the pressure at any point in the soil can then be obtained by simple arithmetic processes.

FABER'S RESEARCH ON CONTACT PRESSURES

In considering the bending moments and shear stresses in a foundation it is often assumed that the pressure distribution at the

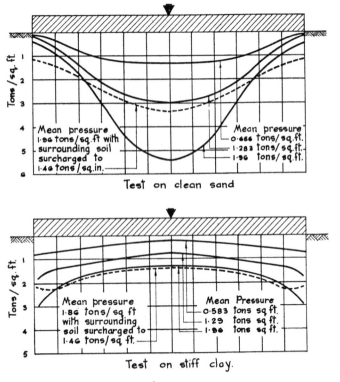

FIG. 7.17. FABER'S CONTACT PRESSURES

surface of contact where the foundation rests on the soil is constant under conditions of uniform loading. This assumption is not strictly true as was shown by Oscar Faber in his paper read before the Institution of Structural Engineers in 1933.[3] Faber designed a special apparatus with which he was able to measure the variations of contact pressure across the undersides of foundations bearing on clean sand and stiff London clay, and summaries of his tests are given in Fig. 7.17.

The tests showed that the pressure under the centre of a foundation bearing on sand was about two and a half times the average pressure, and tailed off nearly to zero at the edges due to the lateral instability of the sand at the free surface: but when the same foundation was surcharged, the edge pressure increased so that the variation of pressure across the foundation was reduced. On clay, the tests showed that the pressures were greatest at the edges of the foundation and became less nearer the centre. Surcharge had no effect on the tests on clay, since truly cohesive soils derive but little strength from internal friction. The results of Faber's tests are to some extent of limited application because his apparatus was of a slightly flexible form, and his soil samples were specially selected so as to be almost truly cohesionless and cohesive. Nevertheless the results are indicative of the form of pressure distribution likely to be developed.

With small foundations the inaccuracies arising from assuming a constant contact pressure may not lead to much trouble; and with large foundations, where the loads are more considerable, the edge effects become relatively less pronounced. Accordingly it is customary not to make any great allowance for variations in contact pressure except in special cases. Contact pressure variations have in any case a very localised effect and according to elastic theories do not influence the stress distribution at depths greater than about the width of the foundation: thus their effect is normally negligible in considering questions of settlement.

ART. 7.7. SETTLEMENT OF SOILS

A knowledge of the probable settlement of the soils under a foundation is just as important as the shear strength. Both require consideration in determining the allowable bearing pressure. The amount a foundation will settle depends on the size and formation of the foundation, as well as on the properties and depth of the various soils in the underlying strata.

The total settlement of soils under load is made up of the sum of three types of deformation:

(i) *elastic deformation* in which vertical movement is accompanied by lateral movements, all of which are recoverable on removal of the load;

(ii) *plastic deformation* in which the soil flows laterally from under the foundation and the movement is not recoverable;

(iii) *consolidation* in which water is squeezed out of the soil, with a consequent change in volume of the soil mass associated with a rearrangement of the soil particles.

In all soils, the greater part of the *elastic* and *plastic deformations* occur as soon as the load is applied; and these two deformations are known collectively as the *immediate settlement*. With cohesionless soils, the immediate settlement generally forms the major part of the total settlement, whereas with cohesive soils it is only a small part.

The *consolidation* of cohesionless soils occurs relatively quickly, because the high permeability enables the water to be expelled rapidly; but the magnitude of the consolidation is generally small. On the other hand cohesive soils, with their much lower permeability, consolidate relatively slowly, generally over a period of several years; and the magnitude of consolidation is considerably greater.

Due to the many uncertain factors involved, settlement estimations are necessarily only approximate. However, they enable a very reasonable indication of the probable total settlement to be gauged, at any rate sufficient for many practical purposes.

FIELD METHODS OF ESTIMATING TOTAL SETTLEMENT

Settlements in cohesionless soils have been related by Terzaghi to the results of Standard Penetration Tests. This is referred to in Article 7.4 – Subsurface Soundings.

A second method well suited for use in cohesive soils is due to Housel, and based on a perimeter shear theory. Housel expressed the allowable bearing pressure for a specified settlement in the form

$$q = \frac{P}{A} n + m \qquad (7.15)$$

where n and m are empirical constants for the soil found by test loads on a number of plates of various sizes of perimeter P and area A.

CALCULATION OF CONSOLIDATION AMOUNT

When load is first applied to a cohesive soil, the load is carried by the pore water under pressure; then as the water is squeezed out, the soil consolidates, and the load is transferred gradually from the pore water to the soil particles, until ultimately the soil particles carry all of the applied load.

The results of the Consolidation Tests described in Article 7.5 may be expressed in the form of a curve as shown in Fig. 7.18a, and if we define m_v, the coefficient of volume change of the soil as the change in unit volume per unit change in pressure, we have

$$m_v = \frac{dh}{dp} \times \frac{1}{h}. \qquad (7.16)$$

Then at any pressure, the slope $\dfrac{dh}{dp}$ of the curve divided by the height of the specimen gives the value of m_v; so that a second curve can be drawn of m_v against p, as shown in Fig. 7.18b.

Referring now to Fig. 7.19a, if the compressible strata (total

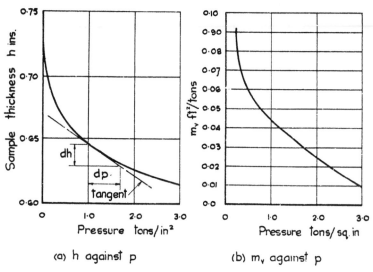

(a) h against p   (b) m_v against p

FIG. 7.18. CONSOLIDATION. GRAPHS FOR COEFFICIENT
OF VOLUME CHANGE

depth z) beneath a foundation be divided into a number of layers (each of height h), and if at any level

p_1 is the stress due to overburden,
p_2 is the stress due to the foundation, and
m_{v12} is the average coefficient of volume change for the pressure change from p_1 to p_2,

then, since equation 7·16 may be rewritten as

$$dh = m_v h \cdot dp,$$

a change in height of any layer under the foundation is given by

$$dh = m_{v12} h \cdot p_2.$$

The total settlement of the compressible strata is, therefore,

$$s = \int_0^z dh = \Sigma m_{v12} h \cdot p_2. \qquad (7.17)$$

This summation is normally carried out in tabular form, with m_{v12} taken from the graph in Fig. 7.18b, and p_2 derived by one of the stress distribution methods described in Article 7.6. The

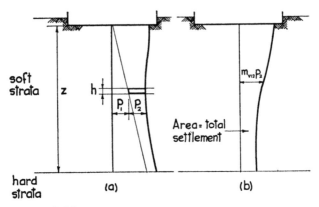

FIG. 7.19. CONSOLIDATION UNDER FOUNDATION

total settlement is depicted graphically by the area of the envelope in Fig. 7.19b.

CALCULATION OF CONSOLIDATION TIME

The *rate* at which consolidation takes place depends on four factors, viz.:

the compressibility of the soil as given by m_v,
the vertical distribution of stress within the soil,
k, the coefficient of permeability of the soil, and
d, the length of the drainage path for the water expelled.

It is assumed here that all drainage takes place vertically, so the length of the drainage path will depend only on the number of escape surfaces available. Thus if the water can escape through both the upper and lower boundaries of the strata, d will be half the thickness of the compressible strata; whereas if one of the boundaries is impermeable, then d will equal the total thickness. Now theoretically the time for complete consolidation is infinite, and so the calculation is made of the time t for only a certain percentage of the total consolidation to take place, frequently 90 per cent. This is known as the degree of consolidation μ so that

$$\mu = \frac{s_t}{s}$$

where s is the total consolidation, and s_t is the amount of consolidation in time t.

It is necessary here to define two further terms used in the analysis, as follows:
the coefficient of consolidation,

$$c_v = \frac{k}{\gamma_w m_v},$$

and the dimensionless time factor,

$$\tau = \frac{c_v t}{d^2}$$

where d is the length of the drainage path.

Consolidation can now be expressed in the form of a differential equation and solutions obtained relating μ to τ. Table 7.2 gives corresponding values of μ and τ for three simplified stress distributions through a soil for two directions of drainage. Thus if drainage for the stress distribution shown in Fig. 7.19a takes place say downwards only towards the boundary where p_2 has its smallest value, the appropriate distribution in Table 7.2 would be as given at (b).

The first stage in the calculation is to determine c_v from the consolidation test results; and from the definition of τ, we have

$$c_v = \frac{\tau d^2}{t}. \tag{7.18}$$

Now at each stage of the consolidation tests, the stress through the thickness of the specimen is sensibly uniform, so that the relationship between μ and τ is given at (a) of Table 7.2; and for this stress distribution it can be shown that for values of μ up to 0·5, the relationship between μ and τ is roughly parabolic and given by the equation

$$\tau = \frac{\pi}{4} \mu^2.$$

Substituting this value of τ in equation 7.18 we have, for values of μ not greater than 0·5,

$$c_v = \frac{\pi}{4} \frac{\mu^2 d^2}{t}. \tag{7.19}$$

Thus to obtain c_v for any test pressure, we can read from the curve of any stage of a test the time t for 50 per cent compression, and substitute this, with 0·5 for μ, in equation 7.19.

However, in practice, for convenience and to iron out test errors, the following method of working is more usually adopted.

TABLE 7.2. *Relationship between time factor and degree of consolidation (after Capper and Cassie)*

Direction of drainage	Type of stress distribution		
Permeable ↑ ⋮ Impermeable	(a)	(b)	(c)

Degree of consolidation μ	Time factor τ		
	Condition (a)	Condition (b)	Condition (c)
0·1	0·008	0·047	0·003
0·2	0·031	0·100	0·009
0·3	0·071	0·158	0·024
0·4	0·126	0·221	0·048
0·5	0·197	0·294	0·092
0·6	0·287	0·383	0·160
0·7	0·403	0·500	0·271
0·8	0·567	0·665	0·440
0·9	0·848	0·940	0·720

For two-way drainage use $\tau_{(a)}$ for all pressure distributions, and in this case also use d equal to half the height of the compressible strata

Since for any one stage of pressure in the consolidation test, the values of c_v and d are constant, it can be seen from equation 7.19 that the initial relationship between μ and t is parabolic: so that if compression is plotted against $\sqrt{t}$ from the test results, the first part of the graph is a straight line as shown in Fig. 7.20; and if this line is extended to cut the ultimate compression line (where $\mu = 1$) we get the hypothetical compression time t_1. Substituting this value of t_1 in equation 7.19, we have

$$c_v = \frac{\pi}{4}\frac{d^2}{t_1} \qquad (7.20)$$

and c_v can be calculated for each test pressure. For the test, d is taken as half the height of the specimen.

Having now derived c_v by one means or another from the results

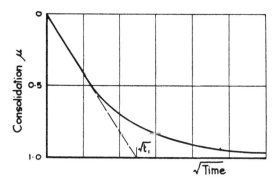

FIG. 7.20. GRAPH FOR THE RATE OF CONSOLIDATION
OF SOILS

of the consolidation tests, we may, for the full-scale foundation, rearrange equation 7.18 to give

$$t = \frac{\tau d^2}{c_v}. \tag{7.21}$$

d of course is already known, and τ may be taken from Table 7.2 for the particular stress and drainage conditions and for any chosen value of μ. c_v is taken for the average value of the stress range in the strata.

ART. 7.8. ULTIMATE BEARING CAPACITIES OF SOILS

The *ultimate bearing capacity* of a soil is the pressure it can just sustain before shear failure occurs. Failure may occur either by the foundation punching its way into the soil, or by the development of a large-scale cartwheel action involving the sympathetic rotation of the adjacent soil mass. Due to the unhomogeneous nature of all soils in the field, punching failure of an isolated base will lead ultimately to a small-scale rotation of the soil, but generally this will be checked by redundant influences in the superstructure: in any case, it is not to be confused with the more dramatic actions associated with cartwheel failure. The study of both forms of failure involves a knowledge of the density of the soil, its cohesion,

248 Reinforced Concrete

and its angle of internal friction. These are all determined from the laboratory tests described in Article 7.5.

PUNCHING FAILURE

Consider the mechanism of failure of the soil at a strip foundation at ground level as shown in Fig. 7.21. The soil in the wedge-

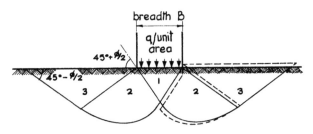

FIG. 7.21. PUNCHING SHEAR FAILURE OF SOIL AT A
STRIP FOUNDATION

shaped active zone 1 moves downwards with the foundation; and the zones of radial shear 2 (assumed to be in a plastic state) shear along radial and logarithmic spiral lines as shown, exerting a lateral thrust on the passive zones 3, which in consequence are displaced sideways and upwards, and break the ground surface.

With plastic clays, which may be regarded as non-frictional (i.e. $\phi = 0$), we have c equal to half the unconfined compressive strength of the soil. Then if the underside of the foundation is smooth, it can be shown mathematically that for a strip foundation at ground level,

$$q_f = 2 \cdot 57 q_u = 5 \cdot 14 c \qquad (7.22)$$

where q_f is the ultimate bearing capacity of the soil, q_u is the unconfined compressive strength of the soil, and c is the cohesion. For foundations below ground level an allowance for the effect of overburden is made by increasing the ultimate bearing capacity by the amount of the overburden pressure. Thus equation 7.22 becomes

$$q_f = 5 \cdot 14 c + \gamma d \qquad (7.23)$$

where γ is the bulk density of the soil, and d is the depth of the foundation below surface level. Terzaghi and Peck[2] have extended the study of this problem to deal with the case where the underside of the foundation is rough, and have shown that with non-frictional soils loaded at ground level

$$q_f = 2 \cdot 85 q_u = 5 \cdot 70 c. \qquad (7.24)$$

With frictional soils the results obtained from the above analysis are generally unsatisfactory. Terzaghi and Peck have pursued an alternative analysis for frictional soils, based on the assumption that the wedge of soil under a rough foundation cannot penetrate the soil unless the pressure exerted by the side of the wedge is equal to the passive pressure of the adjacent soil. They have shown that the ultimate bearing capacity of the soil under a footing

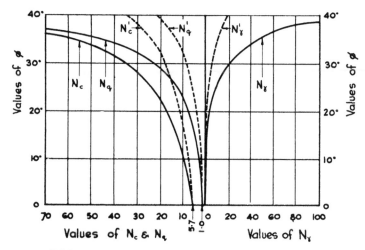

FIG. 7.22. TERZAGHI AND PECK'S BEARING CAPACITY COEFFICIENTS FOR ALL SOILS (PARTICULARLY FRICTIONAL)

(breadth B) at a depth d below surface level can be expressed in the form

$$q_t = cN_c + \gamma dN_q + 0.5\gamma BN_\gamma \qquad (7.25)$$

where γ is the bulk density of the soil, c is the cohesion, and N_c, N_q and N_γ are dimensionless quantities known as *bearing capacity coefficients*, and depend only on the value of ϕ.

The values of the coefficients have been calculated for various ϕ values and are given in Fig. 7.22. On the same basis, but with an eye to experimental results, Terzaghi and Peck have given the ultimate bearing capacity of the soil under a circular foundation as

$$q_t = 1.3cN_c + \gamma dN_q + 0.6\gamma rN_\gamma \qquad (7.26)$$

and under a square foundation as

$$q_t = 1.3cN_c + \gamma dN_q + 0.4\gamma BN_\gamma. \qquad (7.27)$$

Equations 7.25, 7.26, 7.27 apply for foundations on stiff or dense soils. With foundations on soft or loose soil, appreciable settlement is likely to take place before general failure occurs, due to the local breakdown of elemental shear forces within the soil. The ultimate bearing capacity is then given as the point where the settlement curve in Fig. 7.5b dips more steeply, or becomes tangential. Terzaghi and Peck suggest that for soft or loose soils the values of c

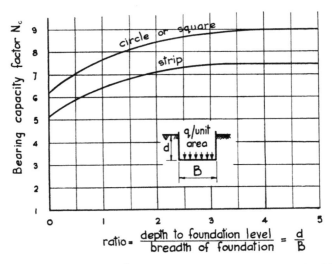

FIG. 7.23. SKEMPTON'S BEARING CAPACITY COEFFICIENTS FOR COHESIVE SOILS (FRICTIONLESS)

and $\tan \phi$ should be reduced by one-third, so that equation 7.25 becomes

$$q_t = \tfrac{2}{3}c \, \mathrm{N'}_c + \gamma d \, \mathrm{N'}_q + 0 \!\cdot\! 5\gamma B \, \mathrm{N'}_\gamma \qquad (7.28)$$

with modified bearing capacity coefficients as shown dotted in Fig. 7.22. Equations 7.26 and 7.27 may similarly be rewritten to suit soft and loose soils.

The ultimate bearing capacity of frictional soils is greatly affected by submergence. The soil density γ should then be replaced in the above equations by the submerged soil density γ_s, and this in some cases will have the effect of roughly halving the ultimate bearing capacity of the soil. When the water level is a depth B below the underside of a foundation (breadth B), it is likely not to affect the soil capacity very much; but when the water stands at any level above this, an appropriate reduction in bearing capacity should be made, varying in accordance with the level of the water table, being the full reduction given by the use of γ_s when the foundation is submerged and tailing off to no reduction when the depth to water level equals B.

With cohesive non-frictional soils, Skempton[4] has shown that the ultimate bearing capacity may reasonably be expressed in the form

$$q_f = cN_c + \gamma d. \tag{7.29}$$

Skempton's values for N_c vary with the depth/breadth ratio of the foundation, and are given in Fig. 7.23 for circular, square and strip foundations. In the case of a rectangular foundation of breadth B and length L, the value of N_c for the rectangle is determined from the value N_c for a square of side B, by the equation

$$N_c \text{ for rectangle} = \left(0\cdot84 + 0\cdot16\,\frac{B}{L}\right) \times N_c \text{ for square.} \tag{7.30}$$

CARTWHEEL FAILURE

Consider the mechanism of failure shown in Fig. 7.24. Failure of

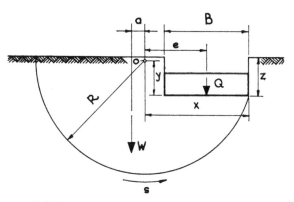

FIG. 7.24. CARTWHEEL FAILURE OF FOUNDATION

this kind is most likely to occur in cohesive frictionless soils where the extent of applied load is considerable, as from a silo or large building, so that shear weaknesses are sought out in the soil at considerable range and depths. Instability of the soil by rotation in this manner arises when the load from the foundation produces a moment about the rotation centre greater than the countermoment due to the weight of overburden and the shear strength of the soil acting along the arc of failure. Analysis of this type of failure was first suggested by Fellenius. A centre of rotation is selected arbitrarily and the ultimate total load Q determined from the consideration that

$$Q = \frac{Wa + slR}{e} \qquad (7.31)$$

where W is the weight of the soil mass within the arc of rotation,
　　　　a is the eccentricity of W about the centre,
　　　　s is the shear strength of the soil,
　　　　l is the length of the arc,
　　　　R is the radius of the arc,
and　　e is the eccentricity of the applied load about the rotation centre.

The process is then repeated trying other centres of rotation until the centre is found which gives the least value for Q.

Consideration of cartwheel failure gives the value for the ultimate bearing capacity of the soil under a continuous foundation at ground level as

$$q_f = 5 \cdot 5c. \qquad (7.32)$$

But where the foundation is square in plan, the ultimate bearing capacity is found in practice to be increased some 20 per cent due to the additional shearing resistance provided at the boundary effects at the two ends of the slip circle.

A check should be made against cartwheel failure in cases where the shear strength of extensive cohesive soil is known to be poor. Variations in the shear strengths of soils in different layers may be taken into account quite easily in making this check.

ART. 7.9. ALLOWABLE BEARING PRESSURES ON SOILS

The *allowable bearing pressure* of a soil is the nett loading intensity it can support, and takes into account the ultimate bearing capacity of the soil (with a suitable load factor), and the amount of settlement acceptable for the problem in hand. The ultimate bearing capacity, and the estimations of settlement, are considered in Articles 7.8 and 7.7 respectively, and the soil properties necessary for evaluating these are found in accordance with the methods described in Articles 7.4, 7.5 and 7.6.

SETTLEMENT CONSIDERATIONS

Except for very small, or very stiff structures, the amount of settlement across the width is likely to vary, depending on the nature of the soil, even though the unit loading intensities throughout are roughly constant. The amount of settlement variation can only be estimated very roughly. For independent or strip foundations on cohesionless soils, the variation is not likely to exceed about 75 per cent of the maximum settlement, and if the intensities of loading and the breadths of foundations are fairly uniform, the variation is not likely to exceed about 50 per cent of the total settlement. Thus for a maximum settlement on sand of 1 in., the variation is unlikely to exceed $\frac{3}{4}$ in. and more probably would be about $\frac{1}{2}$ in. Settlement variations of this order are generally acceptable for most ordinary structures. With a load factor of 3·0 on normal clays, the settlement variation is unlikely to be any more than described above for sand.

Where there are special considerations making settlement variations particularly undesirable, it is possible by judicious arrangement and proportioning of the superstructures and foundations to produce higher intensities of loading at the edges of a structure than at the centre. This will go some way towards preventing the formation of a settlement crater. The settlement of rafts may be reduced by increasing the depth of the foundation so as to increase the amount of overburden removed by excavation. The nett loading intensity is then consequently decreased. If the intensities of loading differ appreciably under various parts of a rafted structure, there may be considerable settlement variations unless the foundations are arranged at different depths so that everywhere the nett loading intensity on the soil is kept sensibly constant. For this reason a deeper basement under the heaviest loaded part of a structure may be an advantage. If the settlements of variously loaded parts of a structure are likely to be materially dissimilar, the different parts should be kept separate from one another by the provision of suitable free joints; and to minimise settlement variations, the more heavily loaded parts should be constructed in advance of the others.

LOAD FACTORS APPLIED TO THE SOIL BEARING CAPACITY

Apart from settlement considerations, the *safe bearing capacity* of a soil is determined by applying a suitable load factor to the ultimate bearing capacity. When the foundations are not near the surface level, it is logical that the load factor be not applied to the overburden pressure. Thus, if the ultimate bearing capacity of a cohesive soil is given by equation 7.29, then the safe bearing capacity will be

$$q_a = \frac{cN_c}{F} + \gamma d \qquad (7.33)$$

where F is the load factor.

Similarly, using equation 7.27, we have

$$q_a = \frac{[1 \cdot 3cN_c + \gamma d(N_q - 1) + 0 \cdot 4\gamma BN_\gamma]}{F} + \gamma d. \qquad (7.34)$$

The load factor against ultimate shear failure is normally taken as 3·0: though for temporary structures, or domestic buildings of not more than two storeys where foundation design loads are seldom realised, it is reasonable with dense and hard soils to reduce this to 2·0.

TABULATED VALUES OF SAFE BEARING CAPACITIES

Many tables have been published giving Safe Bearing Capacities for soils. The wide variations given in any such table, and the wider discrepancies found by comparing tables from different sources, show clearly how misleading the use of such tabulated values could

TABLE 7.3. *Safe bearing capacities for soils*

Types of soils	Safe bearing capacity tons/sq. ft.		Remarks
	Dry	Sub-merged	
Cohesionless soils:			Width of foundation (B) not less than 3 ft. "Dry" means that the ground-water level is at a depth not less than B below the base of the foundation
Compact well-graded sands and gravel-sand mixtures	4–6	2–3	
Loose well-graded sands and gravel-sand mixtures	2–4	1–2	
Compact uniform sands	2–4	1–2	
Loose uniform sands	1–2	½–1	
Cohesive soils :			This group is susceptible to long-term consolidation settlement
Very stiff boulder clays and hard clays with a shaly structure	4–6		
Stiff clays and sandy clays	2–4		
Firm clays and sandy clays	1–2		
Soft clays and silts	½–1		
Very soft clays and silts	.½–nil		To be determined after investigation

be. Indeed with cohesive soils and some silts, the allowable bearing pressure may well depend more on considerations of settlement than on any safe bearing capacity determined by shear considerations. Accordingly in foundation design work, it is better to proceed from field or laboratory tests as described in Articles 7.4 and 7.5, or even to rely on a visual field inspection coupled with experience, rather than to follow blindly the inappropriate values liable to be picked out of any table.

The only use to which tabulated values might reasonably be put, is in the preliminary considerations of design conception at the earliest stages. For this purpose Table 7.3 is included from *Civil Engineering Code of Practice*, No. 4 (1954)[1] which warns that the values given are based solely on consideration of the soil shear strength, and ignore the question of failure from settlement. The values given are based on a load factor of 2·0; and for foundations on clay, if the full design load is likely to be realised it is recommended that the figures given be reduced to two-thirds.

ART. 7.10. STRUCTURAL CONCEPTION OF FOUNDATIONS

MINIMUM DEPTH OF FOUNDATIONS ON SOILS

Where the depth of a foundation is not determined by considerations of bearing capacity or settlement, other factors arise which influence the minimum satisfactory foundation depth.

Foundations in clay constructed near the surface are liable to suffer movements as a result of the shrinkage or swelling of the soil due to variations in moisture content. Normally this effect dies out at a depth of about 3 ft below ground level, but in cases where the slightest movements would be critical it may be necessary to take the foundations down to 5 ft. or 6 ft. depending on the clay and on other topographical features. Most soils suffer the phenomenon of frost heave, particularly saturated silts and sands, chalk and some clays. Frost heave results from the formation of ice in the soil, causing swelling. In temperate zones, foundations about 2 ft. below ground level are generally deep enough to escape the effects of frost, though where the water table is within a foot or two of the surface, it is advisable to keep foundations about 3 ft. down.

Most soils, particularly clays, change in character once they are exposed to the atmosphere; and therefore whatever depth a foundation is constructed, it is important to put down a layer of protective concrete as soon as the excavation has been bottomed up. The bottoms of excavations in sands and gravels should be compacted before covering to make good any loosening done in the process of excavation.

TYPES OF FOUNDATION

The choice of type of foundation depends on the nature of the sub-strata, and on the arrangement of the superstructure. The simplest form of foundation is a continuous strip foundation under a load bearing wall, or an independent base under a single column; and indeed where these types of foundation are suitable, they are often the cheapest arrangement. In cases where column rows are spaced far apart, but the columns in each of the rows are relatively close to one another, an economy in design may be achieved by linking together the column bases in the same row to form a strip foundation, in this way introducing continuity and so reducing the bending moments. Pairs or groups of columns may be considered similarly.

However, where the ground is poor or where the loads from the superstructure are large, the size of individual column foundations may form an appreciable part of the plan area of the building. In such cases, a raft foundation should be considered, again with a view to achieving economy through structural continuity. When the area of individual foundations would occupy more than about half the plan area of the building, it generally pays to provide a raft foundation. Alternatively it may be desirable to provide a raft in order to minimise differential settlement. The differential settlement of raft foundations may be only about half that of individual footings, due to the greater stiffness of the raft and its ability to bridge local weaknesses in the soil. The thickness of solid raft construction may become considerable if the superstructure columns are widely spaced, or if the loadings are high; and it may then be economical to use a cellular or hollow box type of raft. The sides of the raft cells then form beams spanning between the columns or other applied loads, and the top slab normally forms the lowest floor of the building, while the bottom slab rests directly on the soil.

Where the allowable bearing pressure of the soil is very low, there may be a case for constructing a cellular raft of considerable depth to achieve what is known as a *buoyant foundation*. This is constructed at such a depth that the weight of overburden removed by excavation is roughly equal to the total applied load. The nett loading intensity at the underside of the foundation is then approximately nil, or at any rate reduced to a suitably small value. Frequently the use of cellular construction in this way makes for a saving in weight in the foundation (as compared with a solid slab raft); and furthermore the safe bearing capacity at the increased depth may be greater than nearer the surface. With buoyant foundations, it is important to guard against ground-heave – the property of certain soils at the bottoms of excavations to relieve themselves by rising once the overburden has been removed. Ground-heave may be checked to some extent by the use of cut-off piling, or by dewatering. Dewatering needs to be continued until sufficient of the superstructure has been built to hold down any tendency for ground-heave.

If none of the methods described above is satisfactory, then piled foundations or piers need to be considered. Piers are used where a firm strata occurs not more than about 15 ft. down. Beyond this piers generally cease to be economical. Continuous piers enable mechanical equipment to be used for the excavation work, saving time and money; otherwise manual excavation is required for independent pier holes. Pier holes should be large enough for excavation to be carried out in reasonable comfort, even though this may necessitate spacing the piers well apart and bridging between them at ground level with beams or otherwise. Where firm strata is not to be found within a depth of about 15 ft., it becomes necessary to drive piles. Piled foundations are discussed in Chapter 9.

FOUNDATION DESIGN LOADS

Some skill is required in deciding what loads should be taken in designing foundations. For storage structures such as warehouses, silos, and water towers, it is clear that the total loading for which the superstructure has been designed can apply if all parts of the structure are fully loaded simultaneously. But in other structures and certain buildings it may be beyond the realms of probability for the superstructure design loadings ever to occur at all levels at the same time. And an industrial floor may have been designed to allow for some specially heavy machine to be moved to any part of the floor, so that the whole floor has been designed for a very considerable superload; but the one machine clearly cannot stand at all parts of the floor at the same time, and it is reasonable then to design each part of the foundations only to cater for the worst possible loading that can arise from the machine being in any one of the various possible positions.

Where horizontal forces from wind, earthquake or other sources affect the vertical loads on the foundations, full provision must be made in the gross downward loads. If the increase in vertical load due to wind or earthquake is less than 25 per cent, it is customary to ignore it in foundation design; but where the increase exceeds 25 per cent, the foundation should be designed to give a total applied pressure not greater than 25 per cent in excess of the normal allowable bearing pressure. For uplift forces, a minimum factor of stability of 1·5 should be provided, either in the dead weight of the foundation, or by other anchorage devices: the uplift forces in such cases being the nett values calculated by deducting the minimum dead load of the structure from the sum of all applied uplift forces.

Where the underside of a foundation is to come below the level of the water table, an uplift pressure will arise from foundation buoyancy. A basement which will ultimately be held down by the weight of the upper part of the building, may float in its early stage of construction if the containing soil is allowed to become flooded before the superstructure has been erected. This may be prevented by increasing the dead weight of the basement, or by continuing ground-water pumping until sufficient of the superstructure has been built.

ART. 7.11. DESIGN OF REINFORCED CONCRETE FOUNDATIONS

Having determined the loads to be carried, and the safe bearing capacity on the soil, the design of various types of foundations may be carried out by the methods shown in the following examples.

INDEPENDENT BASES

(i) *Square base: concentric load*

Consider the design of an independent base to support a column 24 in. × 24 in. on a soil of 2 tons/sq. ft. safe bearing capacity. Let the load on the column be central and 200 tons.

$$\text{Area of base required} = \frac{200 \text{ tons}}{2 \text{ tons/sq. ft.}} = 100 \text{ sq. ft.},$$

whence size of square base = $\sqrt{100}$ = 10 ft.

The critical section for bending is taken at the face of the column, being XX in Fig. 7.25a. The upward force to the right of XX is

$$4 \text{ ft.} \times 10 \text{ ft.} \times 2 \text{ tons/sq. ft.} = 80 \text{ tons}$$

and the moment at XX is

$$80 \text{ tons} \times 24 \text{ in.} = 1920 \text{ in. tons.}$$

Taking the overall depth of the base as 2 ft. 6 in., we have

$$A_{st} = \frac{M}{p_{st} \times l_a} = \frac{1920 \times 2240}{20,000 \times (0.86 \times 27\frac{1}{2})} = 9.1 \text{ sq. in.}$$

which is given by twelve 1 in. diameter rods (9·4 sq. in.). The reinforcement required in the other direction will clearly be the same. These rods should be spaced evenly across the full width of the section.

$$\frac{M}{bd_1^2} = \frac{1290 \times 2240}{(10 \times 12) \times 27\frac{1}{2}^2} = 47,$$

indicating that the compressive stress in the concrete is quite low.
The critical section for shear is taken to be where the 45° line
from the face of the column cuts the tension reinforcement. This is
because shear failure arises from diagonal tension. The area of base

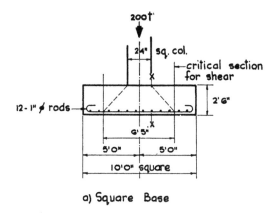

a) Square Base

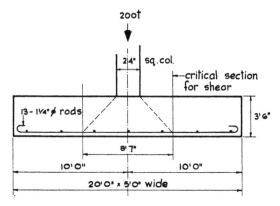

b) Rectangular Base

FIG. 7.25. CONCENTRICALLY-LOADED BASES

pushed through by shear failure would thus be a square of side
24 in. + (2 × 27½ in.) = 79 in. This would be resisted by the shear
strength of the concrete, plus the soil pressure under the area
79 in. square. The latter is $\dfrac{79 \times 79}{144} \times 2$ tons/sq. ft. = 86 tons,
leaving (200 − 86) tons = 114 tons to be taken by the concrete in
shear.

This gives a shear stress in the concrete of

$$q = \frac{Q}{bl_a} = \frac{114 \times 2240}{(4 \times 79) \times 0\cdot86 \times 27\frac{1}{2}} = 35 \text{ lb/sq. in.}$$

This is clearly satisfactory. If the shear stress had exceeded the permissible value for the concrete alone, bent-up bars would be provided.

The critical section for bond stress is taken at XX, Fig. 7.25a. Allowing a value of $16d$ for the hooked end, the bond length is 56 in., more than required to anchor a bar 1 in. diameter.

(ii) *Rectangular base: concentric load*

Where for practical reasons there is not space to provide square bases (for example, to clear other foundations, drains or similar works), a rectangular base may be necessary. Suppose the column in the previous example has its centre-line only 2 ft. 6 in. from another structure. Then width of base is limited to 5 ft., whence length of base $= \dfrac{100}{5} = 20$ ft. (Fig. 7.25b). The critical moment long-way is

$$(9 \text{ ft.} \times 5 \text{ ft.} \times 2 \text{ tons/sq. ft.}) \times 54 \text{ in.} = 4850 \text{ in. tons.}$$

Taking the overall depth of the base as 3 ft. 6 in. we have

$$A_{st} = \frac{4850 \times 2240}{20,000 \times 0\cdot86 \times 39\frac{1}{2}} = 15\cdot9 \text{ sq. in.}$$

which is given by thirteen $1\frac{1}{4}$ in. diameter rods ($16\cdot0$ sq. in.).

$$\frac{M}{bd_1{}^2} = \frac{4850 \times 2240}{(5 \times 12) \times 39\frac{1}{2}{}^2} = 116.$$

For shear, the base is so narrow that the diagonal spread of the load within the base occupies the full width, and only the shear resistance of the concrete between the column and the two ends of the base is available to act. The length of base pushed through by shear failure would be $24 \text{ in} + (2 \times 39\frac{1}{2} \text{ in.}) = 103$ in. Upward pressure on this by soil is

$$\left(\frac{103}{12} \times 5\right) \times 2 \text{ tons/sq. ft.} = 86 \text{ tons,}$$

leaving $(200 - 86)$ tons $= 114$ tons to be taken by the concrete in shear.

This gives a shear stress in the concrete of

$$q = \frac{114 \times 2240}{0.86 \times 39\frac{1}{2} \times (2 \times 60)} = 63 \text{ lb/sq. in.}$$

requiring no special reinforcement.

(iii) *Rectangular base: eccentric load*

In the two previous examples for concentrically loaded columns, the weight of the bases themselves was ignored. The error involved by this is generally less than 5 per cent, taking into account only the *nett* increase in load arising from the *difference* in density between the concrete foundation and the weight of soil it replaces. However, with eccentrically loaded bases where questions of stability arise, the weight of the foundation may assist materially in providing balance, and is normally included in the calculation.

Consider a column 24 in. × 24 in. carrying a load of 100 tons combined with a bending moment in one sense of 1000 in. tons; the safe bearing capacity of the soil being 2 tons/sq. ft. as before. Clearly a rectangular base will be better than a square base for resisting the bending moment. Try a base 14 ft. × 6 ft. × 3 ft 0 in. thick.

Then $W =$ (column load)   100 tons
       (weight of base)   17 tons
                           ───────────
                             117 tons.

$$\text{Eccentricity} = \frac{M}{W} = \frac{1000}{117} = 8\cdot6 \text{ in.} = 0\cdot70 \text{ ft.}$$

$$\text{Max. pressure on soil} = \frac{W}{A}\left(1 + \frac{6e}{d}\right)$$

[e = eccentricity; d = overall dimension of base]

$$= \frac{117}{14 \times 6}\left(1 + \frac{6 \times 0\cdot70}{14}\right)$$

$$= 1\cdot86 \text{ tons/sq. ft.}$$

The critical moment causing bending in the base at XX, Fig. 7.26a, is due to the trapezoidal pressure distribution to the right of XX, less the part of this pressure which is due to the self-weight of the base.

Thus upward force is

6 ft. × 6 ft. × 1·67 tons/sq. ft. (mean) = 60·0 tons
and downward force is
6 ft. × 6 ft. × 0·20 tons/sq. ft.   =   7·2 tons

Difference = 52·8 tons.

The moment at XX due to this force is

$$52\cdot8 \text{ tons} \times 37\cdot8 \text{ in.} = 2000 \text{ in. tons,}$$

whence $\quad A_{st} = \dfrac{2000 \times 2240}{20,000 \times 0\cdot86 \times 33\frac{1}{2}} = 7\cdot80$ sq. in.

which is given by eight $1\frac{1}{8}$ in. diameter rods (8·0 sq. in.).

As a further example of an eccentrically loaded base, consider the same column, with an applied load of only 40 tons and a bending moment of 2400 in. tons. The base will be the same size as before, but will need to be 3 ft. 6 in. thick. Then

$$W = \text{(column load)} \quad 40 \text{ tons}$$
$$ \text{(weight of base)} \quad 20 \text{ tons}$$

60 tons.

Eccentricity $= \dfrac{M}{W} = \dfrac{2400}{60} = 40$ in. $= 3\cdot3$ ft.

This brings the resultant outside the middle-third, so that if tension under the base is not to be relied on, as clearly it cannot, the pressure distribution will be as in Fig. 7.26b, a triangle of length equal to $3 \times 3\cdot7 = 11\cdot1$ ft. The mean pressure over this length is clearly $\dfrac{60 \text{ tons}}{11\cdot1 \times 6} = 0\cdot90$ tons/sq. ft. and the maximum edge pressure is double this equals 1·80 tons/sq. ft. This allows a reasonable margin for stability, since if the moment were increased by 50 per cent to 3600 in. tons, the eccentricity would be 5 ft. giving a maximum edge pressure of 3·33 tons/sq. ft., well within the *ultimate* capacity of the soil.

This example indicates clearly the need to include in the calculations the self-weight of the foundation. Otherwise the eccentricity under normal design conditions would work out as $\dfrac{2400}{40} = 60$ in.

$= 5\cdot0$ ft., giving an edge pressure on the soil of 2·22 tons/sq. ft., as against the 1·80 tons/sq. ft. where the foundation weight is included in the calculation, an increase of 23 per cent.

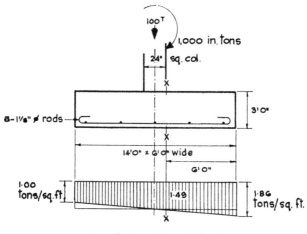

a) Resultant within Middle Third

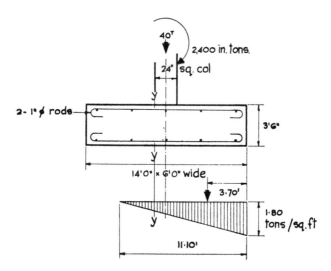

b) Resultant outside Middle Third

FIG. 7.26. ECCENTRICALLY-LOADED BASES

With the pressure distribution shown in Fig. 7.26b the weight of the left part of the base is acting to tail the foundation down, and some top reinforcement is necessary.

M_{YY} due to weight of base $= 8{\cdot}6$ tons $\times$ 36 in. $\quad = 307$ in. tons
less due to soil pressure $= 0{\cdot}25$ ton $\times$ 12 in. $\quad = \quad 3$ in. tons

$$\overline{ 304 \text{ in. tons}}$$

whence $\quad A_{st} = \dfrac{304 \times 2240}{20{,}000 \times 0{\cdot}86 \times 39\frac{1}{2}} = 1{\cdot}00$ sq. in.

which is amply provided by two 1 in. diameter bars in the top of
the base.

CONTINUOUS STRIP FOUNDATIONS

(i) *Carrying a wall*

Where the foundations are of reinforced concrete, there is no
purpose in widening the lower courses of a brick wall into footings
as was common practice in the earlier part of this century.

Consider supporting a 14 in. brick wall bearing 6 tons/foot run
on soil of 1 ton/sq. ft safe bearing capacity. The foundation requires to be 6 ft. wide, as shown in Fig. 7.27a.

The bending moment at XX is

$$\frac{2{\cdot}42^2}{2} \times 12 = 35 \text{ in. tons per ft.}$$

$$A_{st} = \frac{35 \times 2240}{20{,}000 \times 0{\cdot}86 \times 13} = 0{\cdot}35 \text{ sq. in.}$$

which is given by $\frac{1}{2}$ in. diameter rods at 6 in. centres.
The critical shear at SS is $1\frac{1}{3}$ tons per ft., whence

$$q = \frac{Q}{bl_a} = \frac{1\frac{1}{3} \times 2240}{12 \times 0{\cdot}86 \times 13} = 22 \text{ lb/sq. in.}$$

The bond stress is clearly very reasonable.

For convenience in securing the reinforcements in position, a
number of longitudinal rods are normally provided wired up to
form a handy mat.

(ii) *Carrying a row of columns*

Consider a row of columns each 12 in. $\times$ 12 in. carrying 80 tons
and arranged at 12 ft. centres as shown in Fig. 7.27b. Suppose the
soil to be 1 ton/sq. ft. safe bearing capacity. If independent bases
were provided these would need to be 9 ft. 0 in. square, with bending moments in *both* directions of 865 in. tons. Furthermore the
excavation would be tedious in isolated holes, with little opportunity of using mechanical equipment. With a continuous strip

foundation, the width required would be 6 ft. 8 in. with a longitudinal bending-moment of

$$\frac{WL}{16} = \frac{80 \times 12}{16} = 60 \text{ ft. tons} = 720 \text{ in. tons,}$$

and a transverse moment of only 580 in. tons, showing a gross
saving in moment over the independent base of some 25 per cent.
Whereas an independent base 9 ft. square at this bearing

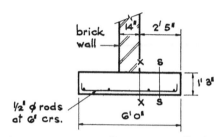

brick
wall

½' ⌀ rods
at 6' crs.

(a) Cross section through wall foundation.

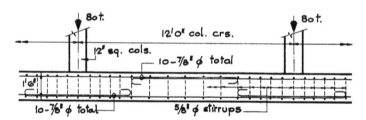

(b) Longitudinal section of foundation
for row of columns.

FIG. 7.27. CONTINUOUS STRIP FOUNDATIONS

pressure would normally be made 2 ft. 0 in. thick, the strip foundation is quite satisfactory if only 1 ft. 6 in. thick due to the
reduced flexibility which arises from continuity. Thus

$$A_{st} \text{ longitudinal} = \frac{720 \times 2240}{20,000 \times 0.86 \times 16} = 5.8 \text{ sq. in.}$$

which is given by five ⅞ in. diameter rods run along the strip, top and bottom, and lapped as necessary to give ten $\frac{7}{8}$ in. rods where required, (6·0 sq. in.).

$$\frac{M}{bd_1{}^2} = \frac{720 \times 2240}{80 \times 16^2} = 79 \text{ only.}$$

A_{st} transverse $= \dfrac{580 \times 2240}{20,000 \times 0·86 \times 16} = 4·7$ sq. in. which is given by sixteen $\frac{5}{8}$ in. diameter rods per bay. If these are arranged in the form of stirrups they act to help support and locate the longitudinal top rods.

The shear force Q is $\left[80 - \left(\dfrac{44 \times 44}{144} \right) \right] = 66·5$ tons,

and the mean stress

$$q = \frac{66·5 \times 2240}{0·86 \times 16 \times (4 \times 44)} = 62 \text{ lb/sq. in.}$$

RAFT FOUNDATION

Consider the design of a foundation for a building having 16 in. square columns at 12 ft. centres in one direction and 14 ft. centres in the other direction. The internal column loads are 80 tons and the safe bearing capacity of the soil is only 0·75 tons/sq. ft. Independent bases under the columns would need to be $\dfrac{80}{0·75} = 107$ sq. ft., more than 10 ft. square – nearly touching one another in fact. Here then clearly is a case for a raft foundation.

The raft is to form the lowest floor of the building, and will carry also a superimposed floor load of 2 cwt/sq. ft. Ignoring the self-weight of the raft, the pressure applied to the soil is therefore

due to columns $\dfrac{80 \text{ tons}}{12 \text{ ft.} \times 14 \text{ ft.}} = 0·48$ tons/sq. ft.

due to floor load 2 cwts $\quad = 0·10$ tons/sq. ft.

_____

$0·58$ tons/sq. ft.

Since the superimposed floor load is transmitted directly to the ground, it causes no bending moment or shear force in the slab and can be ignored in the following calculations. Ground beams are undesirable and the raft is designed as a flat slab. Let us try a raft slab 18 in. thick.

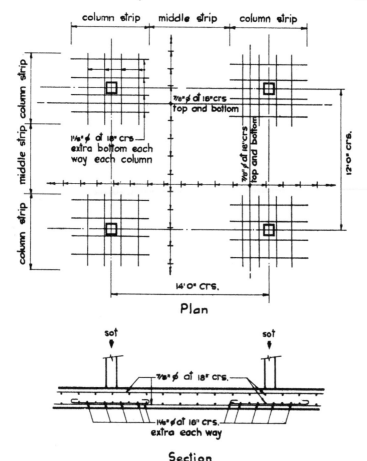

FIG. 7.28. RAFT FOUNDATION

The total bending moment M_0 in any internal panel of the raft in either direction is given empirically (Article 5.5) by

$$M_0 = \frac{wL_2}{10}\left(L_1 - \frac{2D}{3}\right)^2$$

where L_1 is the length of the panel in the direction of span,

$\quad L_2$ is the length of the panel at right angles to the direction of span,

and $\quad D$ is the diameter of the column head and in this case equals 16 in.

The total moment for the slab spanning in the 14 ft. direction is thus

$$M_0 = \frac{0\cdot48 \times 2240 \times 12}{10} \left(14 - \frac{2 \times 1\cdot33}{3}\right)^2 \times 12$$

$$= 2{,}670{,}000 \text{ in. lb.}$$

For the column strip (*see* Table 5.3) we have:

Negative moment $= 0\cdot46 M_o = 0\cdot46 \times 2{,}670{,}000$ in. lb:

$$= \frac{0\cdot46 \times 2{,}670{,}000}{6 \text{ ft.}}$$

$$= 205{,}000 \text{ in. lb/ft.}$$

and positive moment $\quad = \dfrac{0\cdot22 \times 2{,}670{,}000}{6 \text{ ft.}}$

$$= 98{,}000 \text{ in. lb/ft.}$$

While for the middle strip we have:
Negative moment = positive moment

$$= \frac{0\cdot16 \times 2{,}670{,}000}{6 \text{ ft.}} = 71{,}000 \text{ in. lb/ft.}$$

Similarly in the 12 ft. direction, $M_0 = 2{,}240{,}000$ in. lb; and for the column strip the negative moment is 147,000 in. lb/ft. and the positive moment is 70,000 in. lb/ft., while for the middle strip the negative and positive moments are both 51,000 in. lb/ft.

Thus, except directly under the columns, an overall mat of reinforcement in both faces of the slab to resist a moment of 98,000 in. lb/ft. will suffice and is given by

$$A_{\text{st}} = \frac{98{,}000}{20{,}000 \times 0\cdot86 \times 15\tfrac{1}{2}} = 0\cdot37 \text{ sq. in.}$$

which is provided by $\frac{7}{8}$ in. diameter rods at 18 in. centres. The minimum recommended percentage of reinforcement for slabs is 0·15 times the gross cross-sectional area,

$$= \frac{0\cdot15}{100} \times 18 \times 12 = 0\cdot32 \text{ sq. in.,}$$

i.e. 0·16 sq. in. per face and this is amply provided. The maximum negative moment under the column is 205,000 in. lb/ft. and this needs to be increased by 33 per cent over a width equal to one-half the width of the column strip (C.P. 114 (1957)). Thus the area of steel required in the bottom of the slab under the column is

$$A_{st} = \frac{205,000 \times 1\cdot33}{20,000 \times 0\cdot86 \times 15\frac{1}{2}} = 1\cdot02 \text{ sq. in.}$$

less already provided ($\frac{7}{8}$ in. at 18 in. centres) $= 0\cdot40$ sq. in.

Additional reinforcement required $= 0\cdot62$ sq. in.

This is given by additional $1\frac{1}{8}$ in. rods at 18 in. centres extending into the panel a distance equal to one-quarter of the span.

A check for the compressive stress gives

$$\frac{M}{bd_1^2} = \frac{205,000 \times 1\cdot33}{12 \times 15\frac{1}{2}^2} = 95$$

indicating that f_{cb} is well less than 1000 lb/sq. in.

The critical section for shear in flat slab construction is at a distance equal to half the thickness of the slab from the face of the column, i.e. at

$$\frac{16}{2} + \frac{18}{2} = 17 \text{ in. from the column centre-line.}$$

The load transmitted from the column directly to the ground under is therefore

$$0\cdot5 \text{ tons/sq. ft.} \times \frac{34 \text{ in.} \times 34 \text{ in.}}{144} = 4 \text{ tons.}$$

Therefore the shear force at the critical section is $80 - 4 = 76$ tons. The shear perimeter is 4×34 in. $= 136$ in., so that the shear stress $= \dfrac{76 \times 2240}{0\cdot86 \times 15\frac{1}{2} \times 136} = 94$ lb/sq. in. which is satisfactory and no bent-up bars are required.

The arrangement of reinforcement is shown in Fig. 7.28.

BUOYANT FOUNDATION

Consider the case of a building having columns at 20 ft. centres in two directions at right angles, requiring support on poor ground which for a considerable depth is only good for a nett loading intensity of 0·5 tons/sq. ft. The columns carry 400 tons each from the superstructure and the ground floor carries a load of 2 cwt/sq. ft. including its own self-weight. Since the pressure in the ground due to column loads and ground floor is something more than 1 ton/sq. ft. a normal raft slab will not be satisfactory and a buoyant foundation is required.

The depth required for the buoyant foundation is determined so that, in excavating, the relief obtained from the removal of the overburden reduces the nett pressure at the contact surface to 0·5 ton/sq. ft. The bottom slab is then constructed at this level and is loaded through cross walls and beams situated between the raft and ground floor levels, the whole forming a cellular construction as shown in Fig. 7.29. The total pressure on the ground is then

$$\text{from columns} \quad = \frac{400 \text{ tons}}{20 \times 20} \quad = 1\!\cdot\!00 \text{ tons/sq. ft.}$$

$$\text{from ground floor} = \frac{224}{2240} \quad = 0\!\cdot\!10 \text{ tons/sq. ft.}$$

$$\text{Self-weight of foundation (say)} \quad 0\!\cdot\!25 \text{ tons/sq. ft.}$$

$$\overline{}$$

$$1\!\cdot\!35 \text{ tons/sq. ft.}$$

The necessary relief of pressure required due to excavation is therefore $1\!\cdot\!35 - 0\!\cdot\!50 = 0\!\cdot\!85$ tons/sq. ft. Taking 120 lb./sq. ft. for the soil weight, the depth of excavation required to achieve this relief is

$$\frac{0\!\cdot\!85 \times 2240}{120} = 16 \text{ ft.}$$

Consider now an internal bay of the bottom slab. Try a slab thickness of 18 in. As in the previous example, there is no bending or shear produced in the slab by its self-weight. Thus the pressure causing bending is

$$1\!\cdot\!35 - \left(\frac{18}{12} \times \frac{150}{2240}\right) = 1\!\cdot\!25 \text{ tons/sq. ft.}$$

The bending moments in the slab are determined in the manner shown in Article 5.4, whereby from Table 5.2

$$\text{Max. negative moment} = 0\!\cdot\!033 \times 1\!\cdot\!25 \times 2240 \times 20^2 \times 12$$
$$= 444{,}000 \text{ in. lb/ft.}$$

and

$$\text{Max. positive moment} = 0\!\cdot\!025 \times 1\!\cdot\!25 \times 2240 \times 20^2 \times 12$$
$$= 336{,}000 \text{ in. lb/ft.}$$

The reinforcement required under the walls is, therefore,

$$\frac{444{,}000}{20{,}000 \times 0\!\cdot\!86 \times 15\tfrac{1}{2}} = 1\!\cdot\!67 \text{ sq. in.}$$

and the reinforcement required in the top of slab in both directions at mid-span is

$$\frac{336,000}{20,000 \times 0.86 \times 15\tfrac{1}{2}} = 1.26 \text{ sq. in.}$$

A reasonable minimum mat of steel, top and bottom, would be (say) 1 in. diameter rods at 15 in. centres both ways. These can

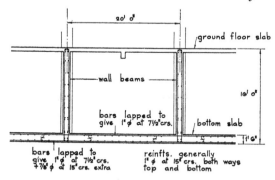

Section

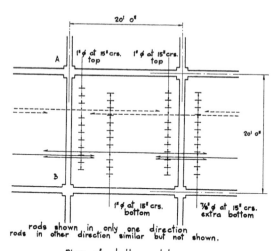

Plan of bottom slab

FIG. 7.29. BUOYANT FOUNDATION

be lapped in the top, at mid-span to give 1 in. at $7\tfrac{1}{2}$ in. centres (1·26 sq. in.), and can be lapped under the walls with additional $\tfrac{7}{8}$ in. rods at 15 in. centres to give a total area of 1·74 sq. in. (see Fig. 7.29). $\dfrac{M}{bd_1{}^2} = \dfrac{444,000}{12 \times 15\tfrac{1}{2}{}^2} = 154$, which is satisfactory.

The critical shear stress in the bottom slab is determined as follows. Assuming that the wall beams are 15 in. wide the critical section for shear is $\dfrac{15}{2} + 15\frac{1}{2} = 23$ in. from the column centre-lines. Thus for the bottom slab panel the shear perimeter

$$= 4 \times \left[20 \text{ ft.} - \left(2 \times \frac{23}{12} \right) \right]$$
$$= 4 \times 17\cdot7 \text{ ft.}$$
$$= 68\cdot7 \text{ ft.}$$

The upward force carried within the shear perimeter

$$= 17\cdot17^2 \times 1\cdot25 = 368 \text{ tons.}$$

Hence shear stress $= \dfrac{368 \times 2240}{0\cdot86 \times 15\frac{1}{2} \times (68\cdot7 \times 12)}$

$$= 75 \text{lb/sq. in.}$$

Consider now the wall beam between columns A and B. This beam is acted on by the pressure of $1\cdot25$ tons/sq. ft. acting upwards, and by the ground floor and self-weight of the beam acting downwards—the difference being reacted by the column loads. The nett upward load causing bending is, therefore,

due to ground

$$= 20 \text{ ft.} \times 10 \text{ ft.} \times 1\cdot25 \qquad\qquad = \qquad\qquad 250 \text{ tons}$$

less due to self

$$= 12\frac{1}{2} \text{ ft.} \times 1\frac{1}{4} \text{ ft.} \times 20 \text{ ft.} \times \frac{150}{2240} = 21 \text{ tons}$$

less due to ground floor

$$= 20 \text{ ft.} \times 10 \text{ ft.} \times \frac{224}{2240} \qquad = 20 \text{ tons} \quad 41 \text{ tons}$$

Therefore nett load on beam $=$ 209 tons.

Thus the end shear $= \dfrac{209}{2} = 104\cdot5$ tons $= 234{,}000$ lb.

and the shear stress $= \dfrac{234{,}000}{0\cdot86 \times 188 \times 15} = 96$ lb/sq. in.

which is a permissible stress on the unreinforced section. Allowing for the breadth of the columns and permissible redistribution of moments, it will be satisfactory to design the wall beams for support and mid-span moments of $\dfrac{WL}{16}$. There again using the nett load on the beam, we have

$$M = 209 \times 2240 \times \frac{20}{16} \times 12 = 7,000,000 \text{ in. lb.}$$

for which

$$\text{required } A_{st} = \frac{7,000,000}{20,000 \times 0.86 \times 188} = 2.16 \text{ sq. in.,}$$

which is given by four $\frac{7}{8}$ in. diameter bars (area $= 2.4$ sq. in.) top and bottom. Additional vertical and horizontal reinforcements are distributed over the height of the wall to minimise shrinkage cracks.

Having now decided on the sizes of the members, the actual ground pressure due to the foundations must be checked back against the assumed pressure:

Self-weight of two wall-beams $= 2 \times 21$ $= 42$ tons

Self-weight of slab $= 20 \times 20 \times 1\frac{1}{2} \times \dfrac{150}{2240} = 40$ tons

Total weight of foundations $= 82$ tons.

Therefore, pressure due to foundations

$$= \frac{82}{20 \times 20} = 0.2 \text{ tons/sq. ft.}$$

which is less than the assumed figure of 0.25 tons/sq. ft.

REFERENCES

1. FOUNDATIONS. *Civil Engineering Code of Practice*, No. 4. Institution of Civil Engineers (1954).
2. TERZAGHI, K. and PECK, R. B. *Soil mechanics in engineering practice.* John Wiley & Sons (1948).
3. FABER, OSCAR. Pressure distribution under bases, and stability of foundations. *The Structural Engineer* (March 1933).
4. SKEMPTON, A. W. *The bearing capacity of clays.* Building Research Congress (Proceedings). Division 1 (1956).
5. BRITISH STANDARD CODE OF PRACTICE, C.P. 2001. *Site investigations.* (1957).
6. TSCHEBOTARIOFF, G. *Soil mechanics, foundations and earth structures.* McGraw-Hill (1956).

CHAPTER 8

RETAINING WALLS

ART. 8.1. INTRODUCTION

IT is everyday knowledge that sound rock, evenly bedded, will stand with a sheer face to considerable heights. This can be seen in artificial form in deep railway cuttings; and there are countless examples in nature – the cliffs of Beachy Head to mention only one. Soils on the other hand cannot stand more steeply than their *natural angle of repose*, as evidenced again in railway work where engineers have cut back to slopes ranging from about 1 in $1\frac{1}{2}$ to 1 in 3 depending on the nature of the soil, and other physical conditions. Where there are practical objections to sloping back in this manner (as for example at basements to city-buildings or where industrial materials like sand and stone have to be stored in limited areas), a wall has to be built to retain all material required to lie above the natural angle of repose.

The main problem in designing *retaining walls* lies in determining the pressures on the back of the wall from the material to be retained, and the capability of the ground in front of the wall and under the base to resist the lateral and vertical forces arising from those pressures. A knowledge of the properties and behaviour of soils is therefore fundamental to the design of retaining walls; and the problem is closely allied to foundation design. Reference should be made to Articles 7.3 to 7.10 inclusive.

In liquid-retaining structures, the applied pressures are directly calculable from the density of the liquid retained, and the head at which this acts. This is because liquids are both frictionless and cohesionless. Soils behave differently. Sand, for example, when dry, acts as a frictional material without cohesion, and has a well-defined angle of repose. If the same sand is now moistened, it develops a certain amount of cohesive strength and its angle of repose increases, somewhat erratically. Further wetting will break down the internal friction forces until the sand slumps and will hardly stand at any angle at all. Clay on the other hand when first exposed in situ stands vertically to considerable depths when reasonably dry, but after a time will subside, depending on its moisture content. And clay, in dry seasons, gives up its moisture to atmosphere with consequent shrinkage, so that at depths less than about four or five feet it may be unreliable as a stop to react the forward movement of a retaining wall.

Thus the lateral pressures from soils can vary very widely depending on the moisture content. If a unit volume of soil is considered at a depth z below the free surface, the lateral pressures can vary from about $30z$ to $90z$ in sands, and from zero to about $90z$ in clays. And within these ranges, the pressures behind retaining walls may vary due to seasonal or other periodic changes. Indeed the construction of the retaining wall itself may cause major changes in the ground conditions – blocking a natural drainage path, or exposing to shrinkage an otherwise stable clay. Similarly the frictional resistance to sliding under the base of a retaining wall is critical of moisture content. This is particularly true of clays, which when dry can be rough and hard, but when wet can be smooth and slippery.

It is clear from the foregoing that no academic mathematical formula can bend sufficiently to accommodate all the practical realities of the matter. The active forces can vary enormously; and the reactive forces must to a large extent be regarded as suspect. Nevertheless, in Articles 8.3 and 8.4 the bones are given of soil theories as related to the determination of active and passive pressures on retaining walls; though refinements to these theories are deliberately omitted, as being in the authors' views far too pernickety in relation to the physical uncertainties arising. Far better that the engineer should review the conditions of the site, and apply his personal judgement and skill, than embark on some lengthy calculations concerning materials and conditions which have no allegiance to or regard for the laws of mathematics.

Considerable effort should be made by the engineer to control site conditions where possible so that his design problem can be more reasonably defined. For example the provision of proper drainage-rubble behind a retaining wall will guard against active water pressure, *provided the rubble is itself properly drained* by weep-holes or otherwise to an effective drainage system. Similarly cut-off drains or other land-drainage provision behind the wall can be used to limit the range of moisture content in the material to be retained. The base on which a retaining wall is built should be scraped clean and dry, and immediately protected by a layer of rough concrete blinding before workmen go in to fix any reinforcements: and a cut-off heel should be provided to prevent drainage from the back seeping under the base to lubricate or soften clay soils. In soils subject to seasonal variations, the base should be kept deep enough to be unaffected. And, in certain circumstances, concrete paving should be provided at high level behind the wall, and at low level in front of the wall: this will help protect the soil from seasonal variations, and can be laid to falls to get rainwater away to drainage gulleys.

ART. 8.2. STABILITY OF RETAINING WALLS

The present article considers the stability of retaining walls which stand free on their own base and are disassociated from other structures. A typical example of such a wall is shown in Fig. 8.1

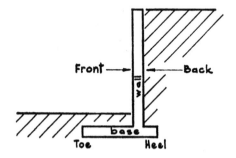

FIG. 8.1. FREE-STANDING RETAINING WALL

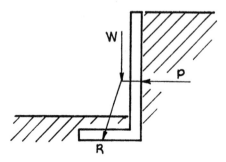

FIG. 8.2. FORCES ACTING ON RETAINING WALL

and comprises, essentially, only a wall and a base. The stability of walls of this form is normally more complicated than walls to basements or deep pits where often the weight of the building over or the active pressures from opposite sides of the structure balance one another by direct compression across walls or floor slabs, so that the risk of failure by general rotation or forward movement is negligible. For convenience we will refer to the main features of free-standing retaining walls as shown in Fig. 8.1: the wall may then be considered as a man with his back to the material being retained and no confusion will arise. Descriptions of actual types of retaining walls, and calculations to determine the necessary con-

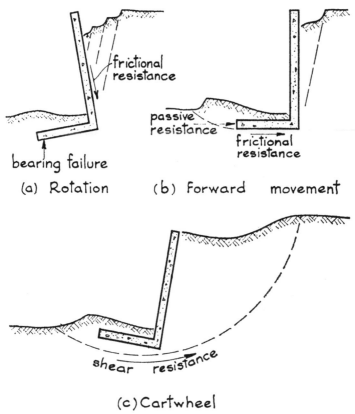

FIG. 8.3. MODES OF FAILURE OF FREE-STANDING
WALLS

crete dimensions and reinforcements are given in Articles 8.6 and 8.7 respectively.

(i) *Failure by rotation*

Referring to Fig. 8.2, the total active thrust P on the back of the wall, combined vectorially with the weight of the wall (and any other vertical loads) W, give a resultant force R. If the line of this resultant falls outside the width of the base, the structure will fail by immediate over-turning. If the line of the resultant comes just within the width of the base, immediate over-turning cannot occur; but the maximum edge pressure resulting from the eccentricity of R may be such as to cause settlements at the toe, which in turn will lead to failure by rotation (*see* Fig. 8.3*a*).

This does not necessarily mean that for stability the resultant has to fall within the middle third of the base (as was shown in Fig. 7.26*b*): but care has to be taken to watch that if a load-factor were applied to the active pressure behind the wall, the resultant R would not then be thrown so near to the edge of the base that the ultimate bearing capacity of the soil would be exceeded. This is important, and not always realised. A factor of safety of 2 or 3 in the strength of the reinforced concrete section, does not necessarily mean that the structure as a whole has a factor of safety against rotation of the same amount: and in view of the difficulty in assessing with any accuracy the pressure on the back of the wall, this point needs careful attention.

For normal design pressures behind the wall, the maximum pressure under the toe should not exceed the allowable bearing pressure of the soil as described in Article 7.9; but in considering over-turning, there should be a stability factor of about $1\frac{1}{2}$, and more in the case of soft clays.

(ii) *Failure by forward movement*

A free-standing retaining wall is prevented from moving forward by a combination of the passive resistance of the soil on the front faces of the base, and the friction on the under-side of the base (*see* Fig. 8.3*b*). With frictional soils, provided a small forward movement of the wall can be tolerated, the full calculated passive resistance can be relied upon. Clays, however, are not reliable, particularly within the depth influenced by seasonal changes, mainly due to shrinkage.

Base friction on frictional soils may be taken as

$$R_{\mathrm{B}} = \mu W$$

where W is the total vertical load, and μ is the friction coefficient, normally taken as $\tan \phi$. On cohesive soils, base friction is taken as

$$R_{\mathrm{B}} = cL_{\mathrm{B}}$$

where L_{B} is the length of base under pressure, and c is the cohesion of the clay up to 1000 lb/sq. ft., beyond which value it should not be taken, as adhesion between the concrete and clay then becomes the limiting factor.

The factor of safety against forward movement of the wall should be not less than 2. This is sometimes achieved by forming a projection below the base of the wall, the projection being best placed near the heel where forward spewing of the soil is prevented by the great bearing pressure nearer the front of the base (*see* Fig. 8.4).

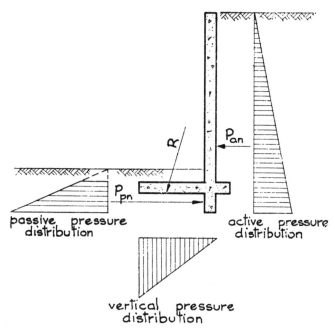

FIG. 8.4. PROJECTING KEY UNDER WALL BASE

(iii) *Cartwheel failure*

Cartwheel failure has already been described in Article 7.8 (*see also* Fig. 8.3c). In the case of retaining walls, the disturbing moment is caused by the weight of soil behind the wall standing above the general level of the soil in front of the wall. A factor of safety of $1\frac{1}{2}$ should be available against failure of this type.

ART. 8.3. ACTIVE PRESSURE ON BACK OF WALL

If the elemental unit of soil in Fig. 8.5a is subjected to principal stresses p_1 vertically (major) and p_3 horizontally, then it is well known that on any plane at an angle θ measured from the major principal plane, the normal stress p and the shear stress s are given by

$$p = \frac{p_1 + p_3}{2} + \frac{p_1 - p_3}{2} \cos 2\theta \qquad (8.1)$$

and $$s = \frac{p_1 - p_3}{2} \sin 2\theta. \qquad (8.2)$$

Now Coulomb's equation for a cohesive and frictional soil is

$$s = c + p \tan \phi \ (see \ \text{equation 7.6}).$$

And in Article 7.5 Coulomb's equation at the condition of failure

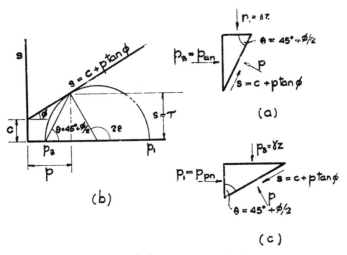

FIG. 8.5. SOIL STRESSES

was represented graphically (Fig. 7.9b) with the "Coulomb line" tangential to the "Mohr circle".

Referring now to Fig. 8.5b, it is simple to demonstrate that the angle of inclination θ of the failure plane is given by

$$\theta = 45 + \frac{\phi}{2}. \qquad (8.3)$$

Then from equations (8.1), (8.2), (7.6), and (8.3), we have

$$p_3 = p_1 \tan^2 \left(45 - \frac{\phi}{2} \right) - 2c \tan \left(45 - \frac{\phi}{2} \right). \qquad (8.4)$$

At depth z, if γ is the density of the soil,

$$p_1 = \gamma z.$$

Thus, at soil failure, the horizontal pressure, known as the normal active pressure, is given by

$$p_{an} = \gamma z \tan^2 \left(45 - \frac{\phi}{2} \right) - 2c \tan \left(45 - \frac{\phi}{2} \right). \qquad (8.5)$$

This formula is generally attributed to Bell[1]. If now the coefficients of active earth pressure are defined as

$$K_a = \tan^2 \left(45 - \frac{\phi}{2} \right) \quad \text{and} \quad K_{ac} = 2 \tan \left(45 - \frac{\phi}{2} \right),$$

Bell's formula becomes

$$p_{an} = \gamma z \, K_a - c K_{ac}. \qquad (8.6)$$

For cohesionless soils ($c = 0$), we have

$$p_{an} = \gamma z \tan^2 \left(45 - \frac{\phi}{2} \right) = \gamma z \left(\frac{1 - \sin \phi}{1 + \sin \phi} \right) \qquad (8.7)$$

which is Rankine's formula[2].
And for frictionless soils ($\phi = 0$, whence $K_a = 1$), we have

$$p_{an} = \gamma z - c K_{ac} = \gamma z - 2c. \qquad (8.8)$$

Thus from theoretical considerations it would appear that the active pressure at any depth can be calculated for all values of c and ϕ. This is indicated diagrammatically in Fig. 8.6. For cohesive soils, the active pressure appears to have negative values

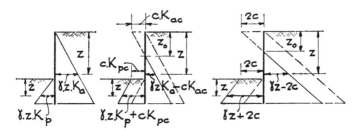

a) Cohesionless   b) c & φ   c) Purely cohesive
   material (c=o)   material   material (φ =o)

FIG. 8.6. ACTIVE AND PASSIVE PRESSURES ON WALLS

down to a depth z_0 where $z_0 = \dfrac{2c}{\gamma} \tan \left(45 - \dfrac{\phi}{2} \right)$: this clearly is theoretical only, and not realistic; and in practice the soil moves away from the wall over the depth z_0 leaving a tension crack.

Obviously no allowance for any tension force can be made in calculations.

Now the soil theory given above only applies in determining the active pressures on the backs of walls where the following conditions are fulfilled:

(i) The wall must be able to yield sufficiently for the soil to reach the condition of failure.

(ii) There must be no friction or adhesion between the soil and the back of the wall.

(iii) The surface of the retained soil must be horizontal back to the plane of rupture.

Frequently these conditions do not apply. Where walls are propped by floors, or tunnel roofs, or where walls are tied back to anchorages, the pressures near the top of the wall will be greater, and the pressures lower down will be less. In cases where the wall as a whole slides forward, the pressure distribution will become more nearly parabolic. And where friction or adhesion develop on the back face of the wall, the amount *and direction* of the total active pressure will change.

SURCHARGE FROM SUPERIMPOSED LOADS

(i) *Uniformly distributed load*

The additional effect of uniformly distributed loads on a horizontal ground surface behind retaining walls is determined most

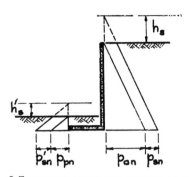

FIG. 8.7. SURCHARGE BY UNIFORM LOAD

conveniently by considering the load as additional soil of equivalent depth h_s (Fig. 8.7) where

$$h_s = \frac{w}{\gamma},$$

and w is the surcharge load per unit area, and γ is the density of the soil.

(ii) *Line load*

An approximate solution for determining the additional effect from a line load behind a retaining wall is due to Terzaghi and Peck[3], and illustrated in Fig. 8.8a. The horizontal force on the wall

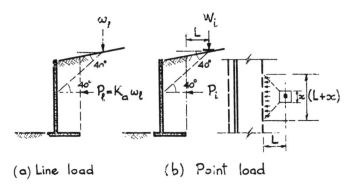

(a) Line load   (b) Point load

FIG. 8.8. SURCHARGE BY LINE- AND POINT-LOADS

due to the line load is assumed to act at a level determined by the 40° line drawn from the position of the load as shown. The intensity of the force per unit length P_1 is then taken as being

$$P_1 = K_a w_1 \qquad (8.9)$$

where $K_a = \tan^2\left(45 - \dfrac{\phi}{2}\right)$,

and w_1 is the intensity of line load per unit length.

(iii) *Point load*

Where point loads are spaced longitudinally behind a retaining wall at intervals sufficient for their effects on the back of the wall not to overlap, it may be assumed that the point load is spread over a width as shown in Fig. 8.8b. The additional effect on the back of the wall, per unit length, is then given by

$$P_1 = K_a \frac{W_1}{(L + x)} \qquad (8.10)$$

where W_1 is the isolated point load,

   L is the distance of the point load from the heel of the
   wall,

   x is the breadth of bearing of the point load.

WALLS SURCHARGED BY SLOPING SURFACE

For cohesionless soils surcharged to an angle β above the
horizontal, Fig. 8.9, the pressure on retaining walls with smooth

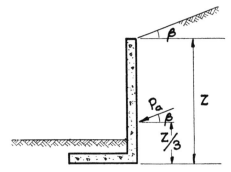

FIG. 8.9. SURCHARGE BY SLOPING FILL

vertical backs, according to Rankine, may be taken as

$$p_a = \gamma z \cos \beta \; \frac{\cos \beta - \sqrt{\cos^2 \beta - \cos^2 \phi}}{\cos \beta + \sqrt{\cos^2 \beta - \cos^2 \phi}}. \qquad (8.11)$$

If the total height of the wall is Z, then the total force on the wall
per unit length of wall is

$$P_a = \frac{\gamma Z^2}{2} \cos \beta \; \frac{\cos \beta - \sqrt{\cos^2 \beta - \cos^2 \phi}}{\cos \beta + \sqrt{\cos^2 \beta - \cos^2 \phi}}. \qquad (8.12)$$

If the wall is not surcharged, this reduces to

$$P_a = \frac{\gamma Z^2}{2} \left(\frac{1 - \sin \phi}{1 + \sin \phi} \right) \qquad (8.13)$$

which is in agreement with equation 8.7.

If β has its maximum possible value of ϕ,

$$P_a = \frac{\gamma Z^2}{2} \cdot \cos \phi. \qquad (8.14)$$

In each case P_a acts at $\dfrac{Z}{3}$ above the base of the wall, and parallel
to the surface of the soil.

For cohesive soils inclined above the horizontal to form a surcharge, the active pressure is most reasonably estimated by using Coulomb's wedge-theory. This graphical method of analysis has already been widely described in textbooks on soil mechanics.

ART. 8.4. PASSIVE RESISTANCE AT FRONT OF WALL

Consider now the elemental unit of soil in Fig. 8.5c, where the principal stresses have been interchanged so that the major stress p_1 is horizontal, and p_3 is vertical. Equations (8.1) and (8.2) still apply; and following the same arguments as before, we have

$$p_1 = p_3 \tan^2\left(45 + \frac{\phi}{2}\right) + 2c \tan\left(45 + \frac{\phi}{2}\right). \qquad (8.15)$$

In this case $p_3 = \gamma z$; and the soil will fail when the horizontal pressure, known as the normal passive pressure, becomes

$$p_{pn} = \gamma z \tan^2\left(45 + \frac{\phi}{2}\right) + 2c \tan\left(45 + \frac{\phi}{2}\right), \qquad (8.16)$$

or
$$p_{pn} = \gamma z K_p + c K_{pc} \qquad (8.17)$$

where $K_p = \tan^2\left(45 + \frac{\phi}{2}\right)$ and $K_{pc} = 2 \tan\left(45 + \frac{\phi}{2}\right)$.

For cohesionless soils $(c = 0)$,

$$p_{pn} = \gamma z \tan^2\left(45 + \frac{\phi}{2}\right) = \gamma z \left(\frac{1 + \sin\phi}{1 - \sin\phi}\right) \qquad (8.18)$$

which again is Rankine's formula.

And for frictionless soils $(\phi = 0$, whence $K_p = 1)$,

$$p_{pn} = \gamma z + c K_{pc} = \gamma z + 2c. \qquad (8.19)$$

Again it would seem that here were formulae to determine the passive pressure at any depth for all values of c and ϕ (see Fig. 8.6). But all too frequently the practical requirements of the theory are not met; and indeed the matter may become critical to such an extent that the theory can be widely misleading, particularly bearing in mind that passive resistance is often sought at relatively small depths where seasonal effects may further militate against calculation, particularly in cases of cohesive soils. Passive resistance may often be less reliable than frictional resistance under the base: and in any event the judgement of the experienced engineer will count for more than any painstaking calculations prepared without an understanding of the limitations of the underlying theory.

ART. 8.5. WATER PRESSURE AND DRAINAGE

The effect ground water has on pressures behind retaining walls is difficult to assess. Consider first the case of a retaining wall constructed tight against an excavated clay face, as may be done with precast-concrete units jacked into position and grouted, or where concrete is cast in situ while vertical timbers are withdrawn.

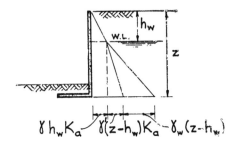

FIG. 8.10. ACTIVE PRESSURE ON WALL FROM
SUBMERGED GRANULAR MATERIAL

Due to the imperviousness of the clay, there should be no water present at the face of the wall, and consequently no water pressure, at any rate below the depth to which the tension crack or seasonal shrinkage may extend. On the other hand, if a very small gap is left, or if the clay is backfilled after the wall has been constructed to a shuttered face, there can be no doubt that any water present will exert its full hydrostatic pressure. But if the water is then between the clay and the wall, the clay cannot press on the wall in addition to the water pressing, so that the design should cater only for the full pressure from the water; except only that where the active pressure from the clay is greater than the hydrostatic pressure from the water, than the clay will push the water away and the design should cater for the full pressure from the clay. In other words in clays where free water is present, the full water pressure should be taken except only where the pressure from the clay is greater: the two need not be taken as additional to one another. It then becomes a moot point as to whether the clay is really home against the wall or not. Is there water pressure, or isn't there? This must remain a matter for the engineer's judgement.

With sands and gravels the problem is less obscure. Clearly

nothing will prevent the water pressure acting against the wall through the soil, and the full value must be allowed for. At the same time, the density of the submerged soil will be reduced by buoyancy so that the active pressure from the soil particles may be less (depending on the particle grading). In some sands however the grading may be such that the soil becomes completely unstable and acts on the wall with the water as a dense fluid. Whichever way

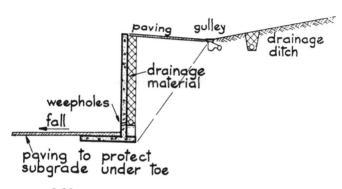

FIG. 8.11. DRAINAGE ARRANGEMENTS AT RETAINING WALL

you look at it, the unit pressure is likely to be of the order of $p = 90z$. The theoretical distribution of pressure on a vertical-backed retaining wall from stable submerged granular materials is indicated in Fig. 8.10. The normal active pressure at any depth z is then given by the equation

$$p_{an} = K_a[\gamma h_w + \gamma'(z - h_w)] + \gamma_w(z - h_w)$$

where h_w is the depth from the free surface to water level,

γ is the normal density of the soil,

γ' is the density of soil when submerged,

γ_w is the density of water,

K_a is $\tan^2\left(45 - \dfrac{\phi}{2}\right)$.

Very high pressures may develop where dry clay is used as back-fill material, consolidated well, but in such a manner that free passage of ground or surface water can cause subsequent swelling of the clay. If the material is then restrained against upward relief, all movement will be directed against the retaining wall, causing considerable forces.

The simplest way of dealing with water-pressure problems is to remove the water! This is not always easy or possible (for example in certain cases of basements or pits); but it is a fact that many retaining walls in existence could have been constructed more cheaply and more successfully if proper attention had been given to the matter of drainage. A layer of drainage-material built up behind the back of the wall will alleviate water-pressure here, and at the same time lower the water table, reducing risk of softening (*see* Fig. 8.11). The water can then be run away, either by a full-length drain discharging at the ends of the wall or into catchpits and manholes, or alternatively through weep-holes. Weep-hole discharge should not be allowed to run on to the ground at the toe of the base, causing softening, but should be collected in channels or pipes, and led away. Below the level of the back-drain (or weep-holes), concrete filling should be provided to prevent water seeping down to foundation level and softening or lubricating the soil under the base.

Where the ground behind a wall rises away for any distance, and considerable quantities of surface water are likely to run down this slope, a drainage ditch should be provided to prevent the water getting into the soil acting behind the retaining wall. It is also advisable, in most circumstances, to protect the soil behind the wall and at the toe by concrete paving as shown in Fig. 8.11.

ART. 8.6. STRUCTURAL ANATOMY OF RETAINING WALLS

Reinforced concrete is well suited to retaining-wall construction. Compared to gravity-walls in mass concrete, reinforced concrete walls of any height are considerably more economical in use of materials; they weigh very much less with obvious advantage in certain circumstances on poor ground; and they occupy less space, which either makes for less excavation cost or alternatively gives greater clear space in front of the wall. In waterlogged ground they have the advantage over masonry of being more nearly impermeable, and certainly less subject to attack from frost and ground-water salts. Vertical shrinkage joints can be provided effectively with modern jointing materials.

Retaining-wall bases have to be taken down to minimum depths to avoid seasonal variations just as any other foundations. Similarly hydrostatic buoyancy requires consideration. And advantage may be taken of nett loading intensities where overburden is removed by excavation in constructing basements and pits. These matters have been explained in Article 7.10.

The simplest retaining-wall structures to design are the sides of pits, trenches or tunnels. Here the walls span as slabs. In small pits, the walls span continuously from corner to corner, with some cantilever action from the base. Where the pits are roofed over, the walls may span in two directions. Trenches cantilever up from the base, but there is normally no problem of forward movement or rotation to consider. Tunnels span vertically between roof and

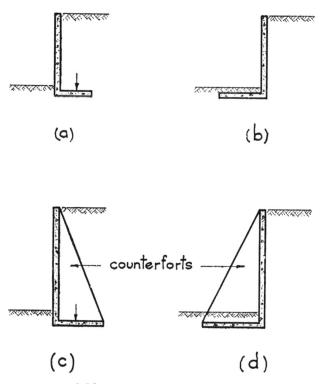

(a) (b)

(c) (d)

FIG. 8.12. FREE-STANDING WALL TYPES

base, with continuity at the supports. Walls to deep basements may cantilever up from foundation level, with or without propping from intermediate floors. Single-storey basement walls are often designed to span either between columns, or between the basement and ground-floor constructions; or if the proportions are suitable, it may be economical to span in both directions. The weight of the building over a basement, or the thrust provided by the lower floors, normally provides adequately against rotation or forward movement.

Free-standing retaining walls are the most difficult to design. Here the problems of rotation and forward movement arise as described in Article 8.2, and a solution can only be found by selecting from experience a structure of suitable arrangement and proportions, and analysing this to determine whether the profile chosen is satisfactory. Design is thus by a process of trial and correction. Basically there are two main types of free-standing walls: with the base behind the wall (*see* Fig. 8.12*a*), or with the base in front of the wall (*see* Fig. 8.12*b*). The type of wall shown in Fig. 8.12*a* requires the shorter base owing to the greater stability given by the weight of material over the base. This arrangement is suitable for material storage (such as coal, sand, stone) but is not so convenient for supporting earthworks because of the greater work involved in excavation and subsequent back-filling. For supporting earthworks the type shown in Fig. 8.12*b* is preferable despite the base generally being required about 25 per cent longer. This is because of savings in excavation cost, and the reduced disturbance to the soil behind the wall.

For high walls the simple slab profile may be elaborated to vertical cantilevered counterforts with the wall slab spanning continuously between (Figs. 8.12*c* and 8.12*d*). This is only a refinement on the simpler types and does not affect the basic principles. With the type of wall which has its base in front, the retained material pushes the slab on to the counterforts. But with the wall with its base behind, the slab has to be tied back to the counterforts: though an advantage of this system is that the wall slab acts as the flange of a T-beam system, moving the neutral axis nearer to the wall with consequent economy of tension-steel in the counterfort. For very high retaining walls it may be an economy to arch the wall slab in plan between the counterforts; this will depend on the relative costs of straight and curved formwork, and the greater interval achieved between counterforts.

ART. 8.7. DESIGN OF REINFORCED CONCRETE RETAINING WALLS

When estimates have been made of the soil pressures and resistances, retaining walls may be designed in the manner shown in the following examples.

WALL WITH BASE IN FRONT

Consider the wall shown in Fig. 8.13*a*, required to retain a 17 ft. height of clay soil surcharged at the back by a superimposed load

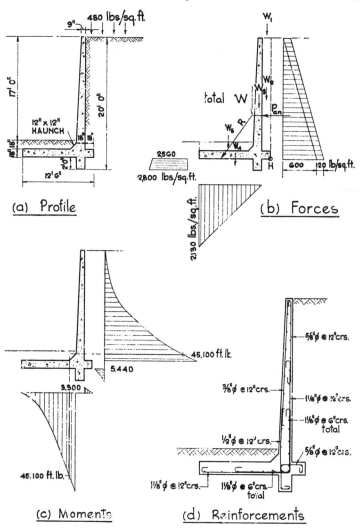

(a) Profile

(b) Forces

(c) Moments

(d) Reinforcements

FIG. 8.13. DESIGN OF WALL WITH BASE IN FRONT

of 480 lb/sq. ft. The soil density γ is 120 lb/cu. ft., and the cohesion c is 1100 lb/sq. ft.

The effect of surcharge is equivalent to additional soil of

$$h_s = \frac{w}{\gamma} = \frac{480}{120} = 4 \text{ ft.}$$

Using equation 8.8 for a frictionless soil, we have the depth of theoretical tension crack

$$z_0 = \frac{2c}{\gamma} = \frac{2 \times 1100}{120} = 18 \cdot 3 \text{ ft.}$$

less surcharge effect $= \quad 4 \cdot 0 \text{ ft.}$

$$14 \cdot 3 \text{ ft.}$$

This implies that there would be active pressure on only the lower $2 \cdot 7$ ft. of the wall back, which is absurd. Accordingly the soil theory must be discarded, and from experience we will design for pressures of

$$p_{\text{an}} = 30z,$$

as shown in Fig. 8.13b. The unit pressure due to surcharge will then be $30 \times 4 = 120$ lb/sq. ft. at all depths.

The maximum bending stresses in the wall will occur at the top of the haunch where

$$p_{\text{an}} \text{ soil} = 30 \times 17\tfrac{1}{2} = 525 \text{ lb/sq. ft.}$$

The total shear down to this level, per foot width of wall is

$$P_{\text{an}} \text{ (soil)} = 525 \times \frac{17\tfrac{1}{2}}{2} = 4600 \text{ lb.}$$

$$\text{(surcharge)} = 120 \times 17\tfrac{1}{2} = 2100 \text{ lb.}$$

$$6700 \text{ lb.}$$

and the bending moment is

$$\left(4600 \times \frac{17\tfrac{1}{2}}{3}\right) + \left(2100 \times \frac{17\tfrac{1}{2}}{2}\right) = 45,100 \text{ ft. lb.}$$

Thus we have $A_{\text{st}} = \dfrac{45,100 \times 12}{20,000 \times (0 \cdot 86 \times 16)} = 1 \cdot 97$ sq. in. which is provided by $1\tfrac{1}{8}$ in. dia. rods at 6 in. centres (area $2 \cdot 0$ sq. in.). $\dfrac{M}{bd_1{}^2} = \dfrac{45,100 \times 12}{12 \times 16^2} = 177$ which is satisfactory for $1:2:4$ concrete. And the shear stress is given by

$$q = \frac{Q}{bl_a} = \frac{6700}{12 \times (0 \cdot 86 \times 16)} = 40 \text{ lb/sq. in.}$$

only. Higher up the wall the bending moments diminish rapidly as shown in Fig. 8.13c, and for economy the reinforcements are part stopped off, and later reduced, as shown.

Considering now the stability of the structure against rotation, and referring to the forces as shown in Fig. 8.13b, we can calculate the anti-clockwise moment applied about H in tabular manner as below:

	Forces		Arm about H	Applied moment
	Vertical	Horizontal		
	lb.	lb.	ft.	ft. lb.
W_1 (surcharge above heel)	720	—	0·75	540
W_2 (soil above heel)	3,330	—	0·75	2,390
W_3 (wall)	3,140	—	2·04	6,400
W_4 (base)	2,810	—	6·25	17,600
W_5 (filling above base)	1,800	—	7·50	13,500
P_{an} (soil)	—	6,000	6·66	40,000
(surcharge)	—	2,400	10·00	24,000
Total forces	11,800	8,400	—	—
Total moment	—	—	—	104,430

The point at which the resultant thrust strikes the underside of the base is therefore distant from H by

$$\bar{x} = \frac{104,430 \text{ ft. lb.}}{11,800 \text{ lb.}} = 8 \cdot 8 \text{ ft.}$$

This is outside the middle third of the base width, so that the pressure distribution will be triangular with a maximum edge pressure of

$$\frac{11,800 \times 2}{3 \times (12 \cdot 5 - 8 \cdot 8)} = 2130 \text{ lb/sq. ft.}$$

From equation (7.24) the ultimate bearing capacity of the soil is

$$q_f = 5 \cdot 70 \times 1100 = 6350 \text{ lb/sq. ft.,}$$

showing a factor of safety of 3. Note also that if the active forces behind the wall had been increased by 50 per cent, the line of thrust would still have been within the base width, and the edge bearing pressure only just in excess of the ultimate. With the undercut heel arrangement, before any rotation could occur, the cohesive strength of the clay behind the wall would bring into play a far greater tailing-down mass of soil than the value taken for W_2, so that the overall stability would be considerably enhanced.

With $\bar{x}$ taken as 8·8 ft. for the normal design pressures on the back of the wall, the bending-moment diagram for the base is as shown in Fig. 8.13c. The maximum bending stresses occur at the edge of the haunch where $M = 45,100$ ft. lb. This is the same value as taken for the wall design; and the same reinforcements can conveniently be carried round, half these being stopped off halfway along the base as shown.

Considering now the matter of the wall sliding forward, we have base friction on the length of base under pressure at the maximum adhesion value of 1000 lb/sq. ft., giving

$$R_B = 1000 \times 11 \cdot 1 \text{ ft.} = 11,100 \text{ lb/ft.}$$

Further resistance is obtained by the provision of a key near the back of the base projecting down 2 ft. Using equation (8.19), we have the average passive pressure on the key as

$$p_{pn} = \gamma z + 2c$$
$$= (120 \times 4) + (2 \times 1100) = 2680 \text{ lb/sq. ft.,}$$

giving the total passive resistance

$$P_{pn} = 2 \times 2680 = 5360 \text{ lb.}$$

Thus the summation of frictional and passive resistances gives $11,100 + 5360 = 16,460$ lb., as against the total active pressure of 8400 lb., showing a factor of safety of 2.

WALL WITH BASE BEHIND

A wall of this type for storage of crushed stone is indicated in Fig. 8.14a. The density of the material γ is taken as 90 lb/cu. ft., the angle of internal friction ϕ as 35°, and the cohesion as zero. To achieve the required capacity of storage, the wall is 25 ft. high; and by mechanical handling methods the stone can be stored to a total height of 30 ft., making for 5 ft. surcharge at the maximum angle of 35° from the top of the wall. In this case it would be proper to design for the full head of 30 ft. of stone, ignoring the small relief due to the sloping stone face at the top of the wall.

Using equation 8.7 for a cohesionless soil, we have

$$p_{an} = \gamma z \left(\frac{1 - \sin \phi}{1 + \sin \phi} \right)$$
$$= 90z \left(\frac{1 - \sin 35°}{1 + \sin 35°} \right)$$
$$= 25z \ (see \text{ Fig. 8.14}b).$$

The unit pressure due to 5 ft. surcharge is therefore

$$25 \times 5 = 125 \text{ lb/sq. ft.}$$

The maximum bending stress in the wall will be at the top of the haunch, where $z = 23\frac{1}{2}$ ft. Then $p_{an} = 25 \times 23\frac{1}{2} = 588$ lb/sq. ft.

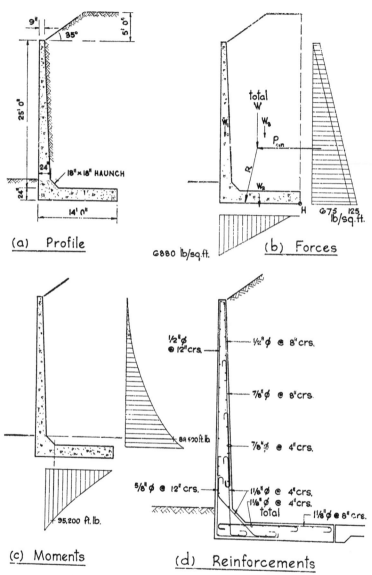

(a) Profile

6880 lb/sq.ft.

(b) Forces

@75 125 lb/sq.ft.

18"×18" HAUNCH

14' 0"

25' 0"

5' 0"

35°

9"

24"

24"

total W

W_1

W_2

W_3

P_{an}

R

H

(c) Moments

88,500 ft.lb

95,200 ft. lb.

(d) Reinforcements

½"φ @ 12"crs.

½"φ @ 8"crs.

⅞"φ @ 8"crs.

⅞"φ @ 4"crs.

5/8"φ @ 12"crs.

1⅛"φ @ 4"crs.

1⅛"φ @ 4"crs.
total

1⅛"φ @ 8"crs.

FIG. 8.14. DESIGN OF WALL WITH BASE BEHIND

295

The total shear down to this level is

$$P_{an} \text{ (stone)} = 588 \times \frac{23\frac{1}{2}}{2} = 6910 \text{ lb.}$$

$$\text{(surcharge)} = 125 \times 23\frac{1}{2} = 2940 \text{ lb.}$$

$$\overline{\underline{\underline{9850 \text{ lb.}}}}$$

and the bending moment is

$$\left(6910 \times \frac{23\frac{1}{2}}{3} \right) + \left(2940 \times \frac{23\frac{1}{2}}{2} \right) = 88,500 \text{ ft. lb.}$$

This requires

$$A_{st} = \frac{88,500 \times 12}{20,000 \times (0 \cdot 86 \times 22)} = 2 \cdot 81 \text{ sq. in.}$$

which is provided by $1\frac{1}{8}$ in. dia. rods at 4 in. centres (area 3·0 sq. in.).

$$\frac{M}{bd_1{}^2} = \frac{88,500}{12 \times 22^2} = 152$$

which is satisfactory.

$$\frac{Q}{bl_a} = \frac{9850}{12 \times (0 \cdot 86 \times 22)} = 43 \text{ lb/sq. in.}$$

The bending-moment diagram for the wall is given in Fig. 8.14c and the reinforcements are reduced higher up the wall to suit.

The stability of the structure against rotation is now considered in tabular manner, using the notation given in Fig. 8.14b:

	Forces		Arm about H	Applied moment
	Vertical	Horizontal		
	lb.	lb.	ft.	ft. lb.
W_1 (wall) . . .	5,140	—	12·8	66,000
W_2 (base) . . .	4,200	—	7	29,400
W_3 (stone above base) .	34,400	—	6	206,400
P_{an} (stone) . . .	—	7,810	10·33	80,800
(surcharge) . .	—	3,120	14·5	45,200
Total forces	43,740	10,930	—	—
Total moment	—	—	—	427,800

The resultant line of thrust cuts the base line a distance from H such that

$$\bar{x} = \frac{427,800}{43,740} = 9\cdot75 \text{ ft.}$$

This is outside the middle third, giving a triangular pressure distribution with a maximum edge pressure of

$$\frac{43,740 \times 2}{3 \times (14\cdot0 - 9\cdot75)} = 6880 \text{ lb/sq. ft.}$$
$$= 3\cdot06 \text{ tons/sq. ft.}$$

This is a reasonable value for a foundation bearing on dry compact sand.

If now we consider the effect of the active forces behind the wall being increased by 50 per cent, the applied moment becomes 490,800 ft. lb. and $\bar{x}$ becomes 11·20 ft. This still lies well within the base width, and increases the pressure under the base to

$$\frac{43,740 \times 2}{3 \times 2\cdot8} = 10,400 \text{ lb/sq. ft.} = 4\cdot65 \text{ tons/sq. ft.}$$

which is modest for dry compact sand in terms of ultimate capacity. Thus the great stability is demonstrated of retaining walls which have their bases behind. The weight of the retained material sitting on the base assists by restraining the resultant line of thrust from becoming too eccentric.

Under normal design conditions, with $\bar{x}$ as 9·75 ft., the bending-moment diagram for the base is as given in Fig. 8.14c. The maximum bending stresses occur at the edge of the haunch where $M = 95,200$ ft. lb., giving

$$A_{st} = \frac{95,200 \times 12}{20,000 \times (0\cdot86 \times 22)} = 3\cdot0 \text{ sq. in.}$$

which is provided by $1\frac{1}{8}$ in. dia. rods at 4 in. centres, as for the wall. Haunch rods of the same size and spacing are provided to carry the tension force round the internal angle, and all the rods at this situation are taken past the points of intersection and given substantial hooked ends to ensure the full bond being developed.

If the sand under the base has an angle of internal friction ϕ of 35°, the ultimate base friction will be

$$R_B = \mu W = \tan \phi \,.\, W$$
$$= 0\cdot7 \times 43,740 = 30,600 \text{ lb.}$$

The total active pressure on the back of the wall is 10,930 lb., showing a factor of safety against sliding of 2·8. Here again we see the merit of this form of construction: the weight of retained material forms a considerable proportion of the value W, thus contributing to the frictional resistance against forward sliding.

WALL WITH BACK COUNTERFORTS

The manner of designing a counterforted retaining wall is most simply demonstrated by introducing counterforts to the structure already designed in the previous example. The proportions here are not as would make for maximum economy, but nevertheless the principle is perfectly clear. The introduction of counterforts enables the wall and base to be made thinner, and there is some saving in weight of reinforcement (*see* Fig. 8.15). The bearing pressure on the ground, and considerations of rotation and sliding remain as before. The counterforts are provided at 17 ft. 6 in. centres.

Owing to the deflection of the wall in spanning 17 ft. 6 in. horizontally, the lowermost 4 ft. of the wall will cantilever up directly from the base.

$$\text{Canti } M = P \times l$$
$$= \left(725 \times \frac{4}{2}\right) \times \left(\frac{4}{3}\right) \times 12$$
$$= 23{,}200 \text{ in. lb.}$$

This requires

$$A_{st} = \frac{23{,}200}{20{,}000 \times (0.86 \times 9)} = 0.15 \text{ sq. in.}$$

which is provided by $\frac{5}{8}$ in. dia. rods at 18 in. centres. Above this level, the wall spans continuously between the counterforts. At 20 ft. depth, $p_{an} = 625$ lb/sq. ft.

$$\text{Moment at counterforts} = \frac{625 \times 17.5^2 \times 12}{12} = 192{,}000 \text{ in. lb.,}$$

requiring $\quad A_{st} = \dfrac{192{,}000}{20{,}000 \times (0.86 \times 10)} = 1.12 \text{ sq. in.}$

which is provided by $\frac{7}{8}$ in. dia. rods at 6 in. centres (area 1.20 sq. in.).

$$\frac{M}{bd_1{}^2} = \frac{192{,}000}{12 \times 10^2} = 160$$

which is satisfactory for 1:2:4 concrete. And the shear stress is

$$\frac{Q}{bl_a} = \frac{5480}{12 \times (0.86 \times 10)} = 53 \text{ lb/sq. in.}$$

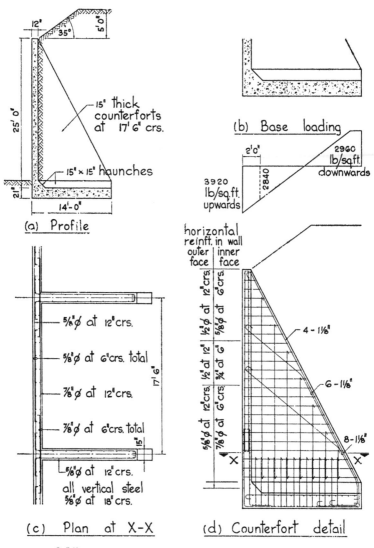

12"
35°
5'0"

25' 0"

15" thick
counterforts
at 17' 6" crs.

15" × 15" haunches

21"

14'-0"

(a) Profile

(b) Base loading

2'0"
2960
lb/sq.ft.
downwards

3920
lb/sq.ft.
upwards

2840

horizontal
reinft. in wall
outer | inner
face | face

5/8"∅ at 12"crs.

5/8"∅ at 6"crs. total

7/8"∅ at 12"crs.

7/8"∅ at 6"crs. total

5/8"∅ at 12"crs.

all vertical steel
5/8"∅ at 18" crs.

17' 6"

15"

5/8"∅ at 12"crs | 1/2 at 6" crs
1/2"∅ at 12" | 5/8"∅ at 6"crs

5/8"∅ at 12"crs | 1/2 at 12" | 3/4" at 6" | 5/8"∅ at 6"crs
7/8"∅ at 6"crs |

X

4 - 1 1/8"

6 - 1 1/8"

8 - 1 1/8"

X

(c) Plan at X-X

(d) Counterfort detail

FIG. 8.15. DESIGN OF WALL WITH BACK COUNTERFORTS

Moment midway between counterforts

$$= \frac{625 \times 17 \cdot 5^2 \times 12}{24} = 96,000 \text{ in. lb.,}$$

requiring A_{st} of 0·56 sq. in., given by $\frac{5}{8}$ in. dia. rods at 6 in. centres (area 0·60 sq. in.). Where the pressure higher up the wall reduces, the amount of reinforcement can be reduced to suit, as shown in Fig. 8.15.

The base-slab spans between the counterforts in much the same way: but the loading applied to the base varies more extremely as will be appreciated from Fig. 8.14b. At the heel the loading on the base slab is downwards due to the tailing-down action of the overlying material, and the unit loading is:

due to stored stone = 30 ft. × 90 lb/cu. ft.   = 2700 lb/sq. ft.
due to base   = 1·75 ft. × 150 lb/cu. ft. = 260 lb/sq. ft.
                                                  _____
                                                  2960 lb/sq. ft.

Near to the toe, however, at a point, say, 12 ft. in from the heel this downward unit loading on the slab is more than balanced by the upward pressure on the underside of the base, and we have to provide for a nett upward loading of 5800 − 2960 = 2840 lb/sq. ft. A suitable arrangement of reinforcement is seen in Fig. 8.15.

The counterfort is designed as a T-beam of varying depth. The lever arm is taken as the distance between the centres of the wall and the tension steel at the back of the counterfort. The calculation may be set out in tabular form as follows:

Depth from top of wall	Applied moment per counterfort	Lever arm	A_{st} required	A_{st} provided
ft.	ft. lb.	ft.	sq. in.	
15	490,000	7	3·5	4 − 1¼ in. dia.
20	1,020,000	9·5	5·4	6 − 1¼ in. dia.
25	1,820,000	12	7·6	8 − 1⅛ in. dia.

The counterfort tension-steel has to be taken well into the base as shown and properly anchored there. Additional anchorage to accommodate the balance of the tailing-down force provided by the overlying stone is given by rods along the length of the counterfort anchored into the base slab.

The maximum force tending to push the wall off the front of the counterfort is

$$625(17\tfrac{1}{2} - 1\tfrac{1}{4}) = 10,150 \text{ lb/ft.},$$

requiring an area of steel of $\dfrac{10,150}{20,000} = 0.51$ sq. in./ft. which is provided by $\tfrac{5}{8}$ in. dia. horizontal rods at 12 in. centres in both faces of the counterfort, anchored well into the wall.

REFERENCES

1. BELL, A. L. The lateral pressure and resistance of clay, and the supporting power of clay foundations. *Proc. Inst. C.E.* Vol. 199, Part I (1914–1915).
2. RANKINE, W. J. M. On the stability of loose earth. *Trans. Royal Soc. London.* Vol. 147 (1857).
3. TERZAGHI, K. and PECK, R. B. *Soil mechanics in engineering practice.* John Wiley & Sons (1948).
4. EARTH RETAINING STRUCTURES. *Civil Engineering Code of Practice No. 2.* Institution of Structural Engineers (1951).

CHAPTER 9

PILING

ART. 9.1. INTRODUCTION

WHEN it is necessary to construct foundations where soil of poor bearing capacity extends to such a depth that the cost and difficulty of sinking piers to a hard stratum would be considerable, pile foundations frequently provide the most reliable and economical solution. For any set of soil conditions, a wide choice is available to the engineer in regard to the type, size, length and number of concrete piles suitable to carry a given concentration of loading. Concrete piles fall into three main types: *precast piles, cast-in-situ driven piles,* and *cast-in-situ bored piles.* Each of these types have certain advantages and disadvantages under different circumstances, and the merits of the various types are discussed in Articles 9.4 and 9.5.

ART. 9.2. SITE INVESTIGATION

As with all foundation problems, the first step will be to carry out a full site investigation.

BORINGS AND SOIL TESTS

Initially a few deep borings will be sunk to ascertain whether in fact piling is necessary or whether the ground conditions are such that any real advantage can be achieved by the provision of piles[1,2,3]. In certain cases, where cohesive material of low shear strength extends to very considerable depths, it may indeed be uneconomical or useless to seek any solution by piling, and resort may more profitably be made to a buoyant foundation of cellular-raft type as described in Article 7.10. A knowledge of the experience of other engineers working in the same area can give a useful guide, as also can observations of settlements of existing structures.

Where preliminary borings indicate that piling gives a suitable answer to the problem, additional borings should then be sunk at close intervals across the site of the work, the borings extending to a depth not less than one and a half times the width of the structure or twice the anticipated depth of the piles, whichever is the greater. Undisturbed samples should be taken from the borings and put through the same series of tests as described in Article 7.5 in order to determine the shear strength of the soil at all depths and also the likely settlement. The shear strength of the soil gives a guide as to the supporting capacity of friction piles, since the side friction of a pile has been found generally to coincide reasonably with the soil shear strength.

SOUNDINGS

Where the results of borings indicate that piles are to be driven through strata to reach end bearing on harder material below, an assessment of the carrying capacity can be obtained by carrying out soundings as described in Article 7.4, though how accurately the assessment of the bearing capacity of a full-size pile can be determined from a small-scale test of this type is arguable. It is essential that such tests be used in conjunction with the findings of adjacent borings which will classify the types of soil being penetrated, since otherwise, for example, the plunger may reach refusal at a large boulder in a clay suggesting that bedrock had been reached, but an adjacent boring would expose this error of interpretation.

TEST PILES

On sites where driven piles are to be used, it is advisable, when funds and conditions permit, to follow up the soil explorations by driving full-size test piles with a drop hammer under works conditions as part of the site investigation. Loading tests on such piles give extremely valuable information and are more reliable than any formula in determining the actual length and number of piles required for the work. Indeed in silts and clays it is only with full-size test piles that any proper assessment of the carrying capacity of the piles can be made. Loading tests on test piles should not be started until after a sufficient interval has elapsed to allow the soil to recover from the disturbance of the pile driving. And the loading tests should be continued over a sufficient period of time to be sure that a true estimate of the settlement of the individual pile is obtained: in sands and gravels this may be a period of only two or three days, but for clays the period may need to be extended over two or three weeks.

Test piles should be sited near to the exploratory borings so that the record of driving can be interpreted in relation to the strata through which the toe of the pile passes. This later gives a guide of the principal strata encountered during the driving of the works piles, and assists in assessing the safe carrying capacity of the works piles; this may be of particular advantage where the test load results show a marked departure from values computed from any piling formula. In driving the test piles it is normal to make chalk marks on the piles at intervals of a foot, so that the number of blows of the monkey falling freely some constant drop-height can be recorded for every foot of penetration of the pile. Similar records can then be kept of the actual works piles for direct comparison.

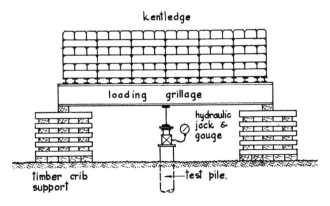

FIG. 9.1. TESTING PILE AGAINST FIXED DEAD
WEIGHT

This reduces the number of loading tests necessary on the works piles. Where the driving records of the test piles and works piles are being used for comparison, the driving should be continued without interruption for the whole length of the pile.

Test loads are normally applied to piles in one of the following ways:

(a) by loading a platform constructed centrally over the pile;
(b) by jacking against a fixed dead weight supported above the pile;
(c) by jacking against a beam which is tailed down to adjacent anchor piles.

Unless the distribution of the load on the platform is carefully controlled, method (a) is liable to produce bending moments in the pile, and at ultimate failure of the pile the platform may tilt and collapse with danger to personnel. The method is not to be recommended. Method (b) – see Fig. 9.1 – is more reliable, provided the kentledge is supported on beams of sufficient length to avoid the ground close to the pile being disturbed. The objection to the method lies in the cost of bringing in the kentledge. Method (c) – see Fig. 9.2 – is the most simple, but is suitable only where a sufficient number of anchor-piles are within ready reach of the pile to be tested. Two or three anchor-piles are required on each side of the pile sometimes making a total of six anchor-piles. With this method it is important that the anchor-piles are at least 5 ft. from the pile, otherwise the test pile may derive support from the upward movement of the ground at the anchor-piles. Both methods (b) and (c) rely on the jack holding its load, and this

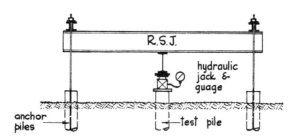

FIG. 9.2. TESTING PILE AGAINST ANCHOR-PILES

necessitates a man standing by continuously for the whole period of the test.

The settlement of a pile during test is normally recorded by a micrometer dial gauge (reading to 1/1000 in.) clamped to a long beam supported from the ground on either side of the area affected by the pile movements. In case the set-up of the dial gauge should be disturbed during the test, it is advisable as a precaution to have a surveyor's level permanently set-up on one side of the pile sighting across to a datum on the other side of the pile and on to a staff with vernier attachment secured above the pile.

The time and money spent in carrying out adequate site investigations and preliminary pile-loading tests is amply repaid. Pile-driving formulae may give some guide in the case of end-bearing piles on sand and ballast, but can be extremely misleading with piles in cohesive strata. Test loads give the surest results for individual piles, and thereafter it is necessary to study soil strengths based on laboratory tests to form any reliable estimate of the total bearing capacity of large groups of piles. Estimates of group settlement and overall stability have also to be based on the known soil properties, and due to the variability of all natural soil formations, all such estimates must necessarily be only approximate[1,2,3].

ART. 9.3. FRICTION AND END-BEARING PILES

The mechanism by which piles are supported falls broadly into two groups: *end-bearing piles* and *friction piles*. In practice, classification of piles into these two groups is by no means clearly marked, and frequently piles are supported by a combination of the two effects. In Fig. 9.3 the driving resistance is shown in blows per foot of penetration at different depths for typical end-bearing and friction piles.

Piles depending mainly on end bearing in cohesionless soils are more amenable to calculation by dynamic formulae, and in many

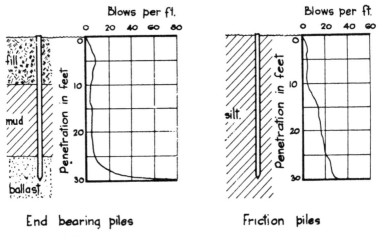

FIG. 9.3. DRIVING RESISTANCE OF END-BEARING PILES
AND FRICTION PILES

cases the capacity of a group of such piles approximates to the sum of the capacities of the individual piles. Nevertheless with groups of piles having end bearing in cohesionless soils a check should be made of the total group load on the general area of pile toes using formulae as in Article 7.8 to see that the strata is not overloaded at toe level, nor at greater depths. Such overloading is possible where piles are arranged close together.

Friction piles rely for their immediate support on the friction force between pile and soil acting on the greater part of the embedded pile length. The carrying capacity of such piles is not readily calculable and tests as described in Article 9.2 are all important. The degree to which any pile is supported by friction can be determined from pull-out tests. Where friction piles occur in

groups, it is important for the piles to be spaced far enough apart
to avoid adjoining piles each striving to derive support by friction
on the same intervening columns of soil: thus if P is the safe load
per individual pile, and q_a is the appropriate safe working pressure
of the ground, then piles driven over an extensive area must be
kept apart so that the minimum spacing, centre to centre, is given
by

$$x_{min} = y_{min} = \sqrt{\frac{P}{q_a}}. \tag{9.1}$$

However, where the pile groups are small in relation to the distance
between the groups, the edge piles of the group can spread their
load outwards depending only on the average shear strength of the
soil and the proximity of adjacent pile groups. Thus, if

   z is the penetration of the piles,
   s is the average shearing resistance of the ground, and
   L is the effective perimeter of the pile group,
then the capacity of the group

$$= \Sigma x \cdot \Sigma y \cdot q_a + zLs. \tag{9.2}$$

It is clear from the above that, in the limit, friction piles closely
spaced over an extensive area are far less effective than individual
friction piles, and can do little more than transfer the applied load
to a lower level in the ground. If the bearing capacity of the ground
does not increase materially with depth, then unless the depth z is
appreciable in relation to the area of the foundation (or L in
equation (9.2)) there can be little benefit derived from side spread
at the edge of the group, and consequently little advantage is
gained by piling. Thus a reduced number of long friction piles has
greater benefit than a large number of short friction piles; the
limit being when the piles become spaced further apart than
required by equation (9.1). Furthermore experience has shown that
the settlement of foundations of given area carrying the same load
decreases as the length of the piles increases even though fewer
piles are required to carry the load. An exception arises in regard
to the reduced unit load per friction pile in groups when driven
into some types of loose sands and silt, where the nature of the
ground can be so improved by the disturbance of driving that the
unit load per pile can in certain circumstances be enhanced.

It is important that adjacent friction piles always be driven to
the same depth even though the number of blows for the last foot
of penetration of individual piles may differ.

A warning is necessary where piles are driven through soft sensitive clay to derive end bearing on a harder strata beneath. The clay

is disturbed by the action of the pile driving, and its internal shear strength is reduced allowing it to settle slowly under its own weight. The piles interfere with this process of settlement and the downward movement of the clay is resisted by skin friction on the sides of the piles. Only a small degree of consolidation of the clay is necessary for the whole weight of the clay to be transferred on to the piles. This action is sometimes referred to as negative skin friction. To cater for this phenomenon the piles should be driven closer together than would otherwise be needed, thus reducing the

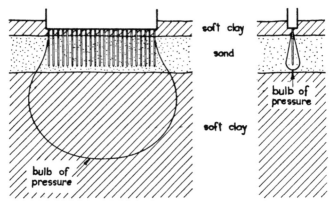

FIG. 9.4. STRESS-BULBS IN GROUND UNDER SINGLE PILE
AND UNDER LARGE PILE-GROUP

amount of clay hanging on each pile. In certain cases the use of bored piles may enable the foundation to be installed without the clay being sufficiently disturbed to cause appreciable negative skin friction.

With all large groups of piles, consideration should be given to settlement effects arising from a large combined bulb of stress extending to a greater depth than the smaller bulbs from individual piles. Settlement measurements taken on single test piles should therefore be regarded with caution in just the same way as small load-bearing tests can mislead. This is made clear by reference to Fig. 9.4. The need for laboratory consolidation tests on soils at considerable depths below toe piles is therefore clear.

Whether piles are end-bearing or friction type they should for ease of driving be spaced, centre to centre, roughly three times the pile diameter: otherwise it may be impossible for some of the piles to penetrate to the depth or strata required. Generally piles in groups should be driven starting at the centre and working outwards, though there may be objections to this when working close to river walls and in similar circumstances.

ART. 9.4. PILES PRECAST

Precast piles have advantages over other types of piles as follows. The concrete and reinforcements are placed in a clean mould above ground and can be properly vibrated and allowed to set and cure under carefully controlled conditions, so that a pile of known strength and construction is assured. In sulphate-bearing grounds, sound dense concrete is essential to prevent the ingress of water into the piles. Whether sulphate-resisting cement is used or not, greater likelihood of success against sulphate erosion results from the dense vibrated concrete achieved with precast piles. Furthermore the concrete in cast-in-situ piles is more vulnerable than in precast piles since it is exposed to attack while still green.

Precast piles have the further merit that in the process of driving they carry a dynamic load greatly in excess of the static working load they have to support subsequently. Consequently if any pile is faulty in manufacture, this will show up in driving: in other words each pile is pre-tested by driving and so has a high degree of reliability. Precast piles are not subject to certain risks associated with cast-in-situ piles such as necking or adulteration of the concrete by mud or silt.

Disadvantages of precast piles include the following. No matter what preliminary survey work may have been carried out, it is never possible to predetermine the exact lengths for the various piles. The strata across many sites varies considerably and not always uniformly: so that three or four test piles on a site 300 ft. square could fail to give any indication of a patch where the piles might need to be considerably longer or shorter. Where piles are required longer than originally cast, there is delay and expense in lengthening the piles and redriving: and this can be troublesome, because the soil sometimes adheres firmly to the pile during the period of rest while the lengthening is effected, causing great difficulty subsequently when starting to redrive. The alternative of dollying the pile down, and later excavating to extend the pile to foundation level, is usually expensive and not very satisfactory. Where the piles have been cast too long there is, of course, considerable waste attached to cutting off and discarding the surplus.

A further objection to precast piles arises as the lengths become considerable. For practical reasons increased lengths necessitate increased cross-sectional areas and the piles, in consequence, become heavy and cumbersome to handle, in which condition they are liable to damage before and during driving. And as the efficiency of the hammer blow is reduced as the ratio of pile-weight to hammer-weight increases, in extreme cases the pile may become so massive as to be uneconomic to drive.

In determining the cross-sectional area and reinforcement required for a precast pile, the following considerations arise:

    (a) transport and handling stresses;
    (b) stresses during driving, from impact and rebound;
    (c) normal working stresses;
    (d) area of side of pile available for friction support;
    (e) bending stresses due to horizontal working forces on the piles.

Items (a) and (b) above have a considerable influence on proportioning precast piles: they have no influence whatever in the case of cast-in-situ piles.

Precast piles must not be too slender in relation to their length, otherwise they buckle in driving and become subject to severe bending stresses in transport and handling. Reasonable minimum sizes for precast piles of various lengths are given in Fig. 9.5. Precast piles about 18 in. square have been driven 80 ft. long and more, but special precautions are then necessary by way of additional reinforcements, special mix and curing of the concrete – using steam – and added care in handling and driving. It is general practice for precast piles to be of square cross-section which facilitates casting and handling. Generous chamfers are advisable on all corners including the head to minimise damage arising from spalling of the concrete during driving.

Reinforcement in precast piles serves principally two functions. Firstly it prevents damage to the piles when they are lifted from their horizontal position in the casting bed and handled prior to the pile being upended. If the steel stress in this condition greatly exceeds 20,000 lb/sq. in., the pile is liable to crack in a manner which would reduce its permanence and good behaviour during driving. In the second place the reinforcement prevents the pile breaking into individual sections immediately following the hammer blow. When the hammer strikes the top of the pile it produces a compression wave which travels down the pile to the toe causing penetration, and this compression wave is then reflected back to the head of the pile; and in the case of a pile approaching refusal the reflected compression wave is of such magnitude that it will cause the hammer to rebound from the pile in just the same way as would occur had the hammer been dropped on to a spring. The tendency of this spring-like release is to cause the upper portion of the pile to fly upwards (just as the original compression wave

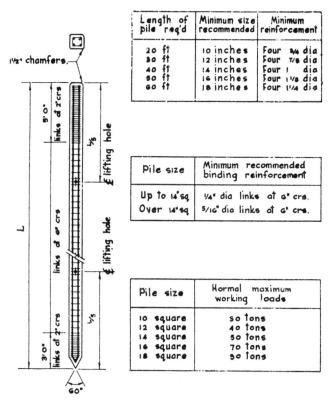

Length of pile req'd	Minimum size recommended	Minimum reinforcement
20 ft	10 inches	Four ¾ dia
30 ft	12 inches	Four ⅞ dia
40 ft	14 inches	Four 1 dia
50 ft	16 inches	Four 1⅛ dia
60 ft	18 inches	Four 1¼ dia

Pile size	Minimum recommended binding reinforcement
Up to 14" sq	¼" dia links at 6" crs.
Over 14" sq	5/16" dia links at 6" crs.

Pile size	Normal maximum working loads
10 square	30 tons
12 square	40 tons
14 square	50 tons
16 square	70 tons
18 square	90 tons

FIG. 9.5. DETAILS OF PRECAST PILES

caused the toe of the pile to penetrate downwards) and longitudinal tensile stresses are produced instantaneously in the pile.

"Foundations"[4], *The Civil Engineering Code of Practice*, No. 4, recommends that the area of main longitudinal reinforcement should be not less than 1¼ per cent of the gross cross-sectional area of the piles for piles of length up to thirty times their least width, 1½ per cent for lengths thirty to forty times, and 2 per cent for lengths over forty times the least width. The reinforcements indicated in Fig. 9.5 comply with this recommendation. Additional reinforcements may be required in the case of longer piles in order to resist handling stresses. In this regard precautions have to be taken to ensure that piles are only lifted at certain specified points. This is achieved by providing lifting holes through the piles, normally at the fifth points, so that the piles are lifted only in the manner indicated in Fig. 9.6(a): other methods of slinging should not be permitted. If then the dead weight of the pile is W and the

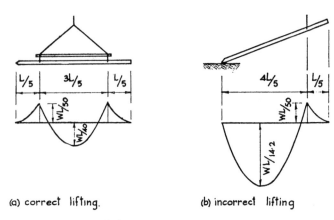

(a) correct lifting.   (b) incorrect lifting

FIG. 9.6. LIFTING PRECAST PILES

length of the pile is L, the cantilever moment is $\dfrac{WL}{50}$ and the mid-span moment is $\dfrac{WL}{40}$.

It is clear however from Fig. 9.6(b) that if the piles are incorrectly lifted by slinging from one point only, the bending moment in the pile is $\dfrac{WL}{14\cdot2}$ which is 2·8 times as severe as the condition shown in Fig. 9.6(a). Therefore the decision has to be made as to whether the cost of supervising the work to ensure all piles being lifted in accordance with Fig. 9.6(a) is greater or less than the cost of providing reinforcement to cover the condition of Fig. 9.6(b). It is perhaps fair comment that longer piles are normally handled with greater respect than short piles. Whatever restrictions are made on the method of lifting the piles, the same restrictions are to be observed when the piles are stacked or transported: Fig. 9.7(b) shows this, where three sets of packs have been provided, clearly giving a bending moment in the lowest pile of $\dfrac{WL}{1\cdot6}$.

Binding reinforcement in precast piles takes the form of links not less than $\frac{1}{4}$ in. at 6 in. centres for piles up to 14 in. square and not less than $\frac{5}{16}$ in. at 6 in. centres for larger piles. The links are closed up to 2 in. centres for the 3 ft. at the toe and 5 ft. at the head. To ensure that the links are kept taught, diagonal cross spreaders are used at about 6 ft. centres.

The concrete for precast piles should be not less than $1:1\cdot6:$ $3\cdot2$ nominal mix. At the head of the pile for a length of 5 ft. the amount of cement in the mix should then be doubled: and arrangements should be made for this enriched mix to be placed continuously with the placing of the remainder of the concrete for

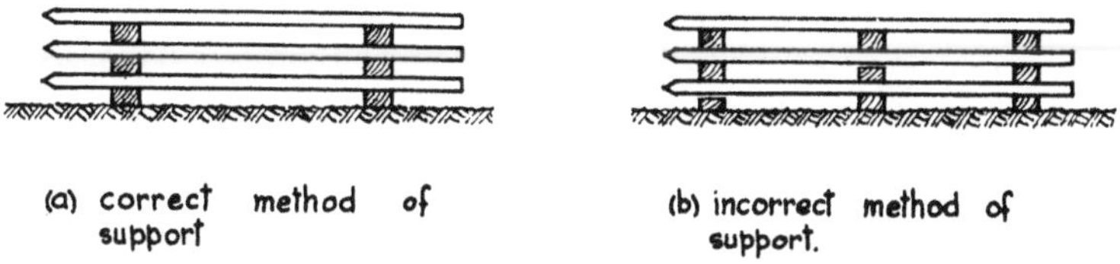

(a) correct method of support

(b) incorrect method of support.

FIG. 9.7. STACKING PRECAST PILES

the pile. Working stresses in piles in position should not exceed permissible stresses given in C.P. 114 (1957), but these stresses may be exceeded by 100 per cent or more during driving. Experience is that piles where embedded in the ground do not fail by lateral buckling even though the ground may be extremely soft. Subject to the supporting capacity of the ground, maximum working loads on piles should not normally exceed the values given in Fig. 9.5. Generally a period of 28 days is allowed between the dates of casting and driving precast piles of Ordinary Portland cement. *The Civil Engineering Code of Practice*, No. 4, recommends that this period be reduced to 10 days where Rapid-Hardening Portland cement is used and to 2 days where High Alumina cement is used.

The capacity of piles to resist lateral working loads is a matter requiring considerable judgement. It will depend upon the passive lateral resistance of the ground, the degree of fixity of the piles where they join with the superstructure, and the overall stiffness of the latter. Groups of piles may act as one unit stiffened by the soil entrapped between the piles.

In early days it was customary to provide precast piles with cast-iron shoes attached to the pile by four rods secured to the point of the shoe and extending up into the concrete. To some extent this was a relic of the days of timber piles, where such a shoe had the main object of preventing the timber from splitting. But where driving concrete piles on clay, marl or sand there seems little case for such shoes; though in strata where boulders or other hard obstructions are likely to be encountered heavy cast-iron shoes are desirable.

ART. 9.5. PILES CAST IN SITU

The advantages of cast-in-situ piles include the following. They can be constructed as soon as the piling plant and construction materials reach the site, without the delay associated with the manufacture and curing of precast piles. Nor is it necessary to know with any accuracy the length of pile required – the steel tube or other former used in constructing the pile can be sunk to the required level and then concreted to the actual length required. This makes for considerable economy. With friction piles, the rougher surface of the cast-in-situ pile gives enhanced bond with the ground.

Bored piles have the merit of causing no vibration to the surrounding ground or adjacent structures. Further they require no tall or heavy plant, and consequently lend themselves well to sites where access is restricted. On jobs where only a few piles are required, the importation of the simpler equipment makes for considerable economy.

The disadvantages of cast-in-situ piles include the following. With many forms, the cage of reinforcement lowered into the ground prior to the pouring of concrete may be eccentric in the hole so that only meagre cover of concrete is provided. This applies particularly in the case of long piles where visual control some 50 or 60 ft. below the ground becomes practically impossible especially while the concrete is being poured. Any check on the infiltration of mud or silt into the concrete is difficult to observe; and with pile types involving the withdrawal of a steel tube as the concrete is placed, there is risk of soft strata pressing into the body of the pile causing necking, i.e. the formation of a length of pile of reduced diameter. Cases are certainly known of piles supposedly 18 in. diameter having necks of only about 8 in. diameter; though there are techniques for guarding against this.

A further objection held against driven cast-in-situ piles is that where they are constructed close together, say 3 ft. 6 in. centres in both directions, the concreting of one pile may be followed immediately by the driving of its neighbour. This may displace the ground sideways and upwards, tending to bend or shear the first pile and lift the upper part of it. Cracks due to this have been found extending right through cast-in-situ piles.

With bored piles there is no dynamic consolidation of the ground through which the pile is constructed, or at toe level. And since there is no driving record, there is no proving evidence of the pile's carrying capacity, and considerable reliance has to be placed on the findings of preliminary soil explorations.

Cast-in-situ piles are of many types, and mostly controlled by specialist firms who carry suitable plant and workmen experienced in their own method. The types here described cover a wide range; but there are other excellent systems each with their own special advantages in appropriate conditions.

SIMPLEX STANDARD DRIVEN PILE

The Simplex Standard Pile (Fig. 9.8) is formed by driving into the ground a hollow steel tube, usually 16 in. or 18 in. diameter, fitted with a double-rimmed cast-iron shoe. The joint between the

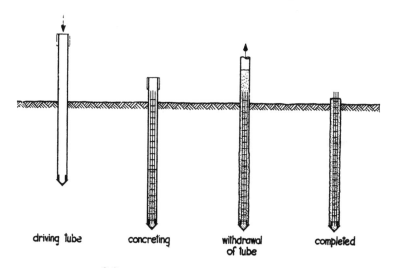

driving tube   concreting   withdrawal   completed
                                           of tube

FIG. 9.8. SIMPLEX STANDARD CAST-IN-SITU
DRIVEN PILE

tube and the shoe is made watertight by the insertion of packing. The tube is driven by drop hammer until the required penetration or set is obtained. The hammer and drive cap are then lifted away, a cage of reinforcement lowered into the tube, and the tube then filled with concrete to the required level. The tube is then slowly withdrawn. During the process of withdrawal, the concrete being fairly fluid issues from the lower end of the tube, and is forced downwards and outwards by its own weight, filling the hole made by the tube together with any voids in the sides of the hole which may have been formed through displacement of stones or boulders during the driving operation.

FRANKI DRIVEN PILE

The Franki Driven Pile (Fig. 9.9) is formed by driving an open-ended steel tube into the ground, using a long cylindrical internal hammer which acts on a temporary gravel-plug jammed inside the bottom of the tube by the hammering action. The gravel-plug technique serves the double function of effectively sealing the

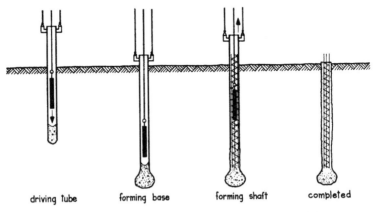

driving tube   forming base   forming shaft   completed

FIG. 9.9. FRANKI CAST-IN-SITU DRIVEN PILE

tube against entry of water or soil during driving, and enables the drop of the hammer to be greater than would be practicable if the hammer were applied to the head of the tube. Nevertheless it is claimed that the greater drop causes but little vibration because the energy is aimed directly to the lowest part of the pile. When the tube has been driven to the required depth, it is anchored by means of cables, and the plug broken and forced out of the bottom of the tube by further hammering. Concrete is then placed in the tube and hammered out of the bottom to form a bulbous base, more and more concrete being added and hammering continued until virtual refusal is obtained. A cage of reinforcement is then lowered into the tube, and further concrete placed stage by stage, the internal hammer, acting inside the cage, consolidating this concrete.

WEST'S SHELL DRIVEN PILE

The West's Driven Pile (Fig. 9.10) is formed by threading rein-forced concrete tubes on to a steel mandrel with a solid concrete shoe. The joints between the shells are made with steel bands which fit into recesses at the ends of the shells, the joint being treated with a bituminous mastic sealer. The whole of the tube, mandrel and shoe is driven into the ground until the required set is obtained. The hammer blow is delivered direct on to the shoe by means of the steel mandrel: simultaneously a cushioned blow is applied to the shells, through a special arrangement on the mandrel driving head, causing the shells to follow the shoe. When the required set is obtained the steel mandrel is withdrawn, leaving the

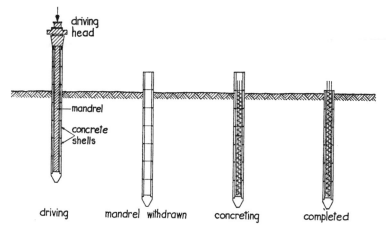

FIG. 9.10. WEST'S SHELL DRIVEN PILE

shoe and reinforced tube in position. A cage of reinforcements is then lowered into the hollow pile formed by the concrete shells, and the pile filled with concrete. The advantages of this method are that there is no risk of variation in cross-section of the pile, and there is no risk of contamination of the core concrete due to impurities in the ground as the cast-in-situ core is protected by the precast shell.

PRESSURE PILING BORED PILE

Pressure Piles (Fig. 9.11) are formed by first making a boring, the hole being lined with steel tubes as the boring proceeds. The steel tubes are sunk in the ground either by their own weight or by lightly tapping with a boring tool. When the steel tube has been sunk to the required level a small amount of concrete is lowered into the tube and consolidated by dropping a ram on it. A cage of reinforcements is then lowered into the tube followed by the first batch of concrete. The lining tubes are then extracted and the concrete consolidated by admitting compressed air to the tubes or by vibrating, further concrete being added and the tubes withdrawn until the pile is concreted up to the required finished level. The air pressure forces the concrete down the casing and presses it

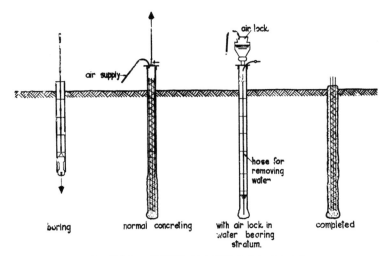

air lock

air supply

hose for removing water

boring   normal concreting   with air lock in   completed
                                      water bearing
                                      stratum.

FIG. 9.11. PRESSURE PILING BORED PILE

into the interstices of the bottom part of the bored hole from which the casing as been withdrawn. In this way the actual diameter of the pile is made to exceed that of the casing itself. Care has to be taken that the bottom of the casing is well below the top of the concrete in the tube so that there is no possibility of soil or subsoil water getting into the space to be occupied by the concrete.

When the pile finishes in a water-bearing stratum, an airlock is fixed on the top of the lining tubes, and after the water has been blown out by means of compressed air acting through a pipe lowered into the tube, concrete is placed through the airlock and the procedure continued as before.

PRESTCORE BORED PILE

The special feature claimed of the Prestcore Bored Pile (Fig. 9.12) is that it is essentially a precast concrete pile. It is installed by first boring a hole in the ground, and lining this with a steel casing. A core pile consisting of precast concrete sections assembled on a central steel pipe is then lowered into the hole and the casing withdrawn, if necessary by jacking it off the core pile inside.

During the withdrawal of the casing the core pile is grouted under pressure, all ground water being expelled and the pile enclosed in a thick skin of grout. Continuous reinforcing rods are provided as required and where necessary an enlarged base can be formed to increase the bearing capacity of the pile.

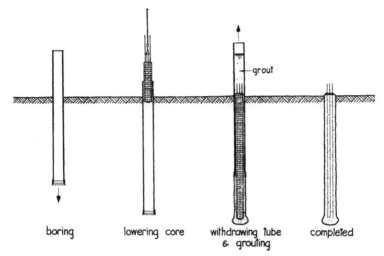

boring   lowering core   withdrawing tube   completed
                                          & grouting

FIG. 9.12. PRESTCORE BORED PILE

ART. 9.6. DRIVING OF PILES

Piles are normally driven using one of four hammer types:

    (a) drop hammer;
    (b) single-acting steam hammer;
    (c) double-acting steam hammer;
    (d) diesel-operated hammer.

With the simple drop hammer, the weight is lifted by a cable run over a pulley at the top of the pile frame and extended back to a winch. Some skill is required of the winch operator to gauge correctly the height of lift for each drop. In order to speed the work, the operator may tend to snatch the weight too soon at the instant of first impact, so that the full force of the blow is spoilt: this manner of operation gives a false set, and interferes with calculation of the carrying capacity of the piles. Single-acting steam hammers operate by the steam raising the movable mass of the hammer which then drops under the influence of gravity. Double-acting steam hammers operate by the steam raising the movable mass and then being switched to give additional energy to the ram in its downward path. In kick-atomised diesel-operated hammers, the ram functions as a piston within an outer cylinder. The ram falling within the cylinder causes compression and impact, resulting in the atomisation of a controlled quantity of diesel fuel injected by a pump which is also operated by the falling ram. The explosion throws up the ram in readiness for the next drop-cycle, and at the same time adds to the force of impact acting on the pile. Whatever type of hammer is used, it is important that full records of the driving characteristics should be kept for the full length of driving of a number of the works piles.

Reference is made in Article 9.7 to pile-driving formulae. Although such formulae may have limitations in assessing the safe carrying capacity of the piles, they do serve as a guide to suitable weights for hammers. Whatever type of pile is driven, it is better to use a heavy hammer with a small drop rather than a light hammer with a big drop. A few ponderous blows from a heavy hammer will drive a pile far more efficiently than innumerable little smacks from a very light hammer. The need for a heavy hammer is particularly great when driving large precast piles owing to the greater mass to be shifted; furthermore, precast piles are more likely to shatter under the punishment of an endless number of light blows than from a reduced number of heavier blows. If the weight of the hammer is very light in relation to the weight of the pile, the efficiency of the blow becomes so reduced that only a small part of the energy of the falling hammer is devoted actually to moving the pile, and the set of the pile becomes so critical that the limitations of any piling formula become greatly exaggerated. This is illustrated by reference to any of the formulae given in Article 9.7. *The Civil Engineering Code of Practice*, No. 4, recommends that the weight of the hammer should be sufficient to develop a blow efficiency of not less than 30 per cent. It also states that the hammer should be at least half the weight of the pile; further that with precast piles it is desirable to limit the driving stresses by using a hammer with a weight not less than thirty times the weight of one foot of pile. The Code recommends that the weight of the hammer should be sufficient to ensure a final penetration of not less than $\frac{1}{10}$ in. per blow unless rock is reached.

The best drop is usually 48 in. or so, and frequently piles are driven to a set of about $\frac{1}{4}$ in. per blow. In driving concrete piles the head should be protected by providing a steel helmet with soft material packed inside to form a cushion between the helmet and the top of the pile. The object of this is to distribute the pressure from the blow uniformly over the top of the pile, and to transform the blow from a "short sharp shock" to a blow of less intensity but acting over a greater period of time. There is no doubt that without such packing the pile is more likely to be fractured: and the packing does little to reduce the efficiency of the driving. About three inches of compressed sawdust in a bag put inside the helmet before lowering on to the pile is effective, the three inches being measured after the sawdust has been compressed by driving. The same thickness of felt is also suitable, as are many other materials.

ART. 9.7. PILE-DRIVING FORMULAE

Many pile-driving formulae have been proposed over the past hundred years; and it would be beyond the claims of any of the engineers who produced these formulae to pretend that the results given are in any way more than a useful guide. The assumption made in all dynamic formulae is that the bearing capacity of a pile under static conditions bears a direct relationship to its final driving resistance. The fallacy of this is clear when one considers that all pile formulae are based on the set of the pile measured over the last few blows, the effects of driving through the various upper strata being not properly taken into account: yet for friction piles it is the whole of the friction down the length of the pile which acts in supporting the pile. Clearly then dynamic pile formulae cannot present a rational estimate of the bearing capacity of piles where side friction contributes materially to the supporting capacity. Nor can pile-driving formulae have cognisance of the soil conditions below the pile toe or the conditions of general stability.

Dynamic formulae may give reasonable results in sands and gravels which have a high permeability and adjust their moisture content immediately at the time of driving. But with saturated clays and silts having low permeability, the toe resistance may appear higher at the instant of driving than subsequently: yet the frictional resistance down the side of the pile may appear misleadingly low in driving owing to entrapped moisture temporarily lubricating the sides of the pile, and lateral vibrations making an enlarged hole which later will close in.

Dynamic pile formulae are derived from considerations of kinetic energy. In the simplest form, if

R is the ultimate driving resistance of the pile (tons),
W is the weight of the hammer (tons),
h is the height of drop of the hammer (inches),
s is the final set of the pile per blow (inches), then

$Wh = Rs$, giving

$$R = \frac{Wh}{s} \text{ tons.} \tag{9.3}$$

This is the Saunders formula to which a considerable factor of safety has to be applied. Such a simple formula ignores a large number of energy losses which arise in driving a pile. To mention a few of these, are the temporary elastic compression of the pile, helmet, and packing; the temporary compression of the ground; the bouncing of the hammer on the pile; and the production of noise and heat. Were it not for these energy losses, the pile would penetrate into the ground not only the distance s but some further distance c giving the pile resistance as

$$R = \frac{Wh}{s + c} \text{ tons.} \tag{9.4}$$

This is the form adopted in the *Engineering News* formula. Again it can only be used with a large factor of safety owing to the great variability which exists between the results given by the formula and the actual carrying capacity of piles in practice.

In 1925 *Engineering* published A. Hiley's formula[5] which takes the various energy losses into account in a very much more rational way, and is expressed as

$$R = \frac{Wh}{s + c/2} \cdot \eta \text{ tons} \tag{9.5}$$

where η is the efficiency of the blow of the hammer and c is the sum of the temporary elastic compressions in the pile, packings and ground in inches. The temporary compressions in the pile and ground are determined by field measurements, but the temporary compression in the packings is obtained from a table.

In the Hiley formula the value h is reduced from the actual drop height by a factor to allow for the method of release and drag of the operating mechanism. The efficiency η is given as $\frac{W + Pe^2}{W + P}$ where P is the weight of the pile, helmet, etc., and e is the coefficient of restitution of the materials under impact, and can be obtained from a table. The tables referred to are available in *The Civil Engineering Code of Practice*, No. 4[4].

A further piling formula worthy of mention is due to Oscar Faber[6], and expressed

$$R = \frac{W\left(h - \dfrac{d}{7}\right)}{s + xh} \text{ tons} \tag{9.6}$$

where d is the diameter of the pile (inches), and

$$x = \frac{0.66}{\sqrt{\dfrac{3d}{7}}} \times \sqrt{\frac{W}{2d^2}\left(\frac{l}{E} + \frac{l_1}{E_1}\right)} \qquad (9.7)$$

where l and E are respectively the length and elastic modulus of the pile, and l_1 and E_1 are respectively the thickness and elastic modulus of the packing (if any).

Accordingly x takes into account the length and material of the pile, the thickness and material of the packing, and the weight of the hammer in relation to the cross-section of the pile. Frequently x has a value of about 0·02. Thus an approximate form of the Faber formula is

$$R = \frac{Wh}{s + 0.02h} \text{ tons.} \qquad (9.6a)$$

Faber recommended that E be taken as 1800 tons/sq. in. for reinforced concrete, and that E_1 be taken as 12·5 tons/sq. in. for compressed sawdust, 8·0 tons/sq. in. for corrugated cardboard, and 5·5 tons/sq. in. for compressed felt.

In his paper Faber showed that over a lifetime's experience, all results within his knowledge gave closer agreement with his own formula than with any other.

ART. 9.8. FACTOR OF SAFETY ON PILES

The ultimate resistance of piles is best determined by actual tests to failure on selected piles at the site of the works. An assessment of the carrying capacity of other piles on the same site is then made by comparison of their driving records. Under such circumstances it would be normal to apply a factor of safety of about 2 in arriving at the safe working load.

When the test loading of actual piles is precluded and reliance has to be placed on the more refined formulae such as equations (9.5) and (9.6), a factor of $2\frac{1}{2}$ to 3 is appropriate for piles bearing in sand and gravel, but a very much higher factor is desirable for piles in soft or loose strata including some silts and clays owing to such piles evading as yet any known mathematical interpretation. Earlier formulae such as the Saunders formula and the *Engineering News* formula have been used with factors of safety of 6 and 8, and on occasions such factors have proved inadequate.

ART. 9.9. EXAMPLE OF USE OF FABER PILING FORMULA

Consider a pile 45 ft. long, 14 in. × 14 in., driven by a 3-ton hammer dropping 4 ft. on to a helmet containing 3 in. compressed sawdust to reach a set of $\frac{1}{4}$ in. per blow in sand. Here:

$$W = 3 \text{ tons;} \qquad h = 48 \text{ in.;} \qquad s = 0\cdot25 \text{ in.;}$$
$$l = 540 \text{ tons;} \qquad l_1 = 3 \text{ in.;} \qquad d^2 = 196 \text{ sq. in.;}$$
$$E = 1800 \text{ tons/sq. in.;} \qquad E_1 = 12\cdot5 \text{ tons/sq. in.}$$

Then
$$x = \frac{0\cdot66}{\sqrt{\dfrac{3d}{7}}} \times \sqrt{\frac{W}{2d^2}\left(\frac{l}{E} + \frac{l_1}{E_1}\right)} \qquad \text{as (9.7)}$$

$$= \frac{0\cdot66}{\sqrt{\dfrac{3 \times 14}{7}}} \times \sqrt{\frac{3}{2 \times 196}\left(\frac{540}{1800} + \frac{3}{12\cdot5}\right)}$$

$$= 0\cdot017.$$

Thus the ultimate driving resistance is given by

$$R = \frac{W\left(h - \dfrac{d}{7}\right)}{s + xh} \text{ tons} \qquad \text{as (9.6)}$$

$$= \frac{3\left(48 - \dfrac{14}{7}\right)}{0\cdot25 + (0\cdot017 \times 48)} = 130 \text{ tons.}$$

Whence safe load, allowing a factor of safety of $2\frac{1}{2}$, is

$$P = \frac{130}{2\frac{1}{2}} = 52 \text{ tons.}$$

Alternatively, using now the approximate formula,

$$R = \frac{Wh}{s + 0\cdot02h} \text{ tons} \qquad \text{as (9.6a)}$$

$$= \frac{3 \times 48}{0\cdot25 + (0\cdot02 \times 48)} = 120 \text{ tons.}$$

Whence safe load, allowing a factor of safety of $2\frac{1}{2}$, is

$$P = \frac{120}{2\frac{1}{2}} = 48 \text{ tons.}$$

ART. 9.10. PILE CAPS

Normally the load from a single member of a superstructure, e.g. a column, is such as to require support from a group of several piles so that the safe load-carrying capacity per pile is not exceeded.

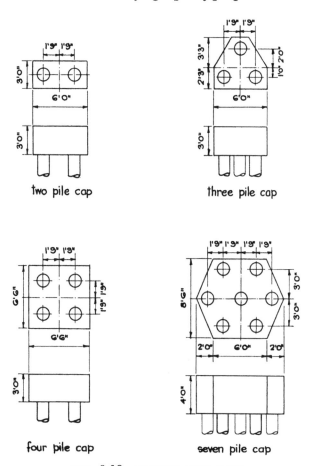

FIG. 9.13. TYPICAL PILE CAPS

It is then necessary to transfer the load from the column on to the piles, so that each pile carries its own share. This is achieved by constructing over the pile group a stiff cap of concrete which spreads the load from the column. Typical arrangements for pile caps are indicated in Fig. 9.13. Pile caps are made a considerable thickness so as to be stiff enough to share the column load equally between each of the piles in the group. The fallacy of this argument is recognised, but it makes for simplicity of work in design, the error being on the side of safety as regards the pile cap. It is the intended load on certain of the individual piles which is initially exceeded, but these piles relieve themselves on to their neighbours as ultimate conditions develop, so that the group

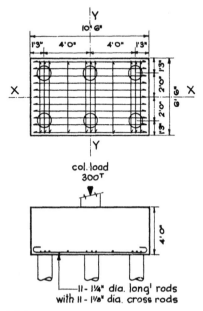

col. load
300ᵀ

11- 1¼" dia. long' rods
with 11 - 1⅛" dia. cross rods

FIG. 9.14. REINFORCEMENT IN SIX-PILE CAP

adjusts itself to a state of equilibrium with all piles carrying reasonable loads.

The first step in the design of a pile cap is to determine the load per pile. Thus with the six-pile cap shown in Fig. 9.14 where the arrangement is symmetrical about both axes, if the applied column load is 300 tons the load per pile is taken as $\frac{300}{6} = 50$ tons. Ignoring the width of the column, the longitudinal moment about YY is due to two piles both acting on an arm of 4 ft., so

$$M_{YY} = (2 \times 50^T) \times 48 \text{ in.} = 4800 \text{ in. tons} = 10,750,000 \text{ in. lb.}$$

Whence

$$A_{st} = \frac{10,750,000}{20,000 \times 0.86 \times 45.5} = 13.8 \text{ sq. in.},$$

requiring 11 — $1\frac{1}{4}$ in. dia. rods.

The shear stress is

$$\frac{(2 \times 50^{\mathrm{T}}) \times 2240}{78 \times 0{\cdot}86 \times 45{\cdot}5} = 74 \text{ lb./sq. in.}$$

requiring no special reinforcement. In cases where the shear stress exceeds the permissible value, bent-up bars are provided as described in Article 3.4.

The cross-moment about XX is similarly calculated and is due to three piles acting at 2 ft., so

$$M_{\mathrm{XX}} = (3 \times 50^{\mathrm{T}}) \times 24 \text{ in.} = 3600 \text{ in. tons} = 8{,}050{,}000 \text{ in. lb.}$$

Whence

$$A_{\mathrm{st}} = \frac{8{,}050{,}000}{20{,}000 \times 0{\cdot}86 \times 44} = 10{\cdot}6 \text{ sq. in.,}$$

requiring 11 — $1\frac{1}{8}$ in. dia. rods.

Note from Fig. 9.14 that the pile cap extends 15 in. beyond the centres of the extreme piles in each direction. This is to enable the

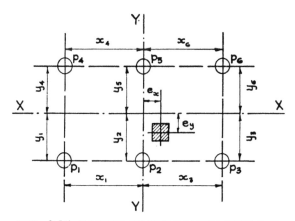

FIG. 9.15. ECCENTRICALLY LOADED PILE CAP

reinforcements to develop the necessary bond; and also incidentally allows some tolerance should the piles be constructed out of their true positions.

Figure 9.15 shows the case where a pile cap is eccentrically loaded about two axes. Eccentricity can arise from errors in the setting-out of the piles, or from fixing moments at the foot of the superstructure column. For example if the load W in a column is

300 tons and this is combined with a fixing moment M equal to 1200 in. tons, the equivalent eccentricity of the load is

$$e = \frac{M}{W} = \frac{1200 \text{ in. tons}}{300 \text{ tons}} = 4 \text{ in.}$$

The general expression for the load on any pile P_n is

$$P_n = \frac{W}{N} \pm \frac{W e_x x_n}{\Sigma(x_1^2 + x_2^2 + \ldots)} \pm \frac{W e_y y_n}{(y_1^2 + y_2^2 + \ldots)}$$

where W is the column load (tons),

N is the number of piles in the group,

e_x and e_y are the eccentricities of the column load in relation to the group axes (feet),

x_n and y_n are the ordinates of pile P_n (feet),

$x_1, x_2, \ldots y_1, y_2, \ldots$ etc., are the ordinates of the other piles P_1, P_2, etc. (feet).

Thus if x_1, etc., are 4 ft. 0 in., and y_1, etc., are 2 ft. 0 in., as Fig. 9.14, and if e_x is 3 in., and e_y is 4 in., and W is 300 tons, then

$$P_1 = 53{\cdot}6^T, \ P_2 = 58{\cdot}3^T, \ P_3 = 63.0^T,$$
$$P_4 = 37{\cdot}0^T, \ P_5 = 41{\cdot}7^T, \ P_6 = 46{\cdot}4^T.$$

Thus the maximum longitudinal-moment, considering piles P_3 and P_6 is

$$(63{\cdot}0^T + 46{\cdot}4^T) \times 45 \text{ in.} = 4910 \text{ in. tons}$$

and the maximum cross-moment, considering piles P_1, P_2 and P_3 is

$$(53{\cdot}6^T + 58{\cdot}3^T + 63{\cdot}0^T) \times 20 \text{ in.} = 3500 \text{ in. tons.}$$

This shows that small errors in setting-out the pile group increases considerably the load on the worst placed pile, though the moments in the pile cap may not be greatly affected.

A further example of an eccentrically loaded pile cap (for a tall chimney) is given in Article 13.8.

Where a column is carried on a single pile, or on a group of only two piles, it is clearly impossible for any eccentricity to be taken up by the piles in the manner just described. And since some amount of eccentricity is unavoidable due to inaccuracies in the positions taken by the piles during driving, it is necessary to provide tie-beams between adjacent pile caps. The tie-beams need to be capable of developing the full moments resulting from accidental eccentricities, plus the effects of other incidental loads.

REFERENCES

1. PLUMMER, F. L. and DORE, S. M. *Soil mechanics and foundations.* Pitman (1940).
2. TERZAGHI, K. and PECK, R. B. *Soil mechanics in engineering practice.* John Wiley & Sons (1948).
3. TSCHEBOTARIOFF, G. *Soil mechanics, foundations and earth structures.* McGraw-Hill (1956).
4. FOUNDATIONS. *Civil Engineering Code of Practice,* No. 4. Inst. C.E. (1954).
5. HILEY, A. A rational pile-driving formula and its application in piling practice explained. *Engineering* (May 29 and June 12, 1925).
6. FABER, OSCAR. A new piling formula. *Journal Inst. C.E.* March (1947).
7. SAURIN, B. F. The design of reinforced concrete piles, with special reference to the reinforcement. *Journal Inst. C.E.* March (1949).
8. CHELLIS, R. D. *Pile foundations.* McGraw-Hill (1951).
9. MORGAN, H. D. and HASWELL, C. K. The driving and testing of piles. *Proc. Inst. C.E.* Vol. 2. No. 1 (January 1953).

CHAPTER 10

BUNKERS AND SILOS

ART. 10.1. INTRODUCTION

MATERIALS such as coal, cement, broken stone, gravel, clinker, wheat, barley, and beans may be stored in bulk in a number of ways. Where ample space is available, the materials may be stored in heaps to their natural angle of repose: this tends to lead to the materials spreading somewhat untidily, and where protection from weather is required may become costly on that account. The provision of retaining walls defines the extent of storage, and at the same time reduces the floor area required, making for economy in roof coverage and materials-handling equipment. The extraction of materials in such cases may be by means of excavating machinery (which is expensive to operate), or by conveyors at floor level or in tunnels below the floor (either of which are not entirely effective). Examples of the design of retaining walls for materials stored in this manner are given in Article 8.7.

Where fully automatic feed and extraction are desirable (and with the present trend of labour costs, the desire is increasing), materials are stored in taller containers provided with outlet devices at the bottoms. Where gravity feed from the outlets is required the containers are elevated, and generally provided with sloping bottoms: in other circumstances, where pneumatic or air-slide transporting equipment is installed, the containers may sit directly on the ground. Elevated hopper bottoms must be designed to provide an angle which ensures the free flow of the material, and this depends on the angles of friction within the material, and between the material and the bottom surface.

Where the containers are not tall enough to prevent planes of rupture of the stored materials cutting the free surface when the container is filled to the top, the lateral pressures on the walls from cohesionless materials are given approximately by Rankine's formula[1] (equation 8.7); and the total weight of the contents is assumed to be supported on the container floor. But for taller containers, the materials tend more to arch across on to the walls, depending on the friction between the material and the wall (see Fig. 10.1): and in these circumstances the lateral pressures on the walls will be less than the Rankine values: and due to the friction at the walls supporting the arching actions, the floor will only receive a part of the total load of the contents. In cases where the height of the container is sufficient to allow of these arching actions, the container is known as a silo: shallower containers are known as bunkers.

Where silo bins are not required to be of small diameter for reasons of blending, or for control or measurement of deliveries and sales, there is no objection to silos being 25 ft. or 30 ft. diameter,

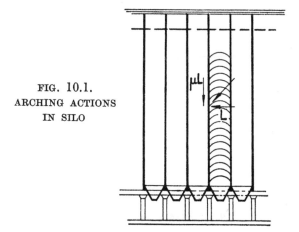

FIG. 10.1.
ARCHING ACTIONS
IN SILO

and the authors have in fact designed silos up to 40 ft. diameter for the storage of cement. In such cases the circular form is usually the most economical. Where, however, the convenience and particular requirements of a miller require different parcels of wheat to be stored separately, with facilities for blending, it may be desirable to divide a battery of silos into many smaller bins; and in these circumstances straight-sided bins, square or rectangular are normally the most economical, sizes generally varying from about 10 ft. square to 14 ft. square. Apart from the millers' convenience, there are certain structural advantages in keeping the bin sizes small in plan while at the same time considerable in height. This is discussed later.

Truly frictional, non-cohesive, materials, stored in circular silos with central inlet and outlet points, behave in accordance with Janssen's theory[2], which may be expressed to give the lateral and vertical pressures in the material at any depth as

$$L = kV = \frac{\gamma r}{\mu}\left(1 - e^{-\frac{\mu k z}{r}}\right) \tag{10.1}$$

where z is any depth below the free surface,

    L is the lateral pressure at depth z,

    V is the vertical pressure at depth z,

    k is a dimensionless ratio $= \dfrac{L}{V}$,

    γ is the density of the contained material,

    r is the hydraulic radius of the container,

    μ is the coefficient of friction of the contained material on the wall.

The ratio k according to Rankine's theory is given by $\dfrac{1 - \sin \phi}{1 + \sin \phi}$ where ϕ is the angle of internal friction of the contained material. The coefficient of friction, μ, is equal to $\tan \delta$, where δ is the angle of friction of the material on the silo wall.

The hydraulic radius, r, is not to be confused with the radius of the silo, and is given by the ratio

$$r = \frac{\text{plan area of silo}}{\text{perimeter of silo}}.$$

For circular silos of diameter D,

$$r = \frac{\pi D^2}{4} \times \frac{1}{\pi D} = \frac{D}{4}.$$

For rectangular silos of sides A and B,

$$r = \frac{A \cdot B}{2(A + B)}$$

which for the special case of a square, where $A = B$, gives

$$r = \frac{A}{4}.$$

In Fig. 10.2 the lateral pressures according to Janssen's theory are shown graphically for reinforced concrete silos of various diameters containing wheat, with $w = 50$ lb/cu ft., $\phi = 28°$, and $\mu = \tan 24°$. The pressures according to Rankine's formula are also shown, and these are of course independent of silo diameter. It will be noted that for all silo diameters, the Janssen curves start off at the same slope as the Rankine line, but fall away at various depths, depending on the silo diameter, becoming asymptopic to maximum pressure values. As the ratio $\dfrac{z}{r}$ increases beyond a certain

value, the term $e^{-\frac{\mu k z}{r}}$ becomes insignificant, and equation 10.1
approximates to

$$L = kV = \frac{\gamma r}{\mu}. \qquad (10.2)$$

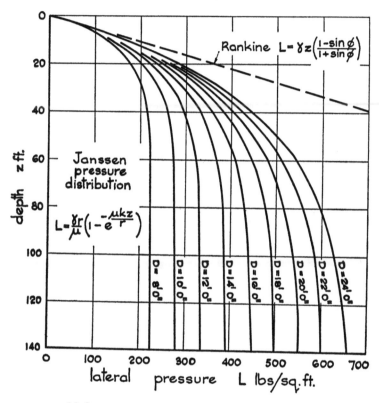

FIG. 10.2. LATERAL PRESSURES IN GRAIN SILOS OF
VARIOUS DIAMETERS (JANSSEN)

$w = 50$ lb/cu. ft., $\phi = 28°$, $\mu = \tan 24°$

With further increase in depth, the lateral pressure then remains
sensibly constant.

Even in tall silos to which the Janssen theory may well apply
lower down, the upper depths of stored material (no matter to
what height the silo is filled) will exert lateral pressures more
nearly in accordance with Rankine's formula. This Rankine condition will apply until sufficient depth is reached for internal arch-

ing effects to develop. It seems that this is not always allowed for.

The structural advantages accruing from silos of small radius and considerable height are now clear by reference to equation 10.2. Such silos enjoy, over a greater proportion of their height, the reduced maximum pressure given by equation 10.2; and the maximum pressure itself is also kept small since L is directly proportional to r. The smaller pressure L acting on the smaller

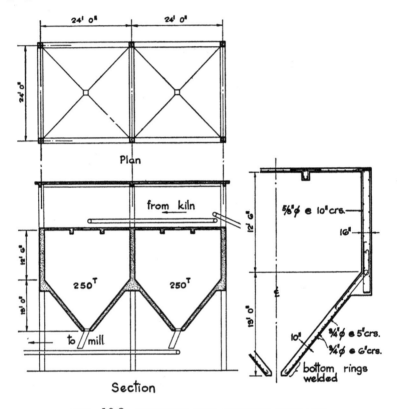

FIG. 10.3. BUNKERS FOR CLINKER STORAGE

sized bin gives obvious structural economies. Furthermore the walls serve to a greater degree their second function as columns, keeping the material load off the silo floor, thus making for economies in the design of the entablature slab. A further practical point arises, in that greater repetition of formwork is achieved; and where sliding forms are used, there is very little difference in cost, once the forms have been set up, whether they travel 60 ft. or 120 ft.

ART. 10.2. DESIGN OF BUNKERS

Figure 10.3 shows an arrangement of bunkers. In this case the bunkers are for storage of clinker at a cement works. The clinker density $\gamma = 85$ lb/cu. ft., and the angle of internal friction ϕ is 30°. The bunkers are elevated to facilitate conveying of material to the grinding mill. Over the bunkers is the Feed Conveyor Floor, which serves to tie the tops of the bunker walls together.

Behaviour of materials in containers of this shape is open to a certain amount of conjecture. There can be no doubt that the material will tend to arch across the pyramid bottom from corner to corner, and it is common experience that on occasions when the outlet-door is opened, the materials jam and have to be released by poking or other means of agitation. The pressures are therefore likely to be least at the centres of the pyramid faces, so that bending will be less than for the uniformly loaded condition. A variety of methods of analysis have been put forward, but perhaps the simplest is as shown in Fig. 10.4(a) where a circle is inscribed in the true elevation of the p⁻⁻ ˙d face. In our case, the circle is of diameter $D = 12·5$ ft., and has its centre at depth $z = 17·5$ ft. from the free surface of the stored material. Accordingly we shall design the pyramid faces for bending on spans of $D = 12·5$ ft., and calculate the pressures on a depth $z = 17·5$ ft.

From Rankine's formula (equation 8.7) the active lateral pressure for clinker is $28·3z$, and since the bunker bottom is at 51° 30′ to the horizontal, we have the normal pressure at 17 ft. 6 in. depth as

$$p_n = (28·3 \times 17·5 \text{ ft.} \times \sin^2 51° 30′)$$
$$+ (85 \times 17·5 \text{ ft.} \times \cos^2 51° 30′) = 890 \text{ lb/sq. ft.}$$

To this has to be added the self-weight of the bottom slab which, allowing for the slope, is

$$120 \times \cos 51° 30′ = \quad 75 \text{ lb/sq. ft.}$$

$$\text{making a total normal pressure} = \underline{\underline{965 \text{ lb/sq. ft.}}}$$

Spanning this in two directions the diameter D, and allowing a little for continuity, we have in each direction

$$M = \frac{p_n D^2}{20} = \frac{965 \times 12\frac{1}{2}^2 \times 12}{20} = 90,000 \text{ in. lb/ft.,}$$

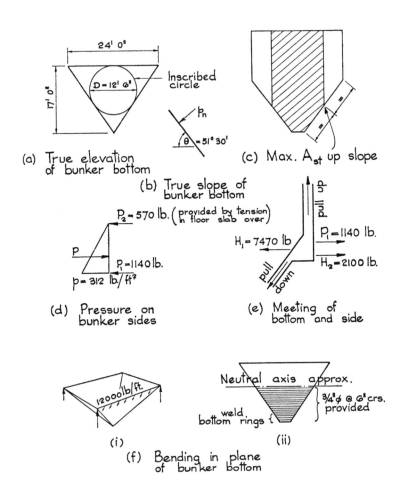

(a) True elevation of bunker bottom

24' 0"

17' 0"

D = 12' 6"

Inscribed circle

(b) True slope of bunker bottom

P_n

$\theta = 51° 30'$

(c) Max. A_{st} up slope

(d) Pressure on bunker sides

$P_2 = 570$ lb. $\left(\begin{smallmatrix}\text{provided by tension}\\\text{in floor slab over}\end{smallmatrix}\right)$

p

$P_1 = 1140$ lb.

$p = 312$ lb./ft²

(e) Meeting of bottom and side

pull up

$H_1 = 7470$ lb

$P_1 = 1140$ lb.

$H_2 = 2100$ lb.

pull down

(f) Bending in plane of bunker bottom

(i)

12000 lb./ft.

(ii)

Neutral axis approx.

bottom rings { weld

¾"φ @ 6" crs. provided

FIG. 10.4. BUNKER DESIGN-DIAGRAMS

requiring $A_{st} = \dfrac{90,000}{20,000 \times (0 \cdot 86 \times 8\frac{1}{2})}$ $= 0 \cdot 62$ sq. in./ft.

Referring to Fig. 10.4(c), this area of steel up the slope has to be increased due to the direct tension at the point midway up the slope resulting from carrying the load shown shaded.

Clinker is 19 ft. × 12 ft. × 12 ft. × 85 ft.

$= 260,000$ lb.

Bunker bottom is

$4 \times 8 \cdot 5$ ft. $\times \dfrac{12}{2}$ ft. $\times 120$ $= 24,500$ lb.

$\overline{284,500 \text{ lb.}}$

Per ft. this is $\dfrac{284,500}{4 \times 12 \text{ ft.}}$ $= 5,930$ lb.

which is taken by force up slope

$\dfrac{5930}{\sin 51° 30'}$, $= 7,550$ lb.

This requires A_{st} due to tension $= \dfrac{7,550}{20,000}$ $= 0 \cdot 38$ sq. in./ft.

Total $A_{st} = \overline{1 \cdot 00 \text{ sq. in./ft.}}$

This is given by $\frac{3}{4}$ in. dia. rods at 5 in. centres (area $= 1 \cdot 05$ sq. in./ft.).

Where the bunker bottom is slung up to the walls, the total tension is equal to the total weight of contents plus self-weight of bottom. These are:

Contents $\left\}\begin{array}{l} 24 \text{ ft.} \times 24 \text{ ft.} \times 12 \text{ ft.} \times 85 \\ \frac{1}{3} \times 24 \text{ ft.} \times 24 \text{ ft.} \times 13 \text{ ft.} \times 85 \end{array}\right.$ $\begin{array}{l}= 585,000 \text{ lb.} \\ = 212,000 \text{ lb.}\end{array}$

Bottom 4×17 ft. $\times \dfrac{24 \text{ ft.}}{2} \times 120$ $= 98,000$ lb.

$\overline{895,000 \text{ lb.}}$

The component of this per foot up the slope is

$$\frac{895,000}{(4 \times 24 \text{ ft.}) \sin 51° 30'} = 12,000 \text{ lb.,}$$

giving a tension stress in $\frac{3}{4}$ in. dia. rods at 5 in. centres of

$$\frac{12,000}{1 \cdot 05} = 11,400 \text{ lb/sq. in.}$$

which is satisfactory.

Considering now the area of steel across the pyramid face, we have M (as before) = 90,000 in. lb/ft. requiring A_{st} (in second

$$\text{layer} = \frac{90,000}{20,000 \times (0.86 \times 7\tfrac{3}{4})} = 0.67 \text{ sq. in./ft.}$$

The horizontal pressure on the adjoining pyramid faces is 28.3×17.5 ft. (mean) = 496 lb/sq. ft., acting on an area $\tfrac{1}{2} \times 24$ ft. $\times 13$ ft. = 156 sq. ft. giving a force of $156 \times 496 = 77,400$ lb. Average tension at corners is therefore

$$\frac{77,400}{2 \times 17 \text{ ft.}} = 2280 \text{ lb/ft.},$$

requiring A_{st} due to tension $= \dfrac{2280}{20,000}$ = 0.12 sq. in./ft.

                 0.79 sq. in./ft.

This is provided by $\tfrac{3}{4}$ in. dia. rods at 6 in. centres (area 0.88 sq. in./ft.).

The horizontal pressure distribution on the bunker sides is as shown in Fig. 10.4(d), varying from zero at the top to

$$p = 28.3 \times 11 \text{ ft.} = 312 \text{ lb/sq. ft.}$$

at the bottom.

Total P per foot width $= 312 \times \dfrac{11 \text{ ft.}}{2} = 1710$ lb., giving bottom and top reactions of $P_1 = 1140$ lb. and $P_2 = 570$ lb. Bending in the side height is $M = 1710 \times 13$ ft. $\times \dfrac{12}{10} = 25,700$ in. lb., re-

quiring $A_{st} = \dfrac{25,700}{20,000 \times 0.86 \times 14}$ = 0.11 sq. in./ft.

(The concrete thickness here is governed by considerations of the bunker side acting as a girder between corner columns, and the detail at the columns themselves.)

Vertical steel is also required to take the vertical tension of 895,000 lb. (*see* page 337).

Per ft., this is $\dfrac{895,000}{4 \times 24 \text{ ft.}} = 9350$ lb.,

requiring $A_{st} = \dfrac{9350}{20,000} = 0.46$ sq. in./ft.,

and with steel in two faces, A_{st} per face   = 0.23 sq. in./ft.

          Total $A_{st} = $ 0.34 sq. in./ft.

This is met by providing ⅝ in. stirrups at 10 in. centres to suit the spacing of the rods in the bunker bottom. The horizontal reinforcements at the top and bottom of the bunker sides are determined from consideration of the sides acting as girders spanning between the corner columns, carrying the self-weight and contents of the bunker, together with other loads from the superstructure, including the floor, walls and roof over, and the incoming conveyor gantry. The calculation for this follows the principles already given in Chapter 4.

In Fig. 10.4(e) the lateral forces are indicated at the corner where the bunker bottom meets the side:

$P_1 = 1140$ lb. as derived above,

H_2 is due to the pressure normal to the bottom face. The total normal pressure per face is

$$\frac{965 \times 24 \text{ ft.} \times 17 \text{ ft.}}{2} = 197,000 \text{ lb.}$$

About one-third of this comes off at top $= 65,700$ lb. Per foot run this is $\frac{65,700}{24 \text{ ft.}} = 2740$ lb. The horizontal component H_2 equals $2740 \sin 51° 30' = 2100$ lb.

H_1 is due to the downward pull in the plane of the bottom and is $12,000 \times \cos 51° 30' = 7470$ lb. and clearly greater than $P_1 + H_2$, so that under the applied loads the corner will try to move inwards. This will be prevented by the passive resistance of the contents.

A further matter needs checking. When the contained material fills the bunker bottom, and stands as steeply as possible without pressing against the upper walls, the pull-down force in the plane of the bottom is unbalanced and has to span to the corner columns (see Fig. 10.4(f)):

M in plane of bottom

$$= \frac{12,000 \times 24^2 \times 12}{8} = 10,400,000 \text{ in. lb.}$$

This moment is taken care of by the bunker bottom acting as an inclined girder, each horizontal rod being stressed in proportion to its distance below the neutral axis of the girder. The rods nearest the outlet are stressed most on this account, but have less duty in regard to bending from normal pressure: and where the rods are stressed to a maximum from normal pressure, they are up to (or above) the neutral axis of the inclined girder. Near to the outlet itself, the individual rods would be too short for the necessary bond to develop, and in this position they need to be welded to form continuous prefabricated rings.

ART. 10.3. DESIGN OF CIRCULAR SILOS

Apart from secondary bending effects, circular silo bins resist the lateral pressures from within by ring tension. The reinforcement is therefore required in the form of continuous circular hoops. In such circumstances, where the whole cross-sections of the walls are in direct tension, it is prudent to work to stresses less than the maximum values given in C.P. 114 (1957), otherwise appreciable cracking may develop. Where the walls are external and exposed to driving rain, these cracks may be wide enough to allow penetration of water sufficient to corrode the reinforcements and also to affect the materials stored in the silo. Cement of course has to be kept scrupulously dry; and wheat, if wetted, becomes sticky and will not flow properly, and in certain conditions starts to germinate. Wheat also, when wetted, swells and produces increased lateral pressures.

A suitable steel stress under these conditions is 15,000 lb/sq. in. This is but little affected by whether the reinforcement is mild steel or high strength steel, because it is the *stretch* of the steel that matters, not the strength: and the elastic moduli of high tensile and mild steels are sensibly alike. A check should also be made that the tensile stress on the equivalent concrete area does not exceed about 175 lb/sq. in. – depending on the mix of concrete. This gives a further guard against likelihood of cracking. (This is discussed more fully in Chapter 12.)

SILOS FOR STORAGE OF GRAIN

Figure 10.5 shows a typical battery of grain silos. Each bin is 20 ft. diameter and 120 ft. high. The grain is fed to the top conveyor-floor by vertical elevator, distributed to the various bins as required by conveyor, and later extracted from the bottoms of the bins by chutes and conveyors below the level of the bin floors. For the present example we will assume that the silo is to be used for the storage of wheat.

The lateral pressures on the bin walls are calculated from Janssen's formula as follows:

$$L = \frac{\gamma r}{\mu} \left(1 - e^{-\frac{\mu k z}{r}} \right)$$
$$= C_1 \left(1 - e^{-C_2 z} \right) \tag{10.3}$$

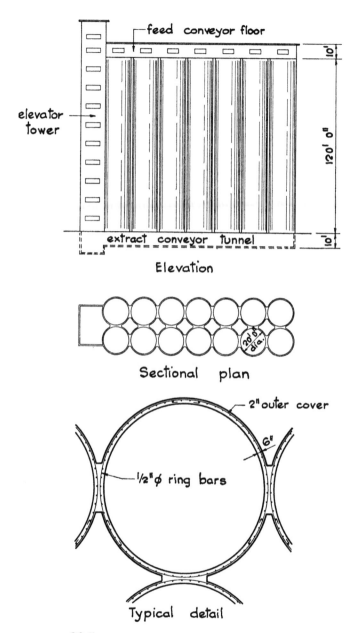

FIG. 10.5. CIRCULAR SILOS FOR GRAIN STORAGE

341

where
$$\gamma = 50 \text{ lb/cu. ft.},$$
$$r = 5 \text{ ft. } (hydraulic \text{ radius}),$$
$$\mu = \tan 24° = 0.445,$$
$$k = \frac{1 - \sin 28°}{1 + \sin 28°} = 0.362,$$
$$C_1 = \frac{\gamma r}{\mu} = \frac{50 \times 5}{0.445} = 562,$$
$$C_2 = \frac{\mu k}{r} = \frac{0.445 \times 0.362}{5} = 0.0322.$$

The calculation is most conveniently set out in tabular manner as below:

z	$C_2 z \ (= x)$	e^{-x}	$1 - e^{-x}$	$L = C_1(1 - e^{-x})$
ft.				lb/sq. ft.
10	0·32	0·726	0·274	154
20	0·64	0·526	0·473	266
30	0·96	0·383	0·617	347
40	1·28	0·275	0·725	405
50	1·61	0·200	0·800	450
60	1·93	0·145	0·855	480
70	2·25	0·105	0·895	503
80	2·57	0·075	0·925	519
100	3·22	0·040	0·960	539
120	3·87	0·021	0·979	550

The ring tension $= L \times R$ where R is the *physical* radius of the bin.

The area of steel required at all heights is then determined in tabular manner as below:

z	L	$T = LR$	A_{st} reqd. at 15,000 lb/sq. in.	A_{st} provided
ft.	lb/sq. ft.	lb/ft.	sq. in./ft.	dia. and crs. (sq. in./ft.)
10	154	1540	0·10	
20	266	2660	0·18	⎫ $\frac{1}{2}$ in. at 9 in. (0·26)
30	347	3470	0·23	⎭
40	405	4050	0·27	⎫ $\frac{1}{2}$ in. at 8 in. (0·30)
50	450	4500	0·30	⎭
60	480	4800	0·32	⎫ $\frac{1}{2}$ in. at 7 in. (0·34)
70	503	5030	0·33	⎭
80	519	5190	0·345	⎫
100	539	5390	0·36	$\frac{1}{2}$ in. at 6 in. (0·39)
120	550	5500	0·365	⎭

It will be seen that for the upper part of the silo the area of steel is not skinned down to the Janssen values: this allows for the material behaving more in accordance with Rankine's theory above the depth where the internal arching actions start. The reinforcements are placed in the outer part of the wall to allow the loads to spread through the concrete on to the steel. Generous cover is provided as protection from weather.

At the bottom of the silo (where the ring tension is greatest), the equivalent area of concrete per foot height of wall

$$A_E = (b \times d) + (m - 1)A_{st}$$
$$= (6 \times 12) + (14 \times 0.39) = 77.5 \text{ sq. in.}$$

Tensile stress in concrete $= \dfrac{5500}{77.5} = 71$ lb/sq. in. which is satisfactory.

The vertical pressure on the silo floor is

$$V = \frac{L}{k} = \frac{550}{0.362} = 1520 \text{ lb/sq. ft.},$$

showing that only the lowermost 30 ft. of grain ($1\frac{1}{2}$ diameters) rest on the floor; the remainder hangs on the walls. Thus total load hanging on walls is determined as:

$$\text{Total contents} \left(\frac{\pi 20^2}{4}\right) \times 120 \times 50 = 1,890,000 \text{ lb.}$$

$$\text{Less on floor} \left(\frac{\pi 20^2}{4}\right) \times 1520 \quad = \quad 480,000 \text{ lb.}$$

$$\overline{ 1,410,000 \text{ lb.},}$$

contributing a compressive stress in the concrete walls of

$$\frac{1,410,000}{\pi \times 20 \text{ ft.} \times 12 \text{ in.} \times 6 \text{ in.}} = 310 \text{ lb/sq. in.}$$

A check on the load on the walls can be made by summing the lateral pressures at all depths, and multiplying these by μ. The average pressure L is 440 lb/sq. ft., and the wall area is

$$\pi 20 \text{ ft.} \times 120 \text{ ft.} = 7550 \text{ sq. ft.}$$

Thus load hanging on wall is approximately

$$7550 \times 440 \times 0.445 = 1,470,000 \text{ lb.},$$

showing reasonable agreement.

SILOS FOR STORAGE OF CEMENT

A typical battery of cement silos is indicated in Fig. 10.6. Cement may be transported by conveyor belt, screw conveyor, and bucket elevator; or pneumatically at all slopes and elevations by pipeline; or by gravity on slides sloping at about 8 degrees from

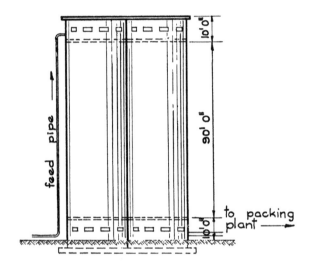

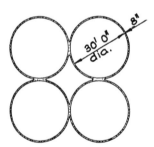

FIG. 10.6. CIRCULAR SILOS FOR CEMENT STORAGE

the horizontal, the cement being aerated from beneath through continuous webbing.

Owing to the erratic cohesive nature of cement, the pressures in cement silos vary widely from values obtained using the Janssen and Rankine formulae (which were never intended for the purpose). So far, very little reliable information has been published on the subject of pressures in cement silos, and no accepted theory has been put forward. Engineers have found that the pressures follow no direct relationship in regard to hydraulic radius and depth: and the authors are quite clear that the depth z_u of cement occurring *below* any particular level under consideration influences the result. The cement below suffers a "cushioning" effect, packing tighter as the head of cement over is increased until it acquires its state of maximum density: and for so long as the "cushioning" goes on, the

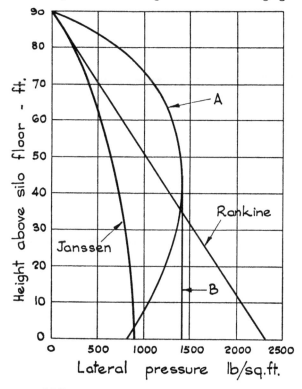

FIG. 10.7. LATERAL PRESSURES IN CEMENT SILO
30 FT. DIAMETER

condition of stress above will be influenced. At depths where the maximum density is reached, so also is the maximum angle of internal friction developed: and indeed compacted cement can stand for considerable heights with a vertical face, whereas shallow lightly sprinkled cement exhibits properties not far removed from those of a fluid. Thus at the top of a silo the lateral pressures build up rapidly, reaching maximum values at some head z dependent on the depth z_u of cement forming the "cushion" under ; and below this level the lateral pressures *decrease* at greater depths down the silo. The plan dimensions of the silo affect not only the arching actions of the material at all depths, but also the rate of change of density with depth, and the influence of the "cushioning" effect. Because extraction of cement occurs from the bottom of the silo, the disturbance on emptying re-creates the "cushion", so that the conditions on emptying follow the same general pattern as on filling: in fact the pressures on emptying are slightly greater than on filling because continuous disturbance enhances the effects previously described. Whether extraction is by mechanical means or by normal low-pressure airslides appears to have very little influence.

It is clear from the above that small-scale tests are useless, and worthwhile results can only be obtained from tests in full-size silos. Until more results are published, engineers are inclined to design cement silos on pressure readings obtained from their own tests and experience rather than on any generalised formulae.

For the silos indicated in Fig. 10.6, the authors made full-scale experiments in a silo belonging to the Associated Portland Cement Manufacturers Limited with assistance from the Cement and Concrete Association. The lateral-pressure distribution for the silos, when full, was found to approximate to the Curve A shown in Fig. 10.7. Allowing for the top level of cement varying as the silo is filled and emptied, it will be wise to design for the pressures shown by Curve B. For interest, in Fig. 10.7, Curves C and D have been indicated based on the erroneous application of Janssen's and Rankine's formulae, taking the popular values of $\gamma = 95$ lb/cu. ft., $\phi = 35°$, $\mu = \tan 35°$. It will be seen that for the upper half of the silo, the true pressures are about double the Rankine values and nearly three times the Janssen values: though when the silo is full, the true values near the bottom are not greatly at variance with the Janssen values.

Setting the design calculation in tabular form, we have as follows:

z	L	$T = LR$	A_{st} reqd. at 15,000 lb/sq. in.	A_{st} provided
ft.	lb/sq. ft.	lb/ft.	sq. in./ft.	dia. and crs. (sq. in./ft.)
10	720	10,800	0·72	⅝ in. at 5 in. (0·72)
20	1100	16,500	1·10	¾ in. at 5 in. (1·07)
30	1300	19,500	1·30	⎫ ⅞ in. at 5 in. (1·44)
40	1400	21,000	1·40	⎭
and deeper				

Where the ring tension is a maximum,

$$A_E = (8 \times 12) + (14 \times 1\cdot44) = 116\cdot2 \text{ sq. in.},$$

giving a tensile stress in the concrete of

$$\frac{21,000}{116\cdot2} = 180 \text{ lb/sq. in.}$$

ART. 10.4. DESIGN OF RECTANGULAR SILOS

In the design of rectangular silos, each wall has to resist direct tension as well as bending. The calculation is most easily done by

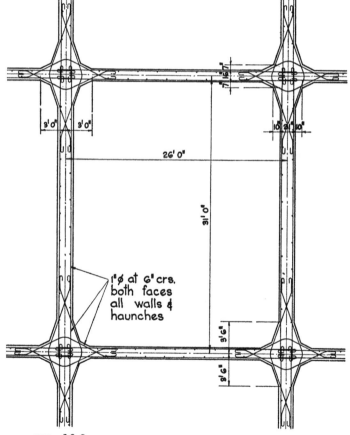

FIG. 10.8. RECTANGULAR SILOS FOR STORAGE OF CRUSHED STONE: SECTIONAL PLAN

selecting a wall thickness from experience, and then determining the areas of reinforcement required. This is shown in the following example of a silo for the storage of crushed limestone. Each bin is 60 ft. high, and has walls at centres of 26 ft. and 31 ft. (*see* Fig. 10.8). We can assume that $\gamma = 100$ lb/cu. ft., $\phi = 35°$, and (for the design of the walls) $\mu = 0.5$.

For the rectangular silo, the hydraulic radius is

$$\frac{29 \cdot 6 \text{ ft.} \times 24 \cdot 3 \text{ ft.}}{2(29 \cdot 6 \text{ ft.} + 24 \cdot 3 \text{ ft.})} = 6 \cdot 7 \text{ ft.}$$

Using Janssen's formula (equation 10.1), we have at depth 55 ft.

$$L = \frac{\gamma r}{\mu} \left(1 - e^{-\frac{\mu k z}{r}} \right)$$

$$= \frac{100 \times 6 \cdot 7}{0 \cdot 5} \left(1 - e^{-\frac{0 \cdot 5 \times 0 \cdot 272 \times 55}{6 \cdot 7}} \right)$$

$$= 900 \text{ lb/sq. ft.}$$

Below this level the walls will cantilever up from the silo floor, and should be reinforced accordingly.

On 26 ft. span:

$$\text{Mid-span } M = \frac{900 \times 26^2 \times 12}{24} = 305,000 \text{ in. lb.,}$$

$$\text{requiring } A_{st} = \frac{305,000}{20,000 \times 0 \cdot 86 \times 14\frac{1}{2}} = 1 \cdot 22 \text{ sq. in.}$$

$$\text{Direct tension} = \frac{31 \text{ ft.}}{2} \times 900 = 14,000 \text{ lb.}$$

Half this will go to offset the compression in the concrete due to bending. We therefore require

$$A_{st} \text{ per face} = \frac{14.000}{2 \times 20,000} = 0 \cdot 35 \text{ sq. in.}$$

$$\overline{ 1 \cdot 57 \text{ sq. in.}}$$

$$\text{Wall-intersection } M = \frac{900 \times 26^2 \times 12}{12} = 610,000 \text{ in. lb.,}$$

$$\text{requiring } A_{st} = \frac{610,000}{20,000 \times 0 \cdot 86 \times 28\frac{1}{2}} = 1 \cdot 24 \text{ sq. in.,}$$

plus direct tension.

On 31 *ft. span:*

Mid-span $M = \dfrac{900 \times 31^2 \times 12}{24} = 432{,}000$ in. lb.,

requiring $A_{st} = \dfrac{432{,}000}{20{,}000 \times 0.86 \times 19\frac{1}{2}}$ $= 1.29$ sq. in.

Direct tension $= \dfrac{26 \text{ ft.}}{2} \times 900 = 11{,}700$ lb.,

requiring $A_{st} = \dfrac{11{,}700}{2 \times 2000}$ $= 0.29$ sq. in.

$\overline{1.58 \text{ sq. in.}}$

Wall-intersection $M = \dfrac{900 \times 31^2 \times 12}{12} = 864{,}000$ in. lb.,

requiring $A_{st} = \dfrac{864{,}000}{20{,}000 \times 0.86 \times 39\frac{1}{2}} = 1.27$ sq. in.

plus direct tension.

All the above requirements are met by providing 1 in. dia. rods at 6 in. centres (area 1·58 sq. in.) in both faces of all walls and haunches as shown in Fig. 10.8. The small out-of-balance between

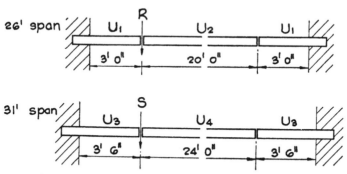

FIG. 10.9. RECTANGULAR SILO: SIMPLE MOMENT ANALOGY

the two wall-intersection moments is taken care of partly by the stiffnesses of the other walls at the intersection, and partly by the action of the contents in the silo. Higher up the silo, the lateral pressures are less, and the wall reinforcements can be reduced to suit.

A rough check on the bending moments derived above may be made as follows. If the haunches at the wall intersections are regarded as held against rotation (due to approximate symmetry) we can consider the 26 ft. and 31 ft. spans as made up of rigid abutments with continuously supported centre spans between, as shown in Fig. 10.9. Then we have:

On 26 ft. span:

$U_2 = 900 \times 20 \text{ ft.} = 18{,}000 \text{ lb.}$

$M \text{ at } R \text{ and at centre} = \dfrac{18{,}000 \times 20 \text{ ft.} \times 12}{16} = \underline{270{,}000 \text{ in. lb.}}$

Shear at $R = 9000 \text{ lb.}$

$U_1 = 900 \times 3 \text{ ft.} = 2700 \text{ lb.}$

$M \text{ at support} = (9000 \times 36 \text{ in.}) + (2700 \times 18 \text{ in.}) + 270{,}000$
$\qquad\qquad = 324{,}000 + 48{,}000 + 270{,}000 = \underline{642{,}000 \text{ in. lb.}}$

On 31 ft. span:

$U_4 = 900 \times 24 \text{ ft.} = 21{,}600 \text{ lb.}$

$M \text{ at } S \text{ and at centre} = \dfrac{21{,}600 \times 24 \text{ ft.} \times 12}{16} = \underline{390{,}000 \text{ in. lb.}}$

Shear at $S = 10{,}800 \text{ lb.}$

$U_3 = 900 \times 3\tfrac{1}{2} \text{ ft.} = 3150 \text{ lb.}$

$M \text{ at support} = (10{,}800 \times 42 \text{ in.}) + (3150 \times 21 \text{ in.}) + 390{,}000$
$\qquad\qquad\qquad\qquad\qquad = \underline{911{,}000 \text{ in. lb.}}$

These all agree reasonably with the values for moments calculated above.

On the 26 ft. span $\qquad \dfrac{M}{bd_1{}^2} = \dfrac{305{,}000}{12 \times 14\tfrac{1}{2}{}^2} = 120$

and the shear stress

$$\frac{Q}{bl_a} = \frac{9000}{12 \times (0 \cdot 86 \times 14\tfrac{1}{2})} = 60 \text{ lb/sq. in.}$$

On the 31 ft. span

$$\frac{M}{bd_1{}^2} = \frac{432{,}000}{12 \times 19\tfrac{1}{2}{}^2} = 95$$

and the shear stress

$$\frac{Q}{bl_a} = \frac{10{,}800}{12 \times (0 \cdot 86 \times 19\tfrac{1}{2})} = 54 \text{ lb/sq. in.}$$

Thus a concrete mixture of 1:2:4 nominal would be adequate. The wall thicknesses chosen are not excessive, considering the jarring effect of crushed rock being dropped and spilt a height of some 60 ft., tending to shatter the internal adhesion forces between the cement and aggregate particles on which the shear strength (tensile) of the concrete depends. The work of constructing a silo like this is kept simple and cheapest by avoiding the introduction of diagonal shear reinforcements. Note also that the reinforcements already provided to resist bending (1 in. dia. at 6 in. centres) are as heavy and as closely spaced as is convenient and practicable.

ART. 10.5. SHUTTERING FOR SILOS

Contractors have shown great ingenuity and skill in forms of shuttering for tall repetitive works such as silos. Common practice now is to use *climbing* or *sliding* shuttering.

With climbing shuttering, the contractor first erects one set of forms (set A), generally about 4 ft. height. These are then filled with concrete, and while this concrete is left to cure, a second set of forms (set B) is erected above. These are then filled with concrete. Set A are now released and erected above set B, and so the process continues like a man climbing a rope, hand over hand.

With sliding shuttering, a special form is prepared, the clear gap between faces for opposite sides of the wall being slightly greater at the bottom than at the top. The whole form is generally about 4 ft. high, and suspended on *yokes* at intervals along the line of the wall of about 5 ft. or 6 ft. The yokes take a bite on vertical *jack-rods* incorporated centrally in the wall. By suitable operation of the jacks, the form travels slowly upwards, fresh concrete being poured in at the top and having time to set and harden sufficiently to stand on its own unsupported when released from the form at the bottom. This technique has been developed and used with great success now for nearly 40 years. Originally the jacks were screw-operated, and the form travelled at a rate of about 6 ft. per day. In the past 10 years hydraulically operated jacks have been used giving a rate of travel of about 20 ft. per day. With the slower system there is always fear in warm weather that the concrete may set sufficiently to jam the forms. This fear does not exist to the same extent with the use of hydraulic jacks; but a fresh difficulty then arises in that the contractor has difficulty in placing the reinforcements sufficiently quickly to keep pace with the rate of slide. Once the slide has started there should be no halt until the total height of the structure has been accomplished, and accordingly close attention to detail is required before and during the process. A number of cases are on record in which the correct quantity of reinforcement was never placed in the structure due to inadequate preparations and supervision: and other cases are known where it was found necessary to abandon the process part way. Nevertheless with proper design and organisation, the authors have used the method on numerous occasions with complete success.

Two practical points in regard to sliding shuttering are worthy of mention. The jack-rods should be left cast into the work and not recovered. Attempts have sometimes been made for supposed economy to operate with jack-rods in sleeves, and later withdraw the rods for subsequent re-use. The outside diameter of the sleeves then becomes a more appreciable part of the total thickness of the wall, making for weakness where shrinkage and other stresses tend to relieve themselves with the consequent appearance of vertical cracks. A second point is that the earlier exposure to atmosphere of the immature concrete calls, in certain circumstances, for special precautions in curing: this is particularly the case in tropical climates, where the authors have seen silos severely cracked on this account alone, even before filling of the silos had started.

REFERENCES

1. RANKINE, W. J. M. On the stability of loose earth. *Trans. Royal Soc. London.* Vol. 147. (1857).
2. JANSSEN, H. Versuche über Getreidedruck in Silozellen. *Zeitschrift des Vereines deutscher Ingenieure.* p. 1045. (1895).

CHAPTER 11

SHELL-CONCRETE ROOFS

ART. 11.1. INTRODUCTION

PERHAPS one of the most beautiful structural shapes in nature is the shell of an egg, combining, as it does, extreme fitness for its purpose with an economy of material and a cleanliness of design which is the envy of engineers and architects. Yet as far as we know the hen manages to construct her shells without a knowledge of mathematics involving double integrals and Fourier series. Consider also an ordinary drinking-straw, manufactured by forming quite thin paper into a tube about 9 in. long $\times$ $\frac{1}{10}$ in. diameter: this tube will certainly carry its own weight over a span equal to its own length; but if the same tube is unrolled and the paper laid flat, it will sag and collapse on only a fraction of the span. These simple examples show what can be achieved when the engineer is given control of his shapes.

Shell concrete is understood to describe concrete construction where the concrete is in the form of a thin curved membrane possessing great strength by virtue of its intrinsic shape. Loads are carried almost entirely by direct forces within the membrane. An early form of construction not dissimilar from shell concrete was the dome, known to the Romans thousands of years ago, and relying for its stability on abutments at the springings which provide vertical support and thrust. With shell concrete construction it is possible to produce satisfactory domes which weigh only a fraction of the weights of the early massive domes. The lightness of shell concrete enables the substructures and foundations to be cheaper. These shell concrete domes can be supported on three or more columns, the lateral thrusts from the dome action being balanced by tensile forces in a ring-beam system at the springing. The ratio of shell thickness to span for reinforced concrete domes has been as little as 1:650, which is less than for an egg shell. Ratios of 1:450 are common.

More general use of shell concrete is made in roofs of arched shape. In this manner large areas can be covered without internal supports. For clear widths of about 150 ft. and over it is found to be economical to provide arched shells spanning direct, with stiffening ribs at about 25 ft. to 35 ft. centres. It is an advantage for the ribs to be on the outer surface of the shell as this eases the use of mobile formwork. An example of this type of shell construction is the aircraft hangar at South Dakota, America, where the arch is 340 ft. wide and 90 ft. to the crown; and the shell thickness varies from 5 in. at the crown to 7 in. at the springing. The stiffening ribs are at 25 ft. centres, and expansion joints are provided at 50 ft. centres, midway between ribs.

For clear spans under about 150 ft. barrel vaulting is used. This, at the present time, remains the most common form of shell concrete. The essential features of a barrel vault are shown in Fig. 11.1.

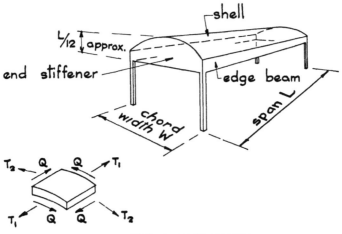

FIG. 11.1. BARREL VAULT

The shell is curved in one direction only. End stiffeners hold the shell to its curved shape and edge beams increase its longitudinal stiffness. This unit is supported on four columns and approximates to a hollow beam of span L. The greater the ratio L/W the nearer the truth does the hollow beam approximation become. As the ratio L/W becomes small some of the load from the shell tends to span directly on to the end stiffeners and the conditions become more complex. Some people hold that the ratio L/W should not exceed 5, and that the most economical value is 2. As with any other structural beam, the barrel vault can be made continuous over the end stiffeners, or it can cantilever. If required the end stiffeners can upstand, leaving the soffite clear for sake of appearance and ease of shuttering.

The main forces acting on any element of the barrel vault shell are T_1 (longitudinal), T_2 (transverse) and Q (shear). On simple spans the T_1 stresses are compressive at the top of the barrel, changing to tension near the bottom and in the edge beam. Thus the edge beam acts as the rib of a T-beam or the tie member of a truss, and would not be stiff enough to support itself if it were not an integral part of the shell construction. The T_2 arching stresses are compressive at the crown but die away at the springing, being absorbed by diagonal tensions in the shell reinforcement and taken back to the end stiffeners. These latter are therefore in tension also. Now the tension in both edge beams and end stiffeners must be balanced by diagonal compression in the shell at each corner. Due to the rise of the barrel these diagonal compressions form thrusts

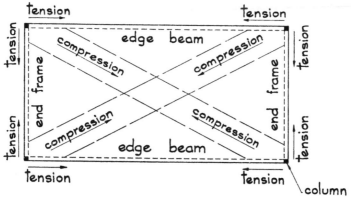

FIG. 11.2. DIAGONAL ARCHING OF BARREL VAULT

of two long diagonal arches (see Fig. 11.2), and it is to a large extent this diagonal arching which interferes with the simple action of the hollow beam and produces the secondary bending moments and torsional couples referred to later. As the L/W ratio decreases, the amount of diagonal arching increases and the departure from the simple hollow beam theory becomes pronounced.

If, in any form of shell, it is assumed that the forces are direct (i.e. no bending moments are taken) and that the shell deforms in a statically determinate manner, then T_1, T_2 and Q can be calculated for any number of chosen points on the shell, using the membrane-theory equations as given in Articles 11.2, 11.3 and 11.4; and the principal stresses at these points can then be determined, and suitable reinforcement chosen. Unfortunately these assumptions are invariably erroneous in practice for normal shell roofs owing to disturbances arising from the edge conditions; the deflections are not statically determinate, and bending moments and torsional couples are set up. Aids to practical design which handle these complications are described in Article 11.2.

The authors are a little surprised by those critics who appreciate only the full-dress and supposedly exact mathematical solutions. Let it be remembered that concrete never behaves elastically, and creep, shrinkage and temperature effects will almost certainly vary the stresses in the steel and in the concrete over a period of time by as much as 50 per cent: also the loads we are to carry, and the strengths of our materials, are not precisely known.

Having calculated the minimum reinforcement required, a mesh is usually provided near the top and bottom of the shell, with additional diagonal bars near the end stiffeners to take the shear (see Fig. 11.4). It could be argued that reinforcement in two directions would suffice without the diagonal bars since reinforcement in two directions must have a component available in every direction, but the use of diagonals puts the steel just where required and makes for economy. The reinforcement is normally not greater than ⅜ in. diameter and therefore small enough to be laid on the formwork without prior bending.

For thin shells the concrete mix used is often $1:1\cdot6:3\cdot2$ with ⅜ in. maximum aggregate size. The concrete is laid in two operations. The first layer is fairly wet and this enables it to flow under the bottom mesh easily. The second layer is stiffer, with about 2 in. slump. The mesh of reinforcement enables slopes of 45° to be concreted conveniently without top shutters. For barrels, it has been found satisfactory to concrete the beams and thickenings first and then follow with the shell itself in strips right over the barrel.

Shuttering is normally supported on tubular steel scaffolding with purlins at roughly 2 ft. centres. Mobile forms have been used where there is much repetition. Where insulation board forms the shuttering it can be left stuck to the underside of the shell. This has not such good insulation value as insulation board secured after concreting. Insulation fixed under the shells does not protect the shells from the sun's rays, but does prevent the shells from shedding their unwanted heat from beneath. Hence, a shell in this condition suffers far more from extremes of temperature than one without insulation. It is thus better to provide insulation in the form of a covering *over* the shells.

The provision of longitudinal expansion joints at about 150 ft. centres is normally desirable. Transverse joints can be spaced further apart where end stiffeners are in the form of curved frames – as opposed to rigid solid panels: transverse joints have then been kept as far apart as 400 ft. with complete success.

To achieve watertightness it is necessary to provide a covering membrane to all shell roofs, and generally bituminous roofing felt answers this requirement perfectly well. The felt keeps the top insulation dry, and also serves to protect the reinforced shells from corrosion (bearing in mind the meagre cover achieved with such thin structures).

Very severe accidental fire tests have been made on barrel vault roofs. One of these was in America, where timber staging caught fire when the concrete was only partly cured. Under the drive of a high wind the staging burned out in less than two hours. Subsequent tests were made on the shell, and apart from local spalling and minor cracks, no major weakness was found.

ART. 11.2. CYLINDRICAL BARRELS

ANALYSIS BY USE OF MEMBRANE THEORY

Where the restraints provided at the edges of the shell are such that they produce no moments or normal shears anywhere in the shell, the analysis can be made in accordance with the membrane theory. The assumptions made in the theory are then as follows:

The forces in the shell are direct (no bending or twisting).
The shell deforms in a statically determinate manner.
The shell is thin compared to the barrel radius.

Then by reference to Fig. 11.3 for a singly curved shell with longitudinal, tangential and radial co-ordinate-axes in directions x, y and z, consider an element of shell of radius R, breadth $R \cdot \delta\theta$ and length δx. Let this be acted on by external forces of X, Y and Z per unit area parallel to x, y and z. The membrane forces will then be:

$T_1 = $ longitudinal force per unit area,
$T_2 = $ transverse force per unit area,
$Q = $ shear force per unit area.

The change of force, over the element, is shown in Fig. 11.3.

Now for equilibrium in the z direction (radially), by resolving forces, we have

$$\left(T_2 + \frac{\partial T_2}{\partial \theta} \cdot \delta\theta \right) \delta x \cdot \sin \frac{\delta\theta}{2} + T_2 \delta x \cdot \sin \frac{\delta\theta}{2} + ZR\delta\theta \cdot \delta x = 0.$$

For small angles $\sin \dfrac{\delta\theta}{2} = \dfrac{\delta\theta}{2}$, so that neglecting second order small quantities, the equation reduces to

$$T_2 = - RZ. \tag{11.1}$$

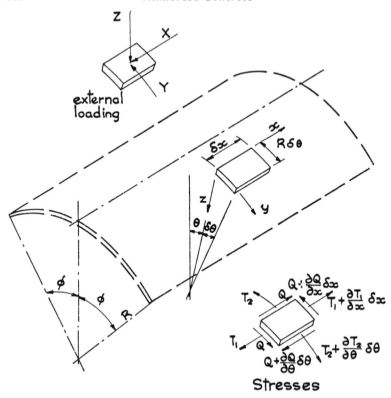

FIG. 11.3. MEMBRANE STRESSES IN CYLINDRICAL
BARREL

For equilibrium in the y direction (tangentially), we have

$$\left(Q + \frac{\partial Q}{\partial x} \cdot \delta x\right) R \cdot \delta\theta - QR \cdot \delta\theta - \left(T_2 + \frac{\partial T_2}{\partial \theta} \cdot \delta\theta\right) \cos\frac{\delta\theta}{2} \cdot \delta x +$$

$$+ T_2 \cos\frac{\delta\theta}{2} \cdot \delta x = \mathrm{Y}R\delta\theta \cdot \delta x.$$

For small angles, $\cos\dfrac{\delta\theta}{2} = 1$, so the equation reduces to

$$- \frac{1}{R} \cdot \frac{\partial T_2}{\partial \theta} + \frac{\partial Q}{\partial x} = \mathrm{Y},$$

and substituting for T_2 from equation 11.1, we have

$$\frac{\partial Q}{\partial x} = \mathrm{Y} - \frac{\partial z}{\partial \theta}. \tag{11.2}$$

For equilibrium in the x direction (longitudinally), we have

$$\left(T_1 + \frac{\partial T_1}{\partial x}\,.\,\delta x\right) R\,.\,\delta\theta - T_1 R\,.\,\delta\theta + Q\,.\,\delta x -$$

$$- \left(Q + \frac{\partial Q}{\partial \theta}\,.\,\delta\theta\right)\delta x = -\,XR\,.\,\delta\theta\,.\,\delta x$$

which reduces to

$$\frac{\partial T_1}{\partial x} - \frac{1}{R}\,.\,\frac{\partial Q}{\partial \theta} = -\,X,$$

and differentiating with respect to x, and substituting from equation 11.2, the equation reduces further to

$$\frac{\partial^2 T_1}{\partial x^2} = -\frac{\partial X}{\partial x} + \frac{1}{R}\left[\frac{\partial Y}{\partial \theta} - \frac{\partial^2 Z}{\partial \theta^2}\right]. \tag{11.3}$$

These are the simplest expressions for T_2, Q and T_1 in terms of x and θ.

For a shell of unit thickness, with loading w per unit surface area, these equations reduce further to:

$$T_2 = -\,Rw\cos\theta, \tag{11.4}$$

$$\frac{\partial Q}{\partial x} = 2w\sin\theta, \tag{11.5}$$

$$\frac{\partial^2 T_1}{\partial x^2} = \frac{2w}{R}\,.\,\cos\theta. \tag{11.6}$$

In similar fashion, by considering strains and rotations, the membrane displacements can be found in terms of the applied forces, and x and θ.

Now in practice the boundary restraints – which depend of course on the characteristics of the particular edge members – never match up to providing exact membrane equilibrium: and the deflection and rotation of the edge-members modify the membrane forces in the shell, and at the same time introduce moments and shear forces normal to the shell surface. The equilibrium of the shell element has then to be reconsidered in terms of forces, displacements and the effects of bending.

At this stage it becomes necessary to introduce simple bending theory relating moments and angular rotations, whereby we arrive at a solution in the form of a differential equation of the eighth order. This has become known as the Equation of Compatibility. The complete solution of this equation is in two parts: the first part is the general solution which yields the membrane stresses, and the second part gives the particular solution related to the boundary conditions at the edge of the shell. The mathematical procedure for the latter is to impose unit forces at the shell boundaries and correlate the effects of these to the restraint provided. This idea came first from Finsterwalder[1], who also produced the first equation of compatibility. Other equations have been published subsequently based on a variety of assumptions and with different simplifications. To date, the most notable are those due to Jenkins[2] and Schorer[3]. Jenkins' equation is the more exact, but is also the more laborious to use: nevertheless it is entirely free from any limitations of application.

Schorer's equation of compatibility is

$$\frac{\partial^8 M_2}{\partial \theta^8} + 4a^2b^2 M_2 = 0 \qquad (11.7)$$

where M_2 is the transverse bending moment, and a function only of θ, and a and b are constants depending on the physical dimensions of the shell.

Analyses have shown that the approximations made by Schorer for shells with L/W ratios not exceeding about 4 have negligible effects on the values obtained for the main forces. Due to the many uncertainties which arise in all reinforced concrete design and construction, Schorer's simple solution is generally acceptable.

DESIGN BY USE OF TABULATED VALUES

Tables have now been prepared to assist engineers in the design of cylindrical barrels. These tables give stress values at points all over the shell surface for an extensive range of shell proportions, with different edge and end conditions. The first such tables *Design of Cylindrical Concrete Shell Roofs*[4] were published by the American Society of Civil Engineers in 1952. The tables are used in two stages: first the internal stresses and the edge forces created by the surface loading on the basis of the membrane theory are read off; then the stresses due to the restraining edge line loading are read off. The American manual is largely the result of work by Parme.

Similarly, tables have been compiled by Rüdiger and Urban under the title *Kreiszylinderschalen*[5] and published by B. G. Teubner Verlagsgesellschaft, Leipzig (1955). These tables appear to be based on a simplified equation proposed by Donnell.

SEMI-EMPIRICAL DESIGN METHOD

A recent development of considerable importance is a paper by Bennett[6], in which the values of T_1, T_2, Q and M_2 for over 200 barrels of various dimensions actually constructed are plotted, and shown to have linear relationships with the shell dimensions and loadings. In other words a sufficient variety of cylindrical barrels have now been constructed for engineers to assess from this wide experience the values of T_1, T_2, Q and M_2 for which they should design. In this manner all symmetrical arrangements of barrels can be designed, except the case of an external barrel where the edge beam is not supported by intermediate columns. The calculations for Bennett's 200 barrels were all based on Schorer's equation, the more recent ones using Tottenham's[7] simplified method. The shells analysed have L/W ratios ranging from 4 to 0·5; shell thicknesses vary between $2\frac{1}{2}$ in. and $3\frac{1}{2}$ in.: and the shell radii vary from 15 ft. to 40 ft.

From the slope of his straight-line graph Bennett shows that if y is the total rise of the barrel, $l_a = 0·72y$; thus, since the longitudinal bending moment is $\dfrac{wL^2}{8}$, we have the total of the tension forces T_1 at mid-span of the barrel as $\dfrac{wL^2}{8 \times 0·72y}$. Similarly Bennett shows that for feather-edged barrels, for example, the shear per half-barrel transmitted to the end stiffening beam is $\Sigma Q = \dfrac{wL}{1·07\theta}$. Relations are also given for M_2 and T_2.

As an example of the use of Bennett's method, consider the feather-edged shell shown in Fig. 11.4.

Span $L = 92$ ft. 6 in.;   Chord $W = 39$ ft. 3 in.;
Radius $R = 27$ ft. 6 in.;   Thickness $= 2\frac{1}{2}$ in.
Unit total loading of shell $= 52$ lb/sq. ft.
Wt. of valley thickening $= 350$ lb/ft.

Now $\theta = \sin^{-1}\left(\dfrac{W}{2R}\right) = \sin^{-1}\left(\dfrac{39·25}{55}\right) = 45° 34'$

whence half arc $= R\theta = 27·5 \times 0·795 = 21·8$ ft.

Rise $y = R(1 - \cos\theta) = 27·5(1 - 0·700) = 8·25$ ft.

Load due to half shell $= 52 \times 21·8 = 1135$ lb/ft.
Load from valley thickening $= 175$ lb/ft.
$\overline{}$
1310 lb/ft.

Then total longitudinal tension T_1 is

$$\frac{wL^2}{8 \times 0·72 \times y} = \frac{1310 \times 92·5^2}{8 \times 0·72 \times 8·25} = 237{,}000 \text{ lb.,}$$

and with steel stressed to 30,000 lb/sq. in., we have

$$A_{\text{st}} = \frac{237,000}{30,000} = 7{\cdot}9 \text{ sq. in.}$$

Total shear force is

$$\Sigma Q = \frac{wL}{1{\cdot}07\theta} = \frac{1310 \times 92{\cdot}5}{1{\cdot}07 \times 0{\cdot}795} = 142,000 \text{ lb.}$$

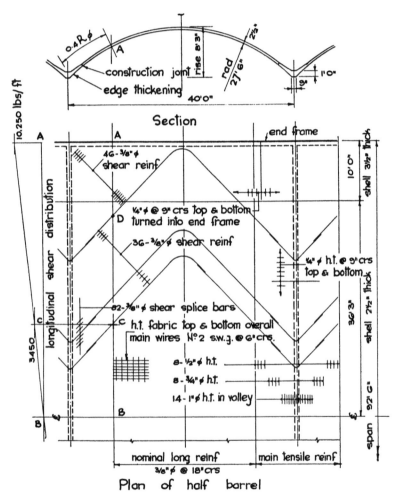

FIG. 11.4. FEATHER-EDGED BARREL

and with steel at 45° stressed to 20,000 lb/sq. in., we have

$$A_{st} = \frac{142,000}{20,000 \times \sqrt{2}} = 5 \cdot 0 \text{ sq. in.}$$

which is provided by 46 bars of $\frac{3}{8}$ in. diameter.

Now the maximum shear force per foot is

$$Q_{max} = 2 \times \frac{\text{total tension}}{L/2}$$

and occurs at the stiffening beam, approximately $0 \cdot 4 R\phi$ up from the springing. This force decreases to zero at mid-span. In our case, we have at A, Fig. 11.4,

$$Q_{max} = 2 \times \frac{237,000}{46 \cdot 25} = 10,250 \text{ lb/ft.}$$

and at B, $\qquad Q = \text{zero.}$

The concrete, unreinforced, will safely carry per foot

$$12 \text{ in.} \times 2\frac{1}{2} \text{ in.} \times 115 \text{ lb/sq. in.} = 3450 \text{ lb.,}$$

so that, from the shear diagram, we see that shear reinforcement is required only between C and A – some 30 ft. Over the length AC the total shear is

$$Q = \left(\frac{10,250 + 3450}{2} \right) \times 30 = 210,000 \text{ lb.,}$$

requiring mild steel reinforcement of

$$A_{st} = \frac{210,000}{20,000 \times \sqrt{2}} = 7 \cdot 4 \text{ sq. in.}$$

Referring again to the figure, it can be seen that $32 - \frac{3}{8}$ in. dia. bars (area $3 \cdot 50$ sq. in.) have already been provided over the length AD as part of the reinforcement for end shear. Therefore additional reinforcement is required over the length DC of

$$7 \cdot 4 - 3 \cdot 5 = 3 \cdot 9 \text{ sq. in.}$$

which is provided by $36 - \frac{3}{8}$ in. dia. bars (area $4 \cdot 0$ sq. in.).

Bennett's curve for M_2 shows that

$$M_2 = \frac{w^2 L}{33,300 \times y} = \frac{1310^2 \times 92 \cdot 5}{33,300 \times 8 \cdot 25} = 580 \text{ ft. lb/ft.}$$

which is satisfactorily carried if we provide a fabric with 2-gauge wires at 6 in. centres arranged in the top and bottom of the shell.

It is interesting to consider the form of the M_2 bending moments. These are shown typically in Fig. 11.5. With four contraflexure points, the shell acts in bending approximately as though it were on free-spans of $\dfrac{W}{4}$: and if the shell were designed for M_2 on this basis, any slight overstressing that might ensue would only lead to

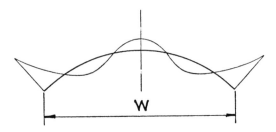

FIG. 11.5. TRANSVERSE BENDING IN BARREL SHELL

plastic adjustment of the bending distribution so that no danger would result. On this basis, then,

$$M_2 = \frac{w}{8} \cdot \left(\frac{W}{4}\right)^2 = \frac{wW^2}{128},$$

giving in our case

$$M_2 = \frac{52 \times 39 \cdot 25^2}{128} = 625 \text{ ft. lb/ft.}$$

which is not so greatly at variance with the value of 580 ft. lb/ft. based on Bennett's method.

ART. 11.3. HYPERBOLIC PARABOLOIDS

Hyperbolic paraboloid shells are curved in two directions in such a manner that the surfaces cannot be developed by simple rotation about one axis. Consider in Fig. 11.6 the plane rectangle OXBY. Let B now be depressed to B'. The warped surface OXB'Y will then be a hyperbolic paraboloid. If each pair of opposite sides is now divided into an equal number of parts, and the corresponding divisions joined by straight lines, it will be seen – from the intersections of the generators – that between O and B', the surface is parabolic – so as to be well suited to resist compressive forces, much in the manner of an arch; and between X and Y the surface, while still parabolic, is inverted – and well suited to take tensile forces, much in the same manner as a catenary. Clearly such a shape has great stiffness and resistance to buckling; and except for secondary bending effects, normal applied loads are carried by direct forces within the thickness of the shell, so that, provided the stresses in the shell are not large, analysis can be made in accordance with the membrane theory. The shears within the shell gather along the four edges, bringing the applied loads to the positions where support is provided.

Apart from the simplicity of analysis, the hyperbolic paraboloid

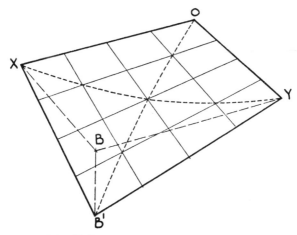

FIG. 11.6. HYPERBOLIC PARABOLOID

has practical merits both in regard to its aesthetic appeal, and also in the fact that – being generated by two systems of straight lines – the whole of the formwork is provided by straight boards, warped only slightly over their length.

Refer now to Fig. 11.7a, where $OX = a$, $OY = b$ and $BB' = h$. In relation to the cartesian axes, x, y and z, the warped surface $OXB'Y$ is then represented by the equation

$$z = \frac{h}{ab} \cdot xy. \tag{11.8}$$

Figure 11.7c shows an element of this surface together with the membrane forces which arise when the element is subjected to load (vertical only) of intensity Z per unit surface area. These forces are most easily determined by projecting the element on to the xy plane as shown in Fig. 11.7b: here the forces indicated are the horizontal components of the inclined membrane forces, together with the change of force occurring over the element: vertical components also occur, but these are omitted from the

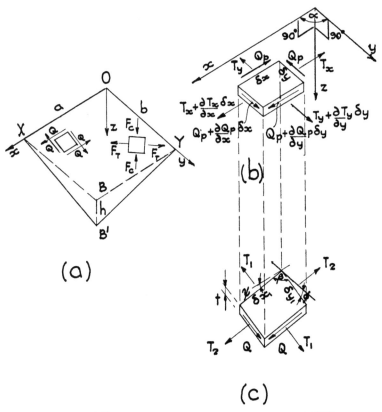

FIG. 11.7. MEMBRANE STRESSES IN HYPERBOLIC PARABOLOID

figure for clarity. For equilibrium, the forces in each direction must balance; and it is easy to demonstrate that:

$$\frac{\partial T_y}{\partial y} + \frac{\partial Q_p}{\partial x} = 0, \tag{11.9}$$

$$\frac{\partial T_x}{\partial x} + \frac{\partial Q_p}{\partial y} = 0, \tag{11.10}$$

and
$$T_x \cdot \frac{\partial^2 z}{\partial x^2} + T_y \cdot \frac{\partial^2 z}{\partial y^2} + 2Q_p \cdot \frac{\partial^2 z}{\partial y \cdot \partial x} = -Z.$$

This latter may be written as

$$\frac{\partial^2 F}{\partial y^2} \cdot \frac{\partial^2 z}{\partial x^2} + \frac{\partial^2 F}{\partial x^2} \cdot \frac{\partial^2 z}{\partial y^2} - 2\frac{\partial^2 F}{\partial x . \partial y} \cdot \frac{\partial^2 z}{\partial x . \partial y} = -Z \qquad (11.11)$$

where the function F is defined by

$$\frac{\partial^2 F}{\partial y^2} = T_x, \quad \frac{\partial^2 F}{\partial x^2} = T_y, \quad \text{and} \quad -\frac{\partial^2 F}{\partial x . \partial y} = Q_p.$$

In the simple case where the load w per unit plan area of shell is uniform, equation 11.11 can be solved without difficulty, since by use of equation 11.8 we have

$$-2\frac{\partial^2 F}{\partial x . \partial y} \cdot \frac{h}{ab} = -w.$$

That is to say

$$Q_p = -\frac{ab}{2h} \cdot w. \qquad (11.12)$$

Also, from equations 11.9, 11.10 and 11.12 it is clear that

$$T_x = T_y = 0$$

and thus

$$T_1 = T_2 = 0.$$

Further, since $\quad Q . \delta y_1 . \cos \psi = Q_p . \delta y,$

we have $\quad\quad\quad\quad Q = Q_p$

whence we have the shear stress

$$q = -\frac{ab}{2ht} \cdot w. \qquad (11.13)$$

Thus the element is in a state of stress where shear forces only act along its edges, parallel to the generatrix (or boundaries of the shell) and no normal forces exist. The principal stresses are therefore of the same magnitude as the shear stress, and occur along lines at 45 degrees to the shell boundaries; the compressions are parallel to OB', and the tensions are parallel to XY. The shears which collect along OX and OY are resisted by these edges acting in tension, and the shears along XB' and YB' are resisted by these edges acting in compression.

Hyperbolic paraboloids with other forms of loading are more difficult to solve: but for many cases – where h is small in relation to a and b – the assumption that the loading is uniform in plan is near enough true for practical purposes.

HYPERBOLIC PARABOLOID UMBRELLA

As a practical example of the design of a hyperbolic paraboloid, consider the umbrella-type shell shown in Fig. 11.8. This comprises four sections, each 24 ft. square, and with one corner depressed

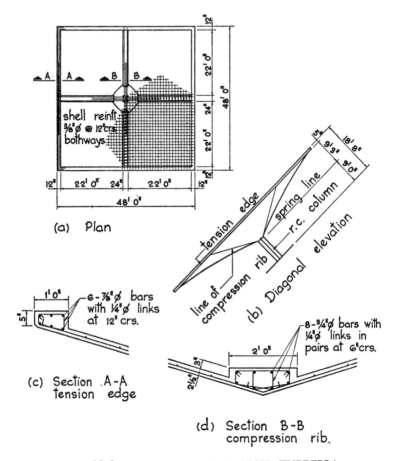

(a) Plan

(b) Diagonal elevation

(c) Section A-A
tension edge

(d) Section B-B
compression rib.

FIG. 11.8. HYPERBOLIC PARABOLOID UMBRELLA

9 ft. 3 in.; and the sections are cast together to form a single unit 48 ft. square, with one depression centred over the single supporting column 24 in. square. The shell is $2\frac{1}{2}$ in. thick, and provided with edge-thickenings along the perimeter to take the tension forces, and four radial-thickenings are provided down the lines of

maximum slope to take the compression forces. The total general loading is as follows:

2½ in. shell on slope	32 lb/sq. ft.
Thickenings	9
Insulation and felt	12
Superimposed load	25
Total:	78 lb/sq. ft. on plan.

Using equation 11.13, we have the shear stress

$$q = \frac{ab}{2ht} \cdot w$$

$$= \frac{24 \text{ ft.} \times 24 \text{ ft.}}{2 \times 111 \text{ in.} \times 2\frac{1}{2} \text{ in.}} \times 78 \text{ lb/sq. ft.}$$

$$= 82 \text{ lb/sq. in.}$$

This stress is also numerically equal to the principal tensile and compressive stresses (due to the absence of normal stresses), and therefore theoretically no reinforcement is required. However, to cope with shrinkage and thermal movements a mesh of ⅜ in. bars at 12 in. centres both ways is provided in the middle of the shell thickness.

At the tension edges, the tension force varies from zero at the umbrella corners to a maximum at the centre-points of each side – at X and Y, Fig. 11.7a – where we have

$$F_{TE} = q \cdot t \cdot a$$
$$= 82 \times 2\frac{1}{2} \times (24 \times 12) \qquad = 59,200 \text{ lb.}$$

In this particular design, allowance had to be made also for edge loads of 100 lb/ft. run all round the shell, and these increased the tension forces by $\qquad$ 9,400 lb.

Therefore total $F_{TE} = $ 68,600 lb.

Area of steel required is therefore

$$\frac{68,600}{20,000} = 3\cdot43 \text{ sq. in.}$$

and this is provided by six ⅞ in. dia. bars (area 3·6 sq. in.).

The weakest part of the umbrella shell is where it is supported on the single column. For this reason an infill thickening-head was provided 8 ft. square, which allows the column bars to be properly anchored so that the necessary continuity can be developed, and assists the compression ribs in taking the cantilever moments which undoubtedly arise in practice. The maximum compressive force in each rib (due to shears from two adjacent quarters of the umbrella) are

$$F_{CR} = 2\,q\,.\,t\,.\,b\,.\sec\gamma$$
$$= 2 \times 82 \times 2\tfrac{1}{2} \times (24 \times 12) \times 1\cdot08 \qquad = 126{,}000 \text{ lb.}$$

To this had to be added the effect of the edge loads
of 100 lb/ft. run $\doteq$ 13,300 lb.

Therefore total $F_{CR} = 139{,}300 \text{ lb.}$

The compression rib near the centre of the umbrella is 24 in. wide by 8 in. deep, and therefore adequately reinforced with eight $\tfrac{3}{4}$ in. dia. bars ($r_p = 1\cdot8$). The rib is restrained from buckling by the shape of the shell with which it is monolithic.

Shear stresses due to vertical loading at the column and at the edge of the thickening-head are calculated in the normal way, and combined with the shell and rib compressions to determine the principal tensile stresses.

The effect on the umbrella shell of the 100 lb/ft. run edge-loads is to modify the membrane stresses and produce bending adjacent to the edges. The only methods of solution of this, available at present, are approximate and unwieldy, and therefore not included here.

While the membrane stresses of such a hyperbolic paraboloid umbrella shell are readily calculable as shown above, there is no practical theory whereby the deflections can be calculated. The matter of deflection of these particular shells was of major significance to the authors in considering a large factory-scheme where the imaginative design of the architects involved the use of 29 such roof shells, each free-standing and with light flexible roofing units between. The requirements of this project prevented the shells from butting up tight against one another, yet the possibility of unbalanced loading was very real (both from edge loads and from snow): and knowledge of the likely overall deflections and rotations was important in order that suitable flexing mechanisms and flashings could be provided along the external lines of curtain-walling, and between the shells themselves. Accordingly by kind co-operation of the Cement and Concrete Association, a model to one-sixth full-size was constructed and tested at their laboratory at Wexham Springs. It is felt that the nature of the test results may be of some general interest, and these are given below, after making the necessary conversions so that the values relate to the full-size shell.

With the uniform load of 25 lb/sq. ft. over the whole area of the shell, and the uniform line-load of 100 lb/ft. run for the full extent of each edge beam, the deflections of the edges were 0·25 in. at the corners and 0·16 in. at the centre-points. These deflections were sufficiently small to be of no practical significance. However, when 15 lb/sq. ft. of the uniform load and 65 lb/ft. run of the edge load were *taken off* one side of the shell, there was a general tilt so that one side rose 2·4 in. while the opposite side fell 2·4 in. About one-third of this was due to the edge loads, and two-thirds due to the snow loads. Of the total tilt it was estimated that about half was due to flexure of the column, and half due to distortion of the hyperbolic paraboloid shell.

ART. 11.4. DOMES

The first shell-concrete roofs were domes – constructed mainly in Germany – and used for covering-in large public spaces. The Market Hall at Leipzig (1929) is a typical example. Domes are still used for roofing large arenas and exhibition buildings, and for covering circular tanks and reservoirs. Formwork for domes is relatively expensive, but the thickness and weight of the concrete and reinforcement can be kept small, even where the spans are very large. The lightness of thin domes can make for inexpensive supporting structures and foundations, especially where arrangements are made in the dome-springing to absorb the lateral thrusts.

Domes are most commonly of spherical form, as shown in Fig. 11.9. Consider the equilibrium of the element, radius R, situated at an angle ϕ from the vertical axis, and subtending an angle $\delta\phi$ at the centre. The horizontal radius of the element about the vertical axis is r_0, and the horizontal angle is $\delta\theta$. Under the action of a vertical external load of uniform intensity w per unit area, it is clear from symmetry of the structure and the loading that there are no shear forces, and that the forces on the *meridian* sides (the sides lying on great circles through the crown) must be equal. The change of force between *parallel* sides (the sides cut by horizontal planes) is shown in the figure.

The lengths of the sides are as follows:

$$\mathrm{DA} = r_0 \cdot \delta\theta; \quad \mathrm{CB} = \left(r_0 + \frac{\partial r_0}{\partial \phi} \cdot \delta\phi\right)\delta\theta; \quad \mathrm{CD} = \mathrm{AB} = R \cdot \delta\phi.$$

Resolving forces radially, we have

$$2T_1R . \delta\phi . \sin\frac{\delta\theta}{2} . \sin\phi + T_2r_0 . \delta\theta . \sin\frac{\delta\phi}{2} +$$

$$+ \left(T_2 + \frac{\partial T_2}{\partial\phi} . \delta\phi\right)\left(r_0 + \frac{\partial r_0}{\partial\phi} . \delta\phi\right)\delta\theta . \sin\frac{\delta\phi}{2} +$$

$$+ w . \cos\phi . r_0R . \delta\theta . \delta\phi = 0.$$

For small angles $\sin\dfrac{\delta\theta}{2} = \dfrac{\delta\theta}{2}$, so that neglecting small quantities and putting $r_0 = R\sin\phi$, we have

$$T_1 + T_2 = - wR\cos\phi. \tag{11.14}$$

Also, it is clear that the total load W (above any level) must

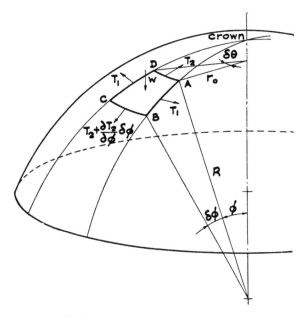

FIG. 11.9. MEMBRANE STRESSES IN DOME

equal the vertical component of the total force T_2 around the periphery. Thus, we have

$$2\pi r_0 T_2 \sin\phi = - W,$$

and, since $r_0 = R \sin \phi$

and $W = 2\pi w \displaystyle\int_0^\phi r_0 R \cdot \delta\phi = 2\pi w R^2 (1 - \cos \phi),$

then we have $T_2 = - \dfrac{wR}{1 + \cos \phi}.$ (11.15)

Substituting now for T_2, equation 11.14 yields

$$T_1 = - wR \left[\cos \phi - \frac{1}{1 + \cos \phi} \right].$$ (11.16)

Clearly, under conditions of uniform loading, the meridian force T_2 is compressive for all values of ϕ from zero to 90 degrees. Further examination shows that the ring force T_1 is compressive in all areas of the dome where ϕ is between zero and 51° 48′, but for values of ϕ greater than 51° 48′ the T_1 force becomes tensile.

The lateral thrusts from the dome can be absorbed in a suitable ring beam provided at the base. This beam can also serve in carrying vertical loads in domes which are supported only at a number of isolated points. The total ring force in the beam due to T_2 is $T_2 \cos \phi \cdot r_0$, and substituting for r_0 and T_2 from above, we have the ring-beam force as

$$wR^2 \frac{\sin \phi \cos \phi}{1 + \cos \phi}.$$

For the proportions adopted for many domes, the T_1 forces at the base are compressive tending to *decrease* the diameter, whereas the T_2 forces at the base are tensile tending to *increase* the diameter. This matter may be resolved by prestressing the ring-beam: though the amount of prestress force required will only achieve balance for one intensity of loading. All other loading intensities will produce effects, local to the edge, that introduce normal shears and bending moments, and at the same time, modify the direct forces determined by the membrane theory – all very much as described in Article 11.2 for Barrels. The problem is then most easily attacked by the use of Geckeler's[8] approximate theory.

REFERENCES

1. FINSTERWALDER, U. Die Querversteiften Zylindrischen Schalenge-wölbe mit Kreissegmentförmigem Querschnitt. *Ingenieur Archiv.* Vol. 4. (1933).

2. JENKINS, R. S. *Theory and Design of Cylindrical Shell Structures.* London, (1947).

3. SCHORER, H. Line Load Action on Thin Cylindrical Shells. *American Society of Civil Engineers – Transactions,* Vol. 101. (1936).

4. DESIGN OF CYLINDRICAL CONCRETE SHELL ROOFS. *American Society of Civil Engineers – Manual of Engineering Practice*, No. 31. (1952).
5. RÜDIGER and URBAN. Kreiszylinderschalen. B. G. Teubner, *Verlagsgesellschaft*. Leipzig, (1955).
6. BENNETT, J. D. *Some Recent Developments in the Design of Reinforced Concrete Shell Roofs*. The Reinforced Concrete Association, London. (November, 1958).
7. TOTTENHAM, H. A Simplified Method of Design for Cylindrical Shell Roofs. *The Structural Engineer*. (June, 1954).
8. GECKELER, J. W. Über die Festigkeit achsensymmetrischer Schalen. *ForschArb. IngWes.*, No. 276. (1926).
9. CANDELA, F. Structural Applications of Hyperbolic·Paraboloidical Shells. *Journal of the American Concrete Institute*. Vol. 26, No. 5. (January, 1955).
10. PARME, A. L. Hyperbolic Paraboloids and other Shells of Double Curvature. *Proceedings of the American Society of Civil Engineers*. Vol. 82. Paper 1057. (September, 1956).

CHAPTER 12

TANKS, WATER TOWERS
AND RESERVOIRS

ART. 12.1. INTRODUCTION

REINFORCED concrete is widely used for water-retaining structures of all kinds, such as tanks, water-towers and reservoirs. In Article 2.4 it is explained that the impermeability of concrete is not absolute: but provided the aggregates are well chosen and the water/cement ratio is successfully controlled, and provided the concrete is mechanically vibrated and subsequently cured in the best manner, concrete can be made near enough impermeable to water for most industrial purposes without the need for a separate waterproof lining. This is not true of liquids of low viscosity and high penetrating power such as petrol; nor is it true of non-aqueous liquids, which will allow the concrete to dry out completely, causing shrinkage cracks. Other liquids, such as certain vegetable oils and sugar solutions, and certain natural waters containing sulphates or other acids, may attack the concrete or corrode the steel, and in such cases it may be necessary to increase the cement content of the mix, or use special cements, or provide some form of lining to protect the concrete. This is discussed in Article 2.2.

Where an impermeable lining is provided, the concrete will in many cases dry out completely, resulting in greater shrinkage movement. This requires careful attention, since the movement at contraction joints may then become sufficient to cause the lining to crack. Linings are liable to require more maintenance than properly constructed unlined structures, and for this reason should be avoided wherever good design will allow.

Chapter 15 describes the application of prestressed concrete techniques to water-retaining structures. The present chapter deals only with structures of normal reinforced concrete.

ART. 12.2. DESIGN CONSIDERATIONS

Movements due to temperature changes, creep and shrinkage of concrete have already been described at length in Chapter 2. These phenomena deserve specially close attention in the design of any water-retaining structure. Cracks and leaks in unsuccessful structures are as frequently attributable to lack of appreciation of shrinkage and thermal movements as they are to stresses resulting from the water pressure. Indeed shrinkage and thermal movements, if restrained, can produce tensile stresses sufficient to cause cracking of the structure even before the water load is applied. Due to the fact that concrete shrinks most when it is allowed to dry out completely, a properly designed tank or reservoir which is kept full of water is less likely to crack than one which is allowed to stand empty. For this reason it is desirable, where circumstances permit, to flood the floors of tanks or reservoirs at an early stage. This not only reduces the amount of shrinkage, but also assists the curing with consequent improvement of the impermeability of the concrete. Slow filling will give reversible shrinkage more time in which to act, so that less stretch will be required of the concrete. Slow filling also allows the concrete more opportunity to relieve itself by creep, thus reducing the tendency of the concrete to crack in tension (direct or bending) due to the water load. In all major water-retaining structures, the specification should stipulate the maximum rate for filling.

Shrinkage and thermal effects can be controlled by the provision of suitable joints for contraction and expansion, and by providing free-sliding surfaces to minimise the restraining friction-forces on the members between these joints. By judicious spacing of the joints, intermediate cracking due to shrinkage and thermal effects can be entirely prevented. The joints themselves require careful consideration in regard to shape and form, as well as the method by which they are sealed to prevent leakage. This is dealt with later (Articles 12.4 and 12.5).

Design calculations for water-retaining structures, to determine the concrete and reinforcement dimensions, are made in two parts. First, to guard against cracking, calculation is made on the elastic behaviour of the full uncracked section, allowance being made for the tensile strength of the concrete. In this calculation, the margin of safety against cracking is considerably less than the factor of safety provided against normal structural failure: and this is reasonable because the formation of cracks, while liable to cause considerable inconvenience and expense, can normally be dealt with by repair-works, whereas the effect of a major structural failure is likely to be more severe. The second calculation is then made to guard against structural failure even if the concrete does crack. This follows the Elastic Theory methods as set out in Chapter 3 (ignoring the tensile strength of the concrete) but working to slightly reduced steel stresses in order to limit the widths of any cracks that may develop. Load-Factor methods of design are not appropriate.

Steels with high yield-points have no advantage over ordinary mild steel in water-retaining structures, since it is essentially the *stretch* that has to be limited. At allowable stresses, the stretch of high-strength steels is approximately the same as that of ordinary mild steel. Nor is much benefit derived from the use of deformed bars, since their special mechanical bond does not become effective until after there has been sufficient movement to allow small cracks to form in the concrete: though some people hold the view that deformed bars give a better distribution of cracking – each crack being limited to a more acceptable width. Saturated concrete shrinks less than dry concrete, and since the bond of the concrete to the reinforcements depends largely on the grip due to shrinkage, it is advisable in water-retaining structures to work to modest bond-stresses. Generous bond-lengths should be provided wherever possible with substantial hooked-ends.

ART. 12.3. BASIS OF DESIGN

In Great Britain, water-retaining structures are designed in accordance with *British Standard Code of Practice*, C.P. 2007 (1960): Design and Construction of Reinforced and Prestressed Concrete Structures for the Storage of Water and other Aqueous Liquids[1]. The complete failure or collapse of any large water-retaining structure could have far-reaching consequences, not only on the owner of the structure, but also on the nearby population and adjoining properties; particularly so when one considers that service-reservoirs are deliberately situated on the highest ground immediately overlooking the towns they supply, and for this reason the Reservoirs (Safety Provisions) Act 1930 requires that large reservoirs are to be designed and supervised only by civil engineers appointed to a panel constituted for the purpose by the Secretary of State.

PREVENTION OF PERCOLATION

Percolation is guarded against by using well chosen aggregates, mechanical vibration, proper curing, and correct water/cement ratio. High water/cement ratios make for porosity and accentuated shrinkage, while low water/cement ratios make for hungry areas and honeycombing, which are worse. The difficulty is best overcome by using a mix sufficiently rich in cement to give good workability while yet keeping the water/cement ratio low. This normally leads to the use of $1:1\cdot6:3\cdot2$ nominal mix except in sections of thickness greater than 18 in., where C.P. 2007 (1960) allows the use of 1:2:4 nominal mix. Alternatively the concrete mix can be designed as described in Article 1.8 – Special concrete mixes. For the nominal mixes, C.P. 2007 (1960) requires works cube-strengths the same as in C.P. 114 (1957) (*see* Table 1.4); but preliminary cube-strengths are required about 11 per cent higher than the values given in Table 1.4. Whatever mix of concrete is used, the grading of the aggregates should be controlled by obtaining the coarse aggregate in two sizes for $\frac{3}{4}$ in. nominal aggregate, and in three sizes for $1\frac{1}{2}$ in. nominal aggregate.

Heavy trowelling of the water-retaining face of the concrete (with a steel trowel) when the concrete is about four hours old assists in giving a dense skin free from surface pores. For this reason it is advantageous to strike vertical formwork to walls about four hours after the concrete has been placed.

PREVENTION OF CRACKING DUE TO SHRINKAGE AND THERMAL
    MOVEMENTS

Shrinkage cracks in slabs are prevented to some extent by the inclusion of a minimum percentage of reinforcement in each of two directions at right angles. On the gross cross-section, C.P. 2007 (1960) recommends this minimum percentage should be 0·3 for plain bars, and 0·25 for deformed bars. To further ensure freedom from shrinkage-cracking, slabs should be concreted in lengths not exceeding 25 ft. for slabs reinforced as described above, and less for slabs with smaller percentages of reinforcement. Generally contraction-joints (as described in Article 12.4) should be provided between adjacent slab-lengths; but at intervals of about 100 ft. the joint should be of expansion-type to accommodate additional thermal movements.

Joints in contiguous walls, floors or roofs of any structure should be arranged to coincide. For example, in a reservoir where the roof-slab is supported on the wall, the movements at each vertical joint in the wall will induce similar movements in the roof, even though a free-sliding joint is provided between the roof-slab and the wall. Sympathetic movements in the roof would then be liable to cause cracking of the roof-slab. This is avoided by providing similar joints in the roof-slab and wall in alignment with one another.

In order that slab units between joints may be as free as possible to take up shrinkage and thermal strains, free-sliding surfaces should be provided. This is particularly important at ground level, where structures are otherwise likely to become continuously keyed to the ground. For floors, the ground should first be covered with a plain concrete screed or under-layer slab at least 3 in. thick, finished to a plane smooth face by polishing with a steel trowel: this is then covered with building paper before the upper-layer slab is laid. For walls, the sides and bottoms of excavations for the foundations should be lined with a 3 in. thickness of concrete cast up to fair-faced shutters set parallel and truly to line, and the concrete face subsequently rubbed smooth with carborundum and painted with bitumen: this then allows the foundations to move without restraint in the direction of their length.

In tall structures such as water-towers there will of course be considerably less restraint; and in protected works such as deep-covered reservoirs, there may be less case for providing expansion joints closer than about 150 ft. intervals.

RESISTANCE TO CRACKING UNDER LOAD

Precautions must be taken in the design to prevent the formation of normal hair-cracks when the structure is filled with water. While such hair-cracks are of no structural consequence, they would of course allow leakage from a water-retaining structure.

TABLE 12.1. *Permissible stresses for Portland cement and Portland Blastfurnace cement concrete in calculations relating to the resistance to cracking*

Nominal mix	Permissible concrete stresses		
	Tension		Shear
	Direct	Due to bending	
	lb/sq. in.	lb/sq. in.	lb/sq. in.
1:1·6:3·2	190	270	280
1:2:4	175	245	250

Calculations are made on the full concrete section, allowing for the tensile strength of the concrete. Allowance is made in the calculations for the effect of steel reinforcements, but under these circumstances the stress in the steel will be low – determined by the concrete stress times the modular ratio (15) – and, in cases of bending, dependent on the position of the steel in relation to the neutral axis. C.P. 2007 (1960) recommends that for nominal mixes the concrete stresses in calculations relating to the resistance to cracking should not exceed the values given in Table 12.1.

Consider now the application of this to the case of a wall in direct tension. With the more usual $1:1\cdot6:3\cdot2$ nominal mix, we have:

Permissible concrete stress $= p_{ct} = 190$ lb/sq. in.
Permissible steel stress $\quad\ = p_{st} = 190 \times 15 = 2850$ lb/sq. in.

In a circular tank 48 ft. diameter, the ring tension at 14 ft. depth $= 62\cdot4 \times 14 \times \dfrac{48}{2} = 21,000$ lb. This is provided for by a wall 8 in. thick with four $\frac{3}{4}$ in. dia. circumferential rods per ft. height of wall. Then:

Tensile resistance due to concrete $= (96 - 1\cdot76)190 = 17,900$ lb.
Tensile resistance due to steel $\quad = 1\cdot76 \times 2850 \quad = \quad 5,000$

$$= 22,900 \text{ lb.}$$

No economy could be gained by increasing this thickness of wall, because the steel area provided is a minimum from strength requirements, since when the tensile strength of the concrete is ignored, we have $f_{st} = \dfrac{21,000}{1\cdot76} = 12,000$ lb/sq. in. (*see* Table 12.2).

On the other hand, if the wall thickness were reduced by only 2 in. the reinforcement would have to be increased to four 1 in. dia. rods per ft. $(r_p = 4\cdot4$ per cent), giving:

Tensile resistance due to concrete $= (72 - 3\cdot14)190 = 13,100$ lb.
Tensile resistance due to steel $\quad = 3\cdot14 \times 2850 \quad = \quad 9,000$

$$= 22,100 \text{ lb.}$$

This saving of 33 per cent concrete has been made at the expense of 80 per cent more steel: no saving in shuttering has been achieved, but the difficulty of placing the concrete and achieving absence of percolation has been increased. Thus it is seen that resistance to cracking determines the economical thickness of concrete within narrow limits.

Consider now the matter of bending. With the permissible stress in bending limited to 270 lb/sq. in.,

since $\qquad\qquad\qquad M_r = fz,$

we have $\qquad\qquad M_r = 270 \cdot \dfrac{bd^2}{6} = 45\,bd^2.$

For this reason, slabs in bending in water-retaining structures have to be considerably thicker than otherwise necessary from strength requirements. The addition of a high percentage of steel gives only a small increase in the section resistance: so that here also the resistance to cracking determines within narrow limits the thickness of concrete required.

In making calculations of resistance to cracking, it is necessary for the structural analysis to be fairly complete and accurate. For

TABLE 12.2. *Permissible tensile stresses in steel rein-forcements for strength calculations relating to water-retaining structures*

Designation of stress			Permissible stress
Members in direct tension			lb/sq. in. 12,000
Members in bending	On liquid-retaining face		12,000
	On face remote from liquid	Members less than 9 in. thick	12,000
		Members 9 in. thick or thicker	18,000*
In shear reinforcement			12,000

* Where deformed bars are used this stress may be increased to 20,000 lb/sq. in.

example, suppose in a rectangular tank there are ample vertical reinforcements in the walls for vertical spanning between the floor and roof. It is certain then that the walls are safe against structural failure. Nevertheless at the vertical corners of the tank there will be horizontal cantilever actions due to deflections of the two adjoining sides. The amounts of the cantilever moments will depend on the proportions of the tank and the thicknesses of the walls. If these cantilever moments are ignored (on the grounds that they cannot influence the structural safety of the tank) vertical cracks are likely to appear in the walls near the corners and these may cause leakage. For reasons such as this, a properly developed understanding of the elastic functioning of monolithic structures is essential for the successful design of water-retaining structures.

STRENGTH CALCULATIONS

The strength of water-retaining structures is determined in accordance with the Elastic Theory as given in Chapter 3. The same permissible concrete stresses are taken as in C.P. 114 (1957) (*see* Table 1.3), but to avoid tension cracks arising from shearing actions, the stress given by $\dfrac{Q}{bl_a}$ is never to exceed the values given in Table 12.1, whatever the reinforcement provided. In order to limit the widths of hair-cracks, C.P. 2007 (1960) recommends the adoption of lower steel stresses than C.P. 114 (1957) – as given in Table 12.2.

The maximum bending-moments to be catered for in water-retaining structures may be less than the values given in Figs. 4.5 and 4.6 since the water load may extend over all spans.

PERMANENCE

Apart from earlier references to the need for dense well-compacted concrete, the life of water-retaining structures is particularly influenced by the adequacy of concrete cover to protect the reinforcing steel. The minimum cover should normally be $1\frac{1}{2}$ in. or the diameter of the bar, whichever is greater, but in the presence of sea-water or waters of specially corrosive nature the minimum cover should be 2 in.

ART. 12.4. TYPES OF JOINTS

Individual engineers differ in their ideas and reasons as to how joints should best be formed. This appears to be the outcome of varied experience. The proposals set out in the present article have been used by the authors over many years with complete success.

RIGID JOINTS

In structures of modest dimensions, or where advantages are obtained from continuous anatomy (as by tying opposite walls together through the floor or roof), or where there are scouring actions from mechanical stirrers or rotating paddles, practical advantages result from the use of rigid joints throughout the structure. A typical detail of such a joint, having the merit of simplicity, is shown in Fig. 12.1, where a continuous rebate is provided with a galvanised steel water-bar cast into the work to obstruct the passage of water trying to seep across the line of the joint. In rigid joints such as this it is important to obtain a good bond between the second- and first-cast concrete. To achieve this, the whole face of the joint is completely removed for a depth of about one-eighth of an inch, by bush hammering or jetting, and a layer of $1\frac{1}{2}$:1 sand cement mortar spread on the first-cast face immediately before the new concrete is placed.

Rigid joints are not satisfactory in larger structures such as reservoirs, where the effects of shrinkage and temperature-change

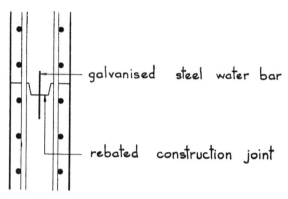

FIG. 12.1. RIGID JOINT – MAINLY FOR SMALL
STRUCTURES

become more significant. It then becomes necessary to provide flexible joints, descriptions of which follow. Indeed, the provision of rigid joints in certain small structures may lead to conditions of high stress, often indeterminate or difficult to analyse, and these could be avoided by the use of flexible joints in appropriate positions.

FLEXIBLE JOINTS

Because flexible joints are intended to allow free movement, no roughening should be made of joint faces, and these should be finished true and even. All second-cast concrete at flexible joints should however be worked into a layer of $1\frac{1}{2}$:1 sand cement mortar spread on to the first-cast face to ensure an adequacy of fine material, and absence of hungry patches.

(i) *Walls*

Figure 12.2a shows how a *horizontal construction joint* would be formed in the wall of a flexible-jointed structure. Horizontal joints in walls of normal height are not provided for the purpose of taking up shrinkage or thermal movements, but are required for convenience in setting the formwork and giving proper access for compaction and supervision of the concreting. Indeed, weaknesses

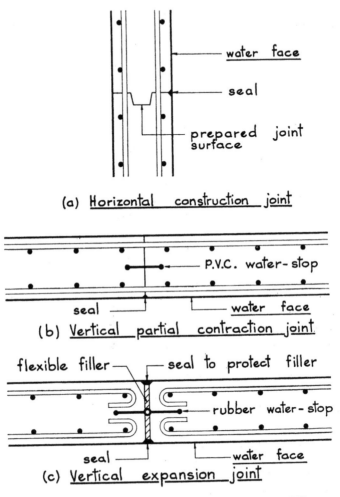

(a) Horizontal construction joint

(b) Vertical partial contraction joint

(c) Vertical expansion joint

FIG. 12.2. WALL JOINTS IN LARGE STRUCTURES

due to shrinkage-movement are caused by these horizontal joints, the lower height of concrete having shrunk before the subsequent height is placed; and for this reason the quicker the wall is built its full height, and the less the number of horizontal joints, the better.

Horizontal joints like this are not in fact required to flex, and indeed may frequently occur where flexing is inadmissible: consequently the face of the joint in this special case should be prepared by removal of all laitence; and care should be taken to keep the rebate parallel-sided and truly in line with the wall, so that the second-placed concrete is resisted as little as possible from shrinking along the line of the first-placed concrete. Water-stops in these horizontal joints are not necessary, but a triangular-fillet seal should be made on the water-face.

Figure 12.2b shows a *vertical partial-contraction joint* in a wall. Joints of this type are spaced at intervals of about 25 ft. and in positions of no shear. The face of the first-placed concrete is left smooth and unkeyed, to allow the joint to open as the concrete on either side shrinks or cools. In this way random cracking is avoided, and all movements are collected at pre-determined positions. With the small amount of movement that occurs, a poly-vinyl-chloride (PVC) water-stop is satisfactory. A triangular-fillet seal is necessary on the water-face. Figure 12.2c shows a *vertical expansion joint* in a wall. Joints of this type are spaced at intervals of about 100 ft. and in positions of no shear. A flexible filler (*see* Article 12.5) about ¾ in. thick is incorporated in the joint, to allow the joint to partially close when the concrete on either side expands. As the movement at expansion joints is considerably greater than at contraction joints, a rubber water-stop is required with a central open-core, capable of accepting far greater movement. The joint needs to be sealed at both faces of the wall, the seal on the water face serving to achieve watertightness, and the seal on the outer face acting to protect the flexible filler and rubber water-stop from deterioration.

(ii) *Floors cast direct on the ground*

Floors bearing directly on the ground are frequently cast in two separate layers, the upper-layer slab being independent of the under-layer slab and free to slide on it. The joints in the upper- and under-layer slabs are staggered one to the other. Typical joint arrangements are shown in Fig. 12.3. The joints in the upper-layer slab are normally *complete-contraction joints* (Fig. 12.3a) with a gap left between adjacent panels. This gap is not filled in until at least seven days after the concreting of the last adjacent slab. The joints in the under-layer slab are *plain butt joints* (Fig. 12.3c), the slab panels being cast alternately, chessboard fashion. Where the upper-layer slabs have 0·3 per cent reinforcement, it is satisfactory for the complete-contraction joints to be spaced at intervals of 25 ft.; but if the under-layer slab is of plain concrete (requiring joints at about 15 ft. centres) it is convenient to arrange the upper-layer slab joints also at 15 ft. centres to simplify the arrangement of staggering.

Figure 12.3*b* shows an expansion joint in an upper-layer slab.

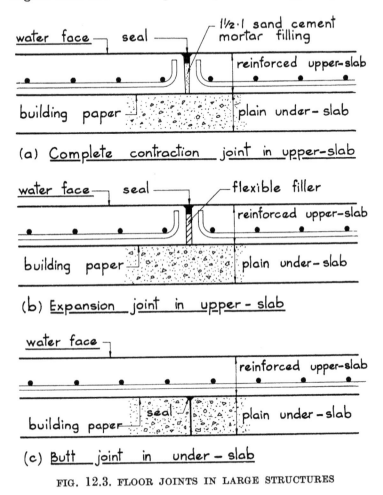

(a) Complete contraction joint in upper-slab

(b) Expansion joint in upper - slab

(c) Butt joint in under - slab

FIG. 12.3. FLOOR JOINTS IN LARGE STRUCTURES

The lower part of the joint is formed by a flexible filler about $\frac{3}{4}$ in. thick, and the top 1 in. run-in with sealing compound.

(iii) *Roofs*

Figure 12.4*a* shows a *partial-contraction joint* in a roof slab. Joints of this type are spaced at intervals of about 25 ft. and at positions of no shear, the panels formed by the joints being cast alternately, chessboard fashion. The top-surface rebate of the joint is later run-in with sealing compound.

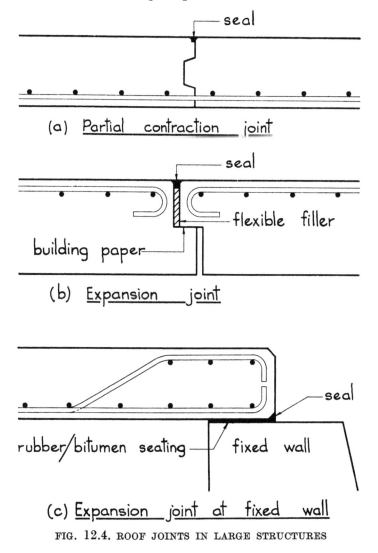

(a) <u>Partial contraction joint</u>

(b) <u>Expansion joint</u>

(c) <u>Expansion joint at fixed wall</u>

FIG. 12.4. ROOF JOINTS IN LARGE STRUCTURES

Figure 12.4b shows an *expansion joint* in a roof. These joints are spaced at intervals of about 100 ft. The stepped arrangement is provided to give support to the $\frac{3}{4}$ in. thick flexible filler. The

horizontal sliding surface at mid-depth of the slab is made as flat
and smooth as possible with an intervening layer of paper, and the
and-surface rebate later run-in with sealing compound. Figure

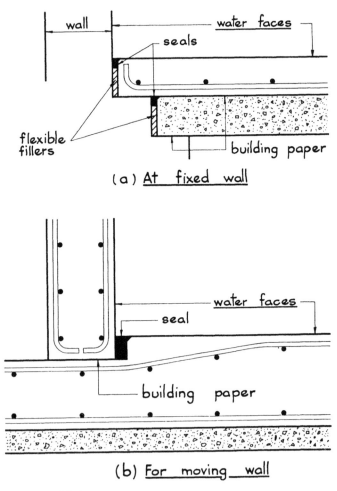

(a) At fixed wall

(b) For moving wall

FIG. 12.5. JOINTS BETWEEN WALLS AND FLOORS

12.4c shows an *expansion joint* where a roof-slab sits on an external
wall. The slab is supported on a continuous rubber-bitumen mat
and the outer edge of the joint sealed from outside.

(iv) *Joints between walls and floors*

Where free-sliding floors adjoin walls built on separate foundations, thermal and shrinkage movements are likely to collect; and it may be desirable to provide an expansion-type joint in both the upper- and under-layer slabs. This is indicated in Fig. 12.5*a*. Freedom for movement is ensured by forming the ledges on the wall foundation as flat and smooth as possible, and interposing the layer of paper. Flexible fillers and seals are required as for other expansion-type joints.

In some designs, as for example cylindrical tanks, there are advantages in providing a sliding-joint, as indicated in Fig. 12.5*b* to achieve complete discontinuity as between the wall and the floor. This goes some way towards preventing vertical cantilever-action in the lower part of the wall. The sliding surface requires careful attention as before. Sealing compound is run into the joint between the wall and the floor.

(v) *General*

The arrangement of joints in water-retaining structures is so vital to the success of the work that all joints should be clearly indicated on the working drawings, and no intermediate joints allowed. All joints in any part of the structure should be arranged in continuous alignment with no side-stepping. This includes vertical joints in walls, with joints in floor and roofs, and all horizontal joints in walls.

In flexible-jointed structures, adjacent slab panels at all *contraction joints* should be cast with time-intervals of at least seven days, so that the first-cast concrete will have done most of its shrinking before the second-cast concrete is poured against it. The only case where the seven-day interval is not beneficial is the *horizontal construction joint* in walls, where it is best to get the wall built up to its full height in the least possible time, as explained earlier.

ART. 12.5. FLEXIBLE JOINTING-MATERIALS

SEALING COMPOUNDS

All sealing compounds have of course to be waterproof. They must also adhere tenaciously to the face of the concrete, and be sufficiently flexible over the required range of temperature to stretch and compress without cracking or excessive extrusion. Also they must be insoluble, non-toxic and taintless. And for practical application they must be readily workable, either by melting for down-hand pouring, or by softening for application to vertical faces. The most satisfactory durable material for the purpose is a rubber-bitumen compound. The surface to which it is applied has first to be dried (in some cases artificially) and then primed to give enhanced adhesion. For down-hand work the compound is heated to a controlled temperature and poured manually. For vertical work on walls the compound is pre-formed at manufacture into continuous strips of square section which are softened by heating and ironed into position.

After the various joints have been sealed, the surface of the concrete should be painted in a band 12 in. or 18 in. wide disposed centrally astride the joint with three coats of bituminous paint, the first coat being thinned down to achieve better penetration into the pores of the concrete. These paint-bands serve the dual function of protecting the rubber-bitumen seals against action from air and water, and also guard against water by-passing the rubber-bitumen seals through any local porosity in the concrete.

FLEXIBLE FILLERS

Flexible fillers at expansion joints require to be readily compressible, yet must be capable of recovery when the joint subsequently re-opens. Otherwise the sealer will not receive proper support, and will be forced by the water-pressure into the crevice left between the filler and the face of the concrete. A suitable filler is composed of cellular cane-fibres, impregnated with bitumen to give protection from rot.

WATER-STOPS

Until fairly recently, engineers used metal water-stops, with varying degrees of success, depending on the amount of movement which occurred at the various joints. Metals suffer fatigue, and in certain circumstances corrosion. They do not key very satisfactorily to the concrete, unless perforated, in which case their purpose is spoiled. The flexible water-stops referred to in this chapter are of poly-vinyl-chloride, or natural or synthetic rubber. These materials can be purchased in considerable lengths in convenient rolls, so that site-joints are generally not necessary. The manufacturers will make up special intersection pieces where required; and even where butt-joints have to be made on the site, the technique is simple and success can be achieved without difficulty.

These flexible water-stops have already been in use in America for about twenty-five years, so that any supposed novelty which may have prejudiced their use in Great Britain has now worn thin. The cross-section of the water-stops provides bulbed ends for casting into the concrete at both sides of the joint, so that the more the joint opens, the tighter the bulbed ends are pulled against the concrete, making for enhanced watertightness. In compacting the concrete, great care is required at these water-stops, since the nuisance to concreting caused by the presence of the water-stops is liable to interfere with successful compaction at the very position where sound dense concrete is especially required.

ART. 12.6. GENERAL CONSIDERATIONS

GROUND-WATER PRESSURES

The arrangements shown in Fig. 12.3 for sealing floor joints are, of course, appropriate only to situations where there is no upward water pressure. Upward pressure would force the seals out of these joints. Other troubles which can arise from upward pressure are bending of the floor, lifting of the floor if not properly tailed-down, or floating of the whole structure. The cost of the work may be so increased in dealing with these matters that it may be preferable to search for another site. Alternatively it may be practicable to relieve upward water pressure by artificial sub-floor drainage.

EXCAVATION

Where the floor of a reservoir or tank is founded directly on the ground, care should be taken in the excavation work to adhere strictly to the depths and dimensions shown on the drawings. Where excavations exceed these dimensions the excess should be filled back with weak concrete, and not with spoil from the excavations; otherwise the floor is likely to settle and crack. Where concrete lining is placed in excavations for wall or column foundations, outside-shuttering must not be used, but vertical timbers withdrawn as the concrete is placed so that the concrete completely fills the gross excavation: in the case of columns this will prevent settlement of the floor; and in the case of walls will ensure minimum lateral take-up before the passive resistance of the ground is developed.

EARTH PRESSURES

Buried tanks and reservoirs should be designed for both maximum earth pressure on the outside of the walls when the reservoir is empty, and full water pressure on the inside of the walls. It is unwise to rely on earth-filling part balancing the water pressure, since the filling may not have been placed at the time of the initial testing of the work for watertightness: furthermore in many materials, particularly clays, under conditions of drought, the soil will shrink away from the structure leaving an appreciable gap.

ROOFS

Roofs of reservoirs to contain potable water are required to be watertight, and have therefore to be designed in accordance with the methods given in Article 12.3. In any case, the underside surfaces of reservoir roofs are permanently dripping from condensation, and corrosion is only prevented by guarding against cracking and working to low steel stresses.

SIMPLICITY OF DESIGN

Simplicity of design tends to make for success of construction: it may also contribute to saving in cost. Elaborate formwork with only a few re-uses is expensive to manufacture and slow and troublesome in the fixing. Continual changes in section and detail of the concrete interrupt the even rhythm of working of the concreting teams and lead to the introduction of patches of poor concrete. Roof slabs of reservoirs and tanks are nowadays frequently of flat-slab construction. This makes for simplicity of work and reduces the height of the columns and walls and hence the cost. Other advantages of flat-slab construction are that efficient roof ventilation is more easily achieved and fewer arrises occur that might lead to the rusting of the reinforcement and spalling of the concrete. Simplicity of reinforcement is likewise desirable. In the example of the reservoir in Fig. 12.12 all the bars in the roof were of the same diameter with generally only two shapes of bending: all the bars in the walls were straight and of only two diameters, and all the bars in the floor were straight and of one diameter.

TESTING FOR WATERTIGHTNESS

A condition of the contract for any water-retaining structure should be for the contractor to demonstrate by test that the work is completely watertight. The structure should be filled at an agreed slow rate to the top-water-level and an accurate record kept of the level over a period of seven to ten days with inspections made throughout that period for evidence of leakage through any parts of the structure.

In the case of reservoirs (except where cylindrical), the water pressure on the walls is normally reacted by the foundations pressing against the natural undisturbed ground. In certain cases the passive resistance of the ground at foundation level may be inadequate to take the full force on the wall until surcharged by the weight of the surrounding bank. For the purpose of testing, however, it may be desirable temporarily to omit the inner part of the bank, constructing only the outer part of the bank as a dumpling. For this reason it is important that the amount of bank to be constructed prior to carrying out the water-test should be clearly defined.

ORDER OF CONSTRUCTION

Contract documents for water-retaining works should state clearly the order in which the various parts of the work are to be constructed. For example, the engineer may require the reservoir roof to be constructed before the corresponding areas of floor, so that the floor is protected from rain or sunshine. Alternatively the engineer may require the whole area of floor to be completed within a certain period of time, and then kept permanently flooded while the remainder of the work is completed. Requirements such as these may be perfectly reasonable, but they must then be made clear in the contract documents.

ART. 12.7. PRACTICAL EXAMPLES OF LIQUID-RETAINING STRUCTURES

The following examples are intended to illustrate various forms of liquid-retaining structures. Brief comments and typical calculations are given, explaining the application of the design-principles set out in Article 12.3. The first four examples are small enough to be satisfactory with *rigid joints*: the two reservoir examples require the use of flexible joints.

Unless fool-proof overflow or other outlet devices are provided, liquid-retaining structures should be designed for standing full to the brim. It is so easy for a works employee to make some stupid mistake by mis-operating a valve or the like so that the water-level rises higher than intended, and the damage caused by once cracking the concrete is irreparable.

CLARIFIER TANK (100 FT. DIA. × 18 FT. DEEP)

The arrangement of the Clarifier Tank is shown in Fig. 12.6. The floor of the tank falls to a central sump. The walls stand freely 16 ft. above the floor haunch. The lower 4 ft. height of wall will cantilever up from the floor, so the maximum ring tension occurs where the head of water is 12 ft.

Ring tension $= 62 \cdot 4 \times 12 \times 50 = 37,500$ lb/ft., so that A_{st} required $= \dfrac{37,500}{12,000} = 3 \cdot 12$ sq. in./ft. This is provided by four 1 in.

dia. rods per ft., i.e. rods 1 in. dia. at 6 in. centres both faces (area 3·14 sq. in.).

A_E (14 in. wall) = (14 × 12) + (14 × 3·14) = 212 sq. in./ft.,

which gives $f_{ct} = \dfrac{37,500}{212} = 177$ lb/sq. in. (as against 190 lb/sq. in. permissible for 1 : 1·6 : 3·2 nominal mix).

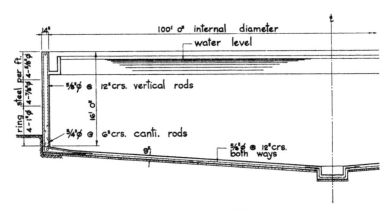

FIG. 12.6. CLARIFIED TANK 100 FT DIAMETER

Thus the full calculations for circumferential steel may be set out in tabular manner as follows:

Depth	Ring tension	A_{st} required at 12,000 lb/sq. in.	Reinforcement provided	
			Rod dia.	A_{st}
ft.	lb/ft.	sq. in./ft.		sq. in./ft.
4½	14,000	1·17	4 – ⅝ in.	1·22
9	28,000	2·34	4 – ⅞ in.	2·40
12	37,500	3·12	4 – 1 in.	3·14

The minimum area for vertical steel in the walls is

$$\frac{0·3}{100} \times 12 \times 14 = 0·51 \text{ sq. in.}$$

This is provided generally by ⅝ in. dia. rods at 12 in. centres in both faces. But at the lowest 4 ft., the cantilever moment is

$$M = 3500 \times 28 \text{ in.} = 100,000 \text{ in. lb/ft.,}$$

requiring $A_{st} = \dfrac{100,000}{12,000 \times 0.86 \times 12} = 0.81$ sq. in./ft.,

which is provided by $\frac{3}{4}$ in. dia. rods at 6 in. centres (area 0·88 sq. in.). Checking this for cracking, we have

$$\frac{M}{bd^2} = \frac{100,000}{12 \times 14^2} = 42.5$$

which is less than 45 and therefore satisfactory.

The floor thickness is made 9 in. to cater for local ground weaknesses. The minimum steel area required is therefore

$$\frac{0.3}{100} \times 12 \times 9 = 0.32 \text{ sq. in./ft.}$$

which is provided by $\frac{5}{8}$ in. dia. rods at 12 in. centres in both directions. The direct tension in the floor from the lower 4 ft. height of wall cantilevering is 3500 lb/ft., giving a steel stress of

$$f_{st} = \frac{3500}{0.3} = 11,700 \text{ lb/sq. in.}$$

(as against 12,000 lb/sq. in. permissible). The stress of the concrete in tension is clearly quite low.

In a tank of this type, in ring tension, the work should be arranged so as to avoid any vertical construction joints in the walls.

CEMENT-SLURRY TANKS (26 FT. DIA. × 35 FT. DEEP)
WITH HOPPER BOTTOMS

The arrangement of twin cement-slurry tanks is shown in Fig. 12.7. The tanks are used for final mixing of chalk/clay slurry prior to the slurry being fed to the kiln (*see* Fig. 1.1). The slurry is agitated in the tanks by compressed air fed in near the bottoms of the hoppers. The weight of the slurry is taken as 100 lb/cu. ft. The foundations in this example were for rock chalk good for 6 tons/sq. ft.

The design considerations for the upper cylindrical portions are much the same as in the previous example.

Ring tension $= 100 \times 35' \times 13' = 45,500$ lb/ft.,

so that A_{st} required $= \dfrac{45,500}{12,000} = 3.8$ sq. in./ft.

This is provided by four $1\frac{1}{8}$ in. dia. rods per ft.

Then A_E (16 in. wall) $= (16 \times 12) + (14 \times 4) = 248$ sq. in.,

giving
$$f_{ct} = \frac{45{,}500}{248} = 183 \text{ lb/sq. in.}$$

The hopper bottoms are similarly in ring tension, the diameter decreasing as the head increases: thus the circumferential reinforcements reduce towards the outlet. Where the hopper bottoms

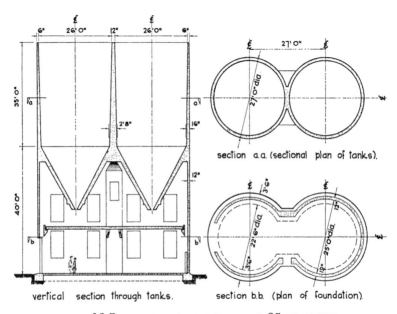

vertical section through tanks.

section b.b. (plan of foundation).

FIG. 12.7. CEMENT SLURRY TANKS 35 FT. DEEP

are slung up to the walls, the total tension is equal to the total weight of contents, plus self-weight of bottom (as for Bunker Design; *see* Article 10.2). Nearer the centre of the hopper bottoms, although the head of liquid is greater, the plan area of liquid to be slung up is less, so that the reinforcements down the slope decrease.

$\frac{1}{4}$-MILLION GALLON WATER-TOWER

Figure 12.8 shows the arrangement of a $\frac{1}{4}$-million gallon water-tower.

The load on the floor of the tank is:

Self-weight of slab (12 in.)   =   150 lb/sq. ft.

Water pressure 62·4 × 21 ft. = 1310 lb/sq. ft.

1460 lb/sq. ft.

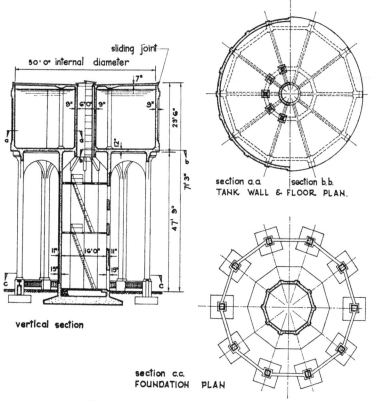

section a.a.   section b.b.

TANK WALL & FLOOR PLAN.

vertical section

section c.c.

FOUNDATION   PLAN

FIG. 12.8. $\frac{1}{4}$-MILLION GALLON WATER-TOWER

The floor slab spans in two directions in panels 14 ft. × 10 ft.

Thus $\dfrac{l_y}{l_x} = \dfrac{14}{10} = 1\cdot 4.$

From Table 5.2 we have moments on 10 ft. span:

− ve $M = 0\cdot054 \times 1460 \times 10^2 \times 12 = 95{,}000$ in. lb/ft.,

+ ve $M = 0\cdot041 \times 1460 \times 10^2 \times 12 = 72{,}000$ in. lb/ft., and moments on 14 ft. span:

$$- \text{ve } M = 0 \cdot 033 \times 1460 \times 10^2 \times 12 = 58{,}000 \text{ in. lb/ft.,}$$
$$+ \text{ve } M = 0 \cdot 025 \times 1460 \times 10^2 \times 12 = 44{,}000 \text{ in. lb/ft.}$$

Allowing for the effect of the width of the supporting beams, the maximum moment to consider is $+ 72{,}000$ in. lb/ft. Trying a 12 in. slab, we have

$$\frac{M}{bd^2} = \frac{72{,}000}{12 \times 12^2} = 42$$

which is less than 45, and therefore satisfactory.

For the circular wall (the pilasters are decorative only) the worst condition of tensile stress is at the bottom of the panels, where the head is 16 ft.

$$\text{Ring tension} = 62 \cdot 4 \times 16 \times 25 = 25{,}000 \text{ lb/ft.,}$$

so that $\quad A_{\text{st}}$ required $= \dfrac{25{,}000}{12{,}000} = 2 \cdot 1$ sq. in./ft.

$$A_E \text{ (9 in. wall)} = (9 \times 12) + (14 \times 2 \cdot 1) = 137 \cdot 4 \text{ sq. in./ft.}$$

which gives $\quad f_{\text{ct}} = \dfrac{25{,}000}{137 \cdot 4} = 182$ lb/sq. in.

(as against 190 lb/sq. in. permissible).

The roof is taken as fixed at the circle of columns, but has a free sliding joint on the outer wall, allowing relative lateral movement so that the ring-tension in the wall is not interfered with. The load on the roof is:

$$\begin{array}{llr}\text{Self-weight of slab (7 in.)} & = & 88 \text{ lb/sq. ft.} \\ \text{Superimposed load} & = & 30 \text{ lb/sq. ft.} \\ \hline & & 118 \text{ lb/sq. ft.} \end{array}$$

Therefore mid-span $M = 118 \times 16^2 \times \dfrac{12}{14} = 25{,}800$ in. lb/ft.

whence $\qquad \dfrac{M}{bd^2} = \dfrac{25{,}800}{12 \times 7^2} = 44.$

The moment on the inner column band is

$$M = 118 \times 16^2 \times \frac{12}{8} = 45{,}000 \text{ in. lb.,}$$

but due to the reduced perimeter at the smaller radius, the moment per ft. is 90,000 in. lb. The slab is haunched down at the columns to 14 in. thickness, giving $\dfrac{M}{bd^2} = \dfrac{90,000}{12 \times 14^2} = 39$. The reinforcement areas are determined as before.

SMALL COVERED RECTANGULAR TANK WITH INTERNAL TIES

Figure 12.9 shows a small holding tank. Requirements were 20 ft. × 20 ft. plan area, × 20 ft. height. A substantial overflow outlet prevents the maximum head exceeding 18 ft. The tank is

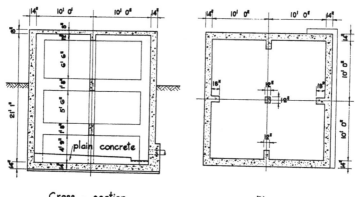

Cross - section   Plan

FIG. 12.9. SMALL RECTANGULAR TANK

too large for the walls economically to span from corner to corner, and accordingly the span was halved by the introduction of vertical counterforts. A central column was required to support the roof, and this is steadied from the counterforts by members which act also to tie-in the counterforts.

With the walls 14 in. thick, the lower 4 ft. will cantilever up from the floor. Immediately above this, the lateral water pressure = 62·4 × 14 = 870 lb/sq. ft.

$$M = 870 \times 10\tfrac{1}{2}^2 \times \frac{12}{12} = 96,000 \text{ in. lb.,}$$

whence tensile stress in concrete $= \dfrac{96,000 \times 6}{12 \times 14^2}$ = 245 lb/sq. in.

Direct tension $T = 5$ ft. × 870 = 4350 lb.

whence tensile stress in concrete $= \dfrac{4350}{12 \times 14}$ = 25 lb/sq. in.

$\overline{ }$ 270 lb/sq. in.

$$A_{st} \text{ bending} \quad = \frac{96,000}{12,000 \times 0.86 \times 12} \qquad = 0.78 \text{ sq. in./ft.}$$

$$A_{st} \text{ direct tension} = \frac{4,350}{2 \times 12,000} \qquad = 0.18 \text{ sq. in./ft.}$$

$$\overline{0.96 \text{ sq. in./ft}}$$

This is provided by $\frac{7}{8}$ in. dia. rods at $7\frac{1}{2}$ in. centres. Higher up the tank the reinforcements are reduced.

The maximum tension in the internal ties is 46,600 lb. (ignoring the cantilever effect of the walls from the floor), whence

$$A_{st} = \frac{46,600}{12,000} = 3.9 \text{ sq. in.}$$

which is provided by six 1 in. dia. rods (area 4.7 sq. in.). With concrete dimensions 20 in. $\times$ 12 in., we have

$$A_E = (20 \times 12) + (14 \times 4.7) = 306 \text{ sq. in.},$$

giving $\qquad f_{ct} = \dfrac{46,600}{306} = 152 \text{ lb/sq. in.}$

The provision here allows a margin for out-of-balance bending.

COVERED TANK 42 FT. SQUARE $\times$ 9 FT. DEEP

Figure 12.10 shows a covered tank 42 ft. square with a small pumping room adjoining. The tank had to be constructed in excavation in ground which flooded seasonally: for this reason the combined weight of the tank and roof-filling were arranged to balance the uplift forces from buoyancy. For the same reason, the floor was made rigid and held down by the loads from the columns; and the wall-thickness was given a margin to meet practical construction difficulties.

The walls were designed to span vertically between the floor and roof, taking into account the worst load-combinations of tank full or empty, ground flooded or dried, and roof loaded or not. The various bending moments for the worst of these conditions are shown in Fig. 12.10.

4-MILLION GALLON COVERED SERVICE-RESERVOIR

Figure 12.11 shows part of a compartmented reservoir designed for construction 300 miles up-country in Africa. In view of the uncertain qualities of labour and supervision, the relatively poor local aggregates, and the unfavourable conditions for curing, the concrete thicknesses are greater than would be provided for first-class work in this country. Further, by comparison with Fig. 12.12 the wall profiles have been kept simple, for ease of formwork and placing of concrete. These are matters requiring careful consideration by engineers designing works to be constructed in

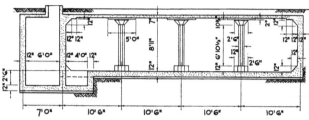

Cross-section through tank.

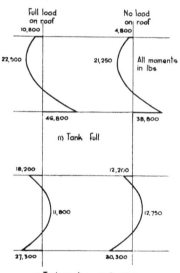

Bending-moments in walls.

FIG. 12.10. TANK 42 FT. SQUARE

under-developed areas far away from home, particularly so with water-retaining structures where percolation through ambitious thin structures could lead to early corrosion, or excessive loss of contents, with possible dangers from softening of the ground.

Generally the foundations bear on laterite at 2 tons/sq. ft., but the wall foundations are keyed against horizontal movement by extensions taken down into sandstone. The roof is composed of

27

four equal units, 106 ft. square, with expansion joints between, each supported on columns independently of the walls to allow complete freedom of movement. The cantilever-moments at the

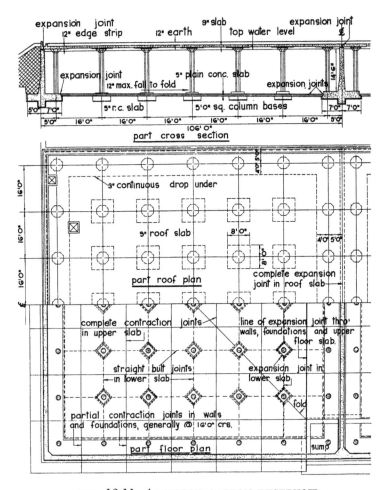

FIG. 12.11. 4-MILLION GALLON RESERVOIR

edges of the roof are arranged to balance the negative moments from the interior panels along the lines of the columns. The loads from the edge columns assist in giving stability to the wall-foundations against overturning, no reliance being placed on the weight of water on the foundation itself, as being too easily nulli-fied in the event of water seeping under the floor, and producing equal and opposite uplift forces from beneath.

5-MILLION GALLON COVERED SERVICE-RESERVOIR

Part of a more elegant reservoir design is shown in Fig. 12.12. The whole reservoir is 215 ft. square internally, and the depth from floor to top-water-level is 18 ft. 6 in. The modest concrete thicknesses have proved a complete success, but only as a result

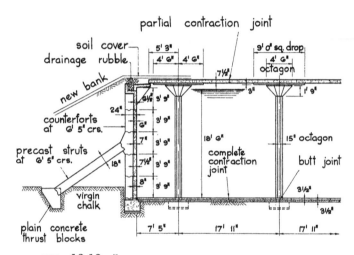

FIG. 12.12. 5-MILLION GALLON RESERVOIR

of first-class workmanship and supervision. The walls vary in thick-ness from 6 in. at the top to 8 in. at the bottom; the floor is in two layers both $3\frac{1}{2}$ in. thick, and the roof slab is $7\frac{1}{2}$ in. thick on spans of 18 ft.

The roof is supported independently of the walls, with a $1\frac{1}{2}$ in. gap between, filled with a rubber-bitumen sealer. Further reduc-tion of stress due to temperature change is achieved by providing two expansion joints through the roof, walls and floor in two direc-tions at right angles, thus dividing the structure into nine separate parts. The reservoir was first filled when the ambient temperature was between 60°F and 75°F, the water temperature being 45°F. The roof edge moved inwards by $\frac{1}{8}$ in. The force necessary to have resisted such a movement against the total cross-sectional area of the roof could not have been provided practically by the thin concrete sections used: and this demonstrates well the need for the movement joints provided. The roof slab is designed for bending moments in accordance with equation 5.5 and Table 5.3, and $\dfrac{M}{bd^2}$ is kept down to 45, taking advantage of the drop thickness where the large negative moments occur at the columns.

The walls span 6 ft. 5 in. between counterforts, and advantage is taken of the width of the counterforts reducing the design-moments to $\dfrac{pl^2}{18}$. The reinforcement is the same on both sides of the wall to resist also the pressure from the external embankments. The counterforts are propped at mid-height by struts inclined at an angle of 35 deg. to the horizontal. The struts were pre-cast, and placed in position before the thrust blocks and counterforts were concreted.

REFERENCES

1. BRITISH STANDARD CODE OF PRACTICE, C.P. 2007. Design and Construction of Reinforced and Prestressed Concrete Structures for the Storage of Water and other Aqueous Liquids. British Standards Institution, (1960).
2. RITCHIE, J. O. C. Water Towers. *The Structural Engineer*. Vol. 35, No. 1. (January, 1957).
3. GRAY, W. S. *Reinforced Concrete Reservoirs and Tanks*. Concrete Publications Limited, (1954).

CHAPTER 13

TALL INDUSTRIAL CHIMNEYS

ART. 13.1. INTRODUCTION

BRICK chimneys have to be stable on a gravity basis. Reinforced concrete chimneys, on the other hand, do not, due to the tensile strength of the reinforcements; and the shell thickness is determined by the compressive stresses on the leeward side. For these reasons, reinforced concrete chimneys can be made slimmer and lighter, with consequent savings in extent and cost of foundations. On congested or valuable sites, or where the ground is of poor bearing capacity, this makes for material advantage. Other merits of reinforced concrete chimneys include low cost, aesthetic satisfaction, and greater permanence (making for reduced long-term maintenance costs).

In some circumstances reinforced-concrete chimneys are built parallel-sided. This simplifies the formwork; but the overturning moments due to wind are then greater, and the overall stability is less than for tapered chimneys. Experienced contractors find no difficulty in constructing chimneys to a taper, and generally these look very much more graceful. The inside taper can, without difficulty, be made less than the outside taper, so that the shell thickness varies as required to suit the stresses. Frequently the rate of change of shell thickness is constant for the total height of the chimney, the shell thickness increasing uniformly from about 5 in. or 6 in. at the top. In order to achieve good cover to the reinforcements, it is not reasonable to go much thinner than this. The technique of forming the taper to tall chimneys is described in Article 13.9. Tapered chimneys have enhanced stability, and require considerably less reinforcement.

Normally it is advisable to provide concrete chimneys with an internal brick lining. The lining protects the concrete from extreme temperature stresses, and serves also as protection against corrosion from sulphuric and sulphurous acids or from other deleterious agents in the flue gases. It is usual to provide a ventilated cavity between the brick lining and the concrete shell in order to increase the overall temperature resistivity, with consequent decrease of temperature difference across the concrete shell itself. The lining is most conveniently carried on reinforced concrete corbel rings sprung from the concrete shell at about 50 ft. intervals up the height of the chimney, each height of lining being tapered to protect the corbel ring carrying the height of brickwork over, while yet being free to allow for expansion movements. Ventilation ports are required in the shell and through the corbel rings; and these details require careful attention. Where gas temperatures exceed about 800°F, and where the gases are of a highly corrosive nature it may be necessary to use extra dense bricks, and special jointing materials.

The downward swirl of smoke and flue gases has been found to corrode the upper few feet of concrete chimneys, and for this reason it is advisable to construct 15 ft. or 20 ft. at the top in 9 in. brickwork suitably banded and capped. Brickwork here is perfectly safe, because tensile stresses cannot occur across horizontal planes in the upper few diameters of the chimney.

At the present time, the design of reinforced concrete chimneys is dealt with most fully in the *Standard Specification for the Design and Construction of Reinforced Concrete Chimneys*, prepared by The American Concrete Institute Committee 505[1]. One of the most important aspects of reinforced concrete chimney design centres on the stresses resulting from temperature effects; and the American Specification accepts that further research is required into this. The present chapter aims to set out the fundamentals of chimney design avoiding the elaboration given in the American Specification.

ART. 13.2. DIMENSIONS OF CHIMNEYS

In regard to height, the Report drawn up in 1932 by the Electricity Commissioners[2] had the following to say:

"The Sub-committee conclude that emissions from chimneys having a height of $2\frac{1}{2}$ times that of the surrounding buildings, plus, where necessary, additional height to compensate for the contour of the adjacent land, will discharge into an air stream which can be depended upon not to come in contact with the earth under normal conditions. The additional height required can be determined with a sufficient degree of accuracy by experiments with a model in a wind tunnel, or by investigations on the site with no-lift balloons."

This statement gives reasonable guidance as far as it goes. But so frequently the local topography influences the matter as much or more than the situation of any surrounding buildings. And a little difficulty can arise in interpreting what are "normal conditions".

The air stream may truly not come into contact with the earth, but what are the conditions necessary to prevent grit, or other bodies in the flue gases falling clear of the air stream?

The minimum internal chimney diameter is determined from consideration of the gas volume to be discharged, and the velocity of the gas. A maximum gas velocity is generally considered to be about 2000 ft. per minute. Where a number of boilers or other items of plant are discharging up one chimney, care has to be taken to guard against the velocity falling too low when a proportion of the plant is out of operation: otherwise the gases in their journey up the chimney may cool below their dewpoint, giving rise to condensation. With certain corrosive gases, the condensate can have extremely deleterious effects on the chimney shell and the lining. Thus where the duty of a chimney is liable to fluctuate, it may be preferable to aim for a high maximum rather than risk a minimum tending too far on the low side. Velocities of 2500 ft. per minute are known.

ART. 13.3. WIND PRESSURES

Circular chimneys in England should normally be designed for wind pressures on their projected areas as given by the curve in Fig. 13.1. American practice is generally similar. The curve is based on a wind velocity of 80 m.p.h. at 30 ft. height, increasing to 120 m.p.h. at 600 ft. height, using Stanton's formula[3]

$$p = k \times 0{\cdot}0032v^2$$

where p is the pressure in lb/sq. ft.,
      v is the wind velocity in m.p.h.,
      k is a form factor (1·0 for square section)
                     (0·67 for circular section).

Figure 13.1 is based on $k = 0{\cdot}67$ for circular chimneys. Chimneys of other sections require the substitution of appropriate k-values. It should be noted that vertical flutings up a circular chimney increase the wind resistance considerably: and tests have shown that the resistance of a 16-sided chimney approximates to that of a square chimney of equal width across flats. For these reasons it is clearly cheapest to adopt chimneys of circular form. Not only is the formwork the most simple, and the quantity and weight of concrete the least, but in addition the applied wind-moments require also the least quantity of steel reinforcement.

It is interesting to note from Fig. 13.2 the general distribution of wind pressure on a tall circular chimney. This record is typical of many such prepared by Dryden and Hill.[4] The positive pressure

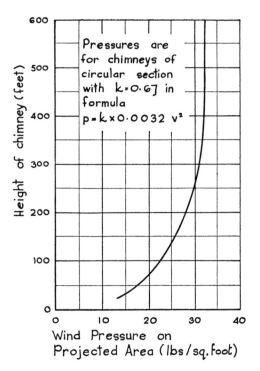

FIG. 13.1. WIND PRESSURES ON TALL CHIMNEYS

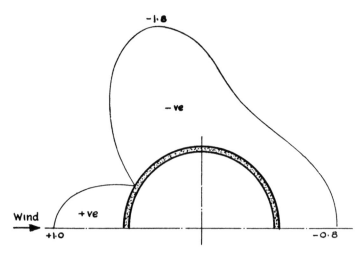

FIG. 13.2. DISTRIBUTION OF WIND PRESSURE ON
CIRCULAR CHIMNEY. (AFTER DRYDEN AND HILL.)

is seen to occur on only a part of the windward front, whereas negative pressures (suction) extend round the whole of the remainder of the chimney, being greatest at the sides, where their intensity is about double the numerical value of the pressure intensity on the front. These side suctions balance one another, and therefore do not influence the general static stability of the chimney: but in chimneys of considerable diameter they may set a limit on the minimum thickness of the shell, causing bending moments in the same sense as the moments caused by the positive pressure on the windward front, and by the temperature gradient across the shell.

ART. 13.4. DESIGN STRESSES

Stresses in the shell of a chimney arise principally from wind-pressure, temperature effects, and the chimney self-weight. Special problems arise with chimneys in earthquake areas.

Stresses arising from oscillation, flutter and temperature effects cannot be determined precisely: and for these and other reasons it is not advisable in chimney design to work to the maximum stresses given in C.P. 114 (1957). On the other hand thick shells should be avoided, owing to their increased weight and cost, and because of the increased temperature gradient which results. It is desirable therefore, in most circumstances, to work with mixes richer than the nominal $1:2:4$ mix.

In calculations based on consideration of self-weight and wind only, the maximum concrete stress should be limited to $0.2 \times u_w$, where u_w is the 28-day works cube strength. However, in calculations when the full effects of temperature have been computed and properly combined with the stresses due to self-weight and wind, experience has shown that maximum stresses up to $0.4 \times u_w$ are generally safe. The latter is not in fact greatly at variance with the maximum value permitted by C.P. 114 (1957) when due allowance is made for wind and age, i.e.

$$0.25 \times u_w \times \underset{\text{(age factor)}}{1.24} \times \underset{\text{(wind allowance)}}{1.25} = 0.39 u_w.$$

But the similarity is by chance, and not the result of parallel reasoning. It will be shown later that stresses in the concrete due to temperature effects can be as great (or greater) than stresses from other considerations: and accordingly there may be arguments (under the 1957 Code) to delay passing hot gases up the chimney until some nine months following completion of the construction work.

Stress in mild steel reinforcements from self-weight and wind only, should be limited to about 12,000 lb/sq. in. However, where the effects of temperature are properly included in the calculations, it is reasonable to go beyond this limit; but owing to effects of oscillation and flutter and consequent fatigue, extreme steel stresses should be avoided in all circumstances.

ART. 13.5. STRESSES DUE TO SELF-WEIGHT AND WIND FORCES

Under the effects of wind, a chimney acts as a vertical cantilever, rigidly supported at its base. Thus, due to wind and self-weight, the chimney is subject to bending and direct compression combined, much as we have already described for rectangular sections in Chapter 3. If we treat the problem by the elastic theory, there are two separate cases to be considered: first, where the eccentricity of the moment is small and no tension develops; and secondly, where the eccentricity produces tension on one side of the member, and the concrete is assumed to have cracked. In the case of a chimney, we are dealing with a hollow member, and it is necessary to evolve fresh formulae. The following applies to chimneys of circular section.

In addition to the normal assumptions of bending (that plane sections remain plane, and all tension is taken by the reinforcement), we shall assume here that the shell is thin in relation to its diameter, and that the reinforcement is situated at the centre of the shell. In practice the reinforcement is placed nearer the outer face, which slightly increases the margin of safety.

The following symbols are used:

A_s is the total area of reinforcement,

   t_s is the thickness of an annular ring of area equal to the reinforcement provided,

   t_c is the equivalent nett thickness of concrete,

   t is the actual shell thickness ($= t_s + t_c$),

   t_E is the equivalent shell thickness ($= mt_s + t_c$),

   r is the mean radius of the shell,

W is the total weight of chimney above any section,

M is the applied moment at that section,

   e is the eccentricity $\left(= \dfrac{M}{W} \right)$.

(i) *Case where tension does not develop*

The simple theory of combined bending and direct load will apply, so that

$$f_{cb} = \frac{W}{A} \pm \frac{M}{z} = W\left(\frac{1}{A} \pm \frac{e}{z}\right).$$

For a hollow circular section

$$A = 2\pi r t_E \quad \text{and} \quad z = \pi r^2 t_E.$$

The limiting case for no tension is when

$$\frac{W}{A} = \frac{M}{z},$$

i.e. when

$$\frac{W}{2\pi r t_E} = \frac{We}{\pi r^2 t_E}$$

or

$$e = \frac{r}{2}.$$

Thus tension will not develop when $\frac{M}{W}$ is less than $\frac{r}{2}$. This is the case with very short chimneys and for the upper parts of tall chimneys.

(ii) *Case where tension does develop*

When e exceeds $\frac{r}{2}$, tension develops on the windward side of the chimney. The calculation then becomes more difficult because of the problem of determining the position of the neutral axis.

Consider Fig. 13.3. Let the tensile stress in the steel be f_{st} and the compressive stress in the concrete be f_{cb}. Then from the stress diagram we have

$$d_n = \frac{2rf_{cb}}{\dfrac{f_{st}}{m} + f_{cb}}.$$

Now $x = r - d_n$,

and if we substitute g for

$$\frac{\dfrac{f_{st}}{m} - f_{cb}}{\dfrac{f_{st}}{m} + f_{cb}},$$

we have $x = rg$.

Furthermore, the angle 2ψ subtended at the centre by the neutral axis is such that

$$\cos \psi = \frac{x}{r} = g.$$

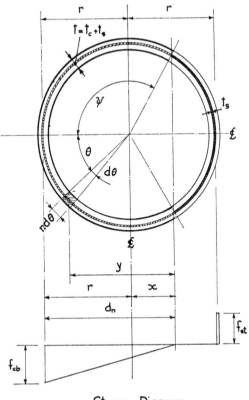

Stress Diagram

FIG. 13.3. STRESS DISTRIBUTION ACROSS CIRCULAR
SHELL WHEN CRACKED

For an elemental length $r \cdot \delta\theta$ on the mean circumference, the total equivalent elemental area is $(t_c + mt_s)r \cdot \delta\theta$; and the mean stress on the element is

$$f_{cb} \cdot \frac{y}{d_n}, \quad \text{or} \quad f_{cb} \cdot \frac{y}{(r-x)}, \quad \text{or} \quad \frac{f_{cb}(\cos\theta - g)}{(1-g)}.$$

Therefore the compressive force on this element is

$$\delta C = \frac{(t_c + mt_s)}{(1-g)} \, r \cdot f_{cb}(\cos\theta - g)\delta\theta$$

and the total compression is

$$C = \frac{2(t_c + mt_s)}{(1 - g)} \cdot r \cdot f_{cb} \int_0^\psi (\cos \theta - g) \cdot d\theta$$

$$= \frac{2(t_c + mt_s)}{(1 - g)} \cdot r \cdot f_{cb} (\sin \psi - g\psi). \tag{13.1}$$

The moment of the compressive element about the neutral axis is $\delta C \cdot y = r(\cos \theta - g)\delta C$, and the total moment due to compression is

$$M_C = \frac{2(t_c + mt_s)}{(1 - g)} \cdot r^2 \cdot f_{cb} \int_0^\psi (\cos \theta - g)^2 \cdot d\theta$$

$$= \frac{2(t_c + mt_s)}{(1 - g)} \cdot r^2 \cdot f_{cb} \left(\frac{\psi}{2} + \frac{\sin 2\psi}{4} - 2g \sin \psi + g^2\psi \right). \tag{13.2}$$

Similarly it can be shown on the tension side that the total tension is

$$T = \frac{2t_s r f_{st}}{(1 - g)} \cdot [\sin (\pi - \psi) + g(\pi - \psi)] \tag{13.3}$$

and the total moment about the neutral axis due to tension is

$$M_T = \frac{2t_s r^2 f_{st}}{(1 + g)} \left[\left(\frac{\pi - \psi}{2} \right) + \frac{\sin 2(\pi - \psi)}{4} + \right.$$

$$\left. + 2g \sin (\pi - \psi) + g^2(\pi - \psi) \right]. \tag{13.4}$$

If now we put $\dfrac{f_{st}}{f_{cb}} = k$, then for any value of k, ψ is determined, and equations 13.1, 13.2, 13.3 and 13.4 reduce to:

$$C = k_1 r f_{cb}(t_c + mt_s), \tag{13.5}$$

$$M_C = k_2 r^2 f_{cb}(t_c + mt_s), \tag{13.6}$$

$$T = k_3 r m f_{cb} t_s, \tag{13.7}$$

$$M_T = k_4 r^2 m f_{cb} t_s, \tag{13.8}$$

where k_1, k_2, k_3 and k_4 are dependent only on the value of k.

If $\bar{y}$ is the distance from the centre of compression to the neutral axis, and l_a is the lever arm between the centre of compression and the centre of tension, we have, by equating moments about the centre of compression,

$$M = W(\bar{y} + x) + T \cdot l_a$$

where

$$l_a = \frac{M_C}{C} + \frac{M_T}{T} = r\left(\frac{k_2}{k_1} + \frac{k_4}{k_3}\right)$$

and

$$\bar{y} = \frac{M_C}{C} = r\frac{k_2}{k_1}.$$

Substituting for T from equation 13.7, and rearranging, we have

$$t_s = \frac{A}{2\pi} \cdot \frac{M}{r^2 f_{cb}} - \frac{B}{2\pi} \cdot \frac{W}{r f_{cb}} \qquad (13.9)$$

where $\quad A = \dfrac{2\pi k_1}{m(k_3 k_2 + k_1 k_4)}, \quad$ and $\quad B = A\left(\dfrac{k_2}{k_1} + g\right),$

and since $A_s = 2\pi r t_s$, we have

$$A_s = A\frac{M}{r f_{cb}} - B\frac{W}{f_{cb}}. \qquad (13.10)$$

Also, equating forces, we have

$$W = C - T$$
$$= r f_{cb}[k_1 t_c + m(k_1 - k_3)t_s]$$

which gives $\qquad t_c = \dfrac{W}{k_1 r f_{cb}} - m\left(\dfrac{k_1 - k_3}{k_1}\right)t_s.$

Substituting for t_s from equation 13.9, and rearranging, we have

$$t_c = k_5 \cdot \frac{W}{r f_{cb}} - k_6 \frac{M}{r^2 f_{cb}}$$

where $\quad k_5 = \dfrac{(2\pi + k_1 B)(k_1 - k_3)m}{2\pi k_1^2}, \quad$ and $\quad k_6 = \dfrac{m(k_1 - k_3)A}{2\pi k_1}.$

Since $t = t_c + t_s$, we have the total shell thickness

$$t = D\frac{M}{r^2 f_{cb}} + F\frac{W}{r f_{cb}} \qquad (13.11)$$

where $\qquad D = \dfrac{A}{2\pi} - k_6, \quad$ and $\quad F = k_5 - \dfrac{B}{2\pi}.$

Thus, by use of equations 13.10 and 13.11, the necessary shell thickness and area of reinforcement can be determined for any ratio of steel and concrete stresses.

This theory is due to Taylor, Glenday and Faber[5] and was first published in 1908. A range of values for A, B, D and F is given in Table 13.1, based on $m = 15$. Intermediate values may be obtained by interpolation.

TABLE 13.1. *Constants A, B, D and F for use in design formulae*

$$A_s = A \frac{M}{rf_{cb}} - B \frac{W}{f_{cb}} \quad \text{and} \quad t = D \frac{M}{r^2 f_{cb}} + F \frac{W}{rf_{cb}}$$

$k = \dfrac{f_{st}}{f_{cb}}$	A	B	D	F
2	2·212	1·292	− 4·700	3·098
4	0·818	0·521	− 1·550	1·381
6	0·456	0·310	− 0·757	0·920
8	0·305	0·216	− 0·420	0·705
10	0·226	0·167	− 0·220	0·600
12	0·178	0·136	− 0·095	0·537
14	0·146	0·114	− 0·012	0·498
16	0·122	0·097	+ 0·050	0·470
18	0·106	0·086	+ 0·100	0·448
20	0·092	0·076	0·140	0·430
22	0·082	0·068	0·175	0·420
24	0·074	0·062	0·205	0·410
26	0·067	0·057	0·230	0·403
28	0·061	0·052	0·255	0·396
30	0·056	0·049	0·275	0·390

The use of equations 13.10 and 13.11 (as they stand) calls for some considerable skill, because the calculation has necessarily to be made on a trial and error basis. Accordingly, to minimise the labour, the authors have rearranged the equations and prepared a design curve as given in Fig. 13.4. Thus equation 13.11 may be rearranged

$$\frac{M}{tr^2 f_{cb}} = \frac{1}{\left(D + \dfrac{F}{e/r}\right)},$$

and since D and F are dependent on the value for k, $\dfrac{M}{tr^2 f_{cb}}$ can be plotted against $\dfrac{e}{r}$ for various values of k. Also, since the steel percentage is $r_p = \dfrac{A_s}{2\pi rt} \times 100$, we have from equations 13.10 and 13.11

$$r_p = \frac{100}{2\pi} \left(\frac{A - \dfrac{B}{e/r}}{D + \dfrac{F}{e/r}} \right),$$

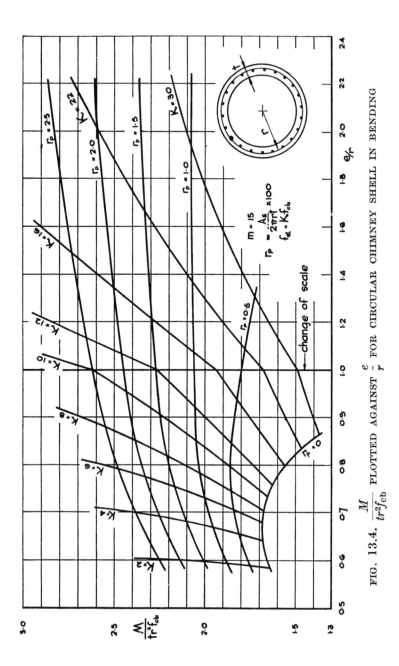

FIG. 13.4. $\dfrac{M}{tr^2 f_{cb}}$ PLOTTED AGAINST $\dfrac{e}{r}$ FOR CIRCULAR CHIMNEY SHELL IN BENDING

so that similarly, for various values of r_p, curves can be drawn of

$$\frac{M}{tr^2 f_{cb}} \text{ against } \frac{e}{r}.$$

Figure 13.4 is thus very similar to Fig. 3.18 for the case of combined bending and compression in rectangular columns.

For any chosen section, $\frac{M}{tr^2 f_{cb}}$ and $\frac{e}{r}$ are known; so that r_p can be read directly from Fig. 13.4, and the reinforcement determined, As a check on f_{st}, k is then read off, and, knowing f_{cb}, we have

$$f_{st} = k . f_{cb}.$$

An example of the use of Fig. 13.4 is given in Article 13.7.

ART. 13.6. STRESSES DUE TO TEMPERATURE

Temperature stresses arise in the shell due to the inner face being hotter than the outer face. The inner face expands more than the outer face, tending to produce curvature; and the restraint of the shell, preventing this curvature, causes the stressing. The degree of curvature, and consequently the intensity of stress. depends on the temperature difference, T_x, across the shell thickness, and the first step is therefore to evaluate this.

The total temperature difference across the brick lining and the concrete shell is shown diagrammatically in Fig. 13.5. There are five stages in the fall of temperature from the flue-gas temperature T_1 to the external-air temperature T_0. These are:

(a) surface effect on the inside of the lining;
(b) temperature drop through the lining;
(c) double surface effect at the cavity, and effect of ventilation of the cavity;
(d) temperature drop through the concrete shell;
(e) surface effect on the outside of the shell.

If now the quantity of heat transferred at (a) be denoted by Q, then the heat transferred at (b) must also be Q; and if $(1 - r)Q$ is the heat taken up the cavity by ventilation, the balance rQ will be the quantity transferred at (c), (d) and (e).

Q can be expressed in the form

$$Q = U(T_1 - T_0) \tag{13.12}$$

where T_i is the flue-gas temperature,
$\quad\quad T_0$ is the external-air temperature,
$\quad\quad U$ is the overall transmission in BTU/sq. ft./°F/hour,
so that

$$\frac{1}{U} = \frac{1}{K_1} + \frac{t_1}{C_1} + r\left(\frac{1}{K_c} + \frac{t}{C_2} + \frac{1}{K_2}\right) \qquad (13.13)$$

where K_1 and K_2 are transmission coefficients for the inner and
$\quad\quad\quad\quad\quad$ outer surfaces of the chimney construction,
$\quad\quad K_c$ is the cavity transmission coefficient,
$\quad\quad C_1$ and C_2 are conductivity coefficients per inch thickness
$\quad\quad\quad\quad\quad$ of the lining and the shell,
$\quad\quad t_1$ and t are the thicknesses in inches of the lining and the
$\quad\quad\quad\quad\quad$ shell.

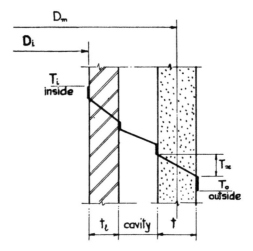

FIG. 13.5. GRADIENT OF TEMPERATURE DROP
THROUGH CHIMNEY CONSTRUCTION

Considering the concrete shell alone,

$$rQ = \frac{C_2}{t} \cdot T_x \qquad (13.14)$$

where T_x is the temperature difference across the shell thickness.
Thus from equations 13.12 and 13.14, we have

$$T_x = r\frac{t}{C_2}(T_1 - T_0)U$$

and, allowing for the increase in shell area due to the shell diameter being greater than the internal diameter of the chimney,

$$T_x = r \frac{D_i}{D_m} \cdot \frac{t}{C_2} (T_i - T_0)U \qquad (13.15)$$

where D_i is the internal diameter of the chimney, and D_m is the mean diameter of the concrete shell.

K_1 varies according to the flue-gas temperature and velocity: 2·0 BTU/sq. ft/°F/hour might be regarded as an average value. K_2 depends on the degree of exposure of the chimney, but an average value is 6·0 BTU/sq. ft/°F/hour. K_c may be taken as $\frac{T_i}{150}$ BTU/sq. ft/°F/hour for a ventilated cavity.

A well ventilated cavity may give r as low as 0·5, but higher values should be taken when the degree of ventilation is in doubt. Conductivity coefficients in BTU/sq. ft/in. thickness/°F/hour may be taken as 8·0 for brickwork (C_1) and 10·0 for concrete (C_2).

Having now determined the temperature gradient across the shell, we may consider in detail how this produces stress. The inner face tries to expand more than the outer face causing curvature, concave on the outer (cooler) face; but due to restraint this curvature cannot develop, and the shell is strained against its tendency to bend outwards, resulting in tension stresses on the outer face and compression stresses on the inner face. Ignoring here stresses from other causes, there will clearly be a neutral surface within the shell thickness, free from strain and stress. And if we assume the concrete cracks, leaving the reinforcement to take the whole of the tension, we have the conditions shown in Fig. 13.6. If T_{x1} is the temperature difference between the reinforcement and the inner surface of the concrete, so that

$$T_{x1} = \frac{d_1}{t} \cdot T_x,$$

then the temperature difference between the neutral surface and the inner surface of the concrete is

$$T_{x1} \cdot \frac{d_n}{d_1}, \quad \text{or} \quad T_{x1} \cdot n_1.$$

Also, since   stress = strain × E,

we have $$f_{cb} = \alpha T_{x1} n_1 E_c, \qquad (13.16)$$

and similarly $$f_{st} = \alpha T_{x1} (1 - n_1) E_s \qquad (13.17)$$

where α = coefficient of expansion of concrete,
       = coefficient of expansion of steel,
       = 0·0000055 per °F temp. diff.

Now n_1 is determined by the percentage of reinforcement (equation 3.8 or Fig. 3.5). Thus the stresses in the concrete and reinforcement due to temperature effects are found using equations

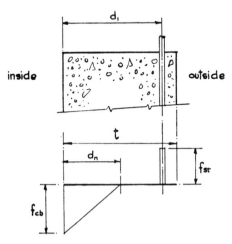

FIG. 13.6. STRESS DISTRIBUTION DUE TO TEMPERATURE
ACROSS SHELL THICKNESS

13.16 and 13.17 – the amount of vertical reinforcement having first been calculated from equation 13.10, and the amount of horizontal reinforcement being selected from experience as a practical minimum.

The example given in Article 13.7 shows the practical application of the above.

ART. 13.7. DESIGN OF CHIMNEY 350 FT. HIGH

Figure 13.7 shows a chimney 350 ft. high above ground-level, tapering uniformly from 26 ft. 6 in. outer diameter at the bottom to 13 ft. 0 in. at the top. The uppermost 15 ft. is in 9 in. brickwork: below this the concrete shell thickness varies uniformly from 5 in. at 335 ft. height, to 8 in. at 95 ft. height, and to 13 in. at the foundation (5 ft. below ground level). The flue floor is at 18 ft. height, and above this the chimney is lined with $4\frac{1}{2}$ in. brickwork in 40 ft. lifts supported on corbel rings from the concrete shell:

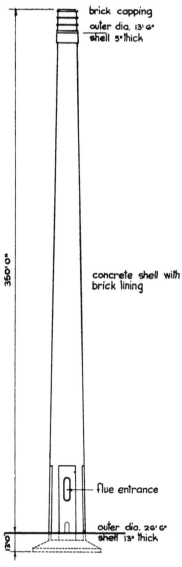

brick capping
outer dia. 13' 6"
shell 5" thick

concrete shell with
brick lining

flue entrance

outer dia. 26' 6"
shell 13" thick

350' 0"

13' 0"

FIG. 13.7. ELEVATION OF 350 FT. HIGH CHIMNEY
(AS CALCULATED IN TABLE ON PAGE 423)

the lining is 4 in. clear of the concrete shell, providing a ventilated cavity. The chimney serves three incoming flues; and vertical brick-baffle walls 30 ft. high are built in star-formation on the flue floor to prevent excessive turbulence between the three gas-streams, and to direct these upwards.

STRESSES DUE TO SELF-WEIGHT AND WIND

The calculations for the combined effects of self-weight and wind are most conveniently set out in tabular manner as shown on page 423. At 95 ft. height above ground level (for example) the detail of the calculation is as follows:

$$\text{Weight, } W = 1{,}804{,}000 \text{ lb.,}$$
$$\text{Moment, } M = 15{,}800{,}000 \text{ ft. lb.}$$

whence
$$e = \frac{M}{W} = 8\cdot75 \text{ ft.}$$

The mean radius of the shell $r = 10\cdot92$ ft.

whence
$$\frac{e}{r} = \frac{8\cdot75}{10\cdot92} = 0\cdot802.$$

If we are to use a concrete with a works cube strength of 3500 lb/sq. in. (say $1:1\cdot6:3\cdot2$ nominal mix), we should work to a design stress of $0\cdot2 \times 3500 = 700$ lb/sq. in. in the concrete. Then

$$\frac{M}{tr^2 f_{cb}} = \frac{15{,}800{,}000 \times 12}{8 \times (10\cdot92 \times 12)^2 \times 700} = 1\cdot97$$

and from Fig. 13.4 we read off $r_p = 0\cdot85$ per cent, which in 8 in. thickness is $\dfrac{0\cdot85 \times 8 \times 12}{100} = 0\cdot81$ sq. in /ft. This is provided by $\frac{3}{4}$ in. dia. rods at 6 in. centres (area $0\cdot88$ sq. in./ft.). Also from Fig. 13.4 we read off $k = 9\cdot2$, whence tensile stress in steel is

$$f_{st} = 9\cdot2 \times 700 = 6450 \text{ lb/sq. in.}$$

Note also the steel stress on the compression side of the chimney is $15 \times 700 = 10{,}500$ lb/sq. in.

At greater heights up the chimney, it will be seen that the reinforcement provided exceeds that required by calculation. Nevertheless the authors consider the quantities given are a practical minimum. It might then be argued that the shell thickness could be reduced to make fuller use of these steel quantities. But for the dimensions of this particular chimney, any reduction in shell thickness would result in oscillations of an undesirably high frequency.

Calculations of stresses in chimney shell due to self-weight and wind forces

Height above ground level (ft.)	Outer diameter (ft.)	Thickness (in.)	Mean radius (ft.)	Total vertical load (lb. × 10⁻³)	Wind moment (ft. lb. × 10⁻³)	e/r (× 10⁻³)	Design stress in concrete (lb./sq. in.)	$\dfrac{M}{t\,r^{2}\,f_{cb}}$	Required r_p (%)	Required A_{st} (sq. in./ft.)	Provided Rod dia. and spacing (in. in.)	Provided A_{st} (sq. in./ft.)	Provided r_p (%)	Actual f_{cb} (lb./sq. in.)	Actual k	Actual f_{st} (lb./sq. in.)
335	13·5	5	6·54	62							½ at 6					
295	15·0	5½	7·27	261	620						½ at 6					
255	16·5	6	8·00	493	1,920	0·486					½ at 6					
215	18·0	6¼	8·73	758	4,020	0·608	700	0·96		0	½ at 6	0·40	0·51	360	2·1	800
175	19·5	7	9·46	1,065	6,950	0·690	700	1·32	0	0	½ at 6	0·40	0·48	505	5·3	2,700
135	21·0	7¼	10·19	1,415	10,900	0·757	700	1·67	0·20	0·18	⅝ at 6	0·60	0·67	620	8·0	5,000
95	22·5	8	10·92	1,804	15,800	0·802	700	1·97	0·85	0·81	¾ at 6	0·88	0·92	680	9·2	6,500
55	24·0	10	11·65	2,239	21,800	0·836	700	1·92	0·70	0·84	¾ at 6	0·88	0·73	690	10·8	7,500
-5	26·5	13	12·70	3,366	33,050	0·775	700	1·88	0·55	0·86	¾ at 6	0·88	0·57	695	9·0	6,300

423

Where flue openings occur in the chimney shell, provided the openings do not form an appreciable part of the total cross-section, it is satisfactory to make up the lost concrete-area by the provision of breast-thickenings at the sides of the openings. The vertical reinforcements then need to be swept clear of the openings into the breasts, the latter being well tied-back to the main structure with horizontal links.

TEMPERATURE STRESSES

Suppose the flue-gas temperature is 750°F. Then pursuing our calculation for the same part of the chimney as before, i.e. at 95 ft. height above ground level, we have (Equation 13.13)

$$\frac{1}{U} = \frac{1}{K_1} + \frac{t_1}{C_1} + r\left(\frac{1}{K_c} + \frac{t}{C_2} + \frac{1}{K_2}\right),$$

and with $t_1 = 4\frac{1}{2}$ in. and $t = 8$ in., and adopting suitable constants, we have

$$\frac{1}{U} = \frac{1}{2} + \frac{4\frac{1}{2}}{8} + \frac{1}{2}\left(\frac{1}{5} + \frac{8}{10} + \frac{1}{6}\right)$$

$$= 1\cdot65$$

whence $U = 0\cdot608.$

Then from equation 13.15, the temperature difference across the concrete shell is

$$T_x = r\,\frac{D_i}{D_m}\cdot\frac{t}{C_2}\,(T_i - T_0)U$$

$$= \frac{1}{2} \times \frac{19\cdot75}{21\cdot8} \times \frac{8}{10}\,(750 - 20)0\cdot608$$

$$= 160°F.$$

(i) *Circumferential stresses*

Allowing 2 in. external cover to the circumferential reinforcements, we have

$$T_{x1} = \frac{5\frac{3}{4}}{8} \times T_x = 115°F.$$

Now with ring rods of $\frac{1}{2}$ in. dia. at 6 in. centres,

$$r_{p1} = \frac{A_{st}}{bd_1} \times 100 = \frac{0\cdot4}{5\frac{3}{4} \times 12} \times 100 = 0\cdot58 \text{ per cent,}$$

so that $n_1 = 0\cdot33.$

Then from equation 13.16

$$f_{cb} = \alpha T_{x1} n_1 E_c$$
$$= 0{\cdot}0000055 \times 115 \times 0{\cdot}33 \times 2{,}000{,}000$$
$$= 420 \text{ lb/sq. in.,}$$

and from Equation 13.17

$$f_{st} = \alpha T_{x1}(1 - n_1)E_s$$
$$= 0{\cdot}0000055 \times 115 \times 0{\cdot}67 \times 30{,}000{,}000$$
$$= 12{,}700 \text{ lb/sq. in.}$$

These stresses are clearly acceptable.

(ii) *Vertical stresses*

The vertical reinforcements come within the circumferential reinforcements, so we have

$$T_{x1} = \frac{5\frac{1}{8}}{8} \times 160 = 103{°}\text{F.}$$

With $\frac{3}{4}$ in. dia. rods at 6 in. centres,

$$r_{p1} = \frac{0{\cdot}88}{5\frac{1}{8} \times 12} \times 100 = 1{\cdot}43 \text{ per cent}$$

whence $n_1 = 0{\cdot}48$.

(a) Compression side of chimney.

The additional compressive stress in the concrete is

$$f_{cb} = 0{\cdot}0000055 \times 103 \times 0{\cdot}48 \times 2{,}000{,}000$$
$$= 545 \text{ lb/sq. in.}$$

With the $1:1{\cdot}6:3{\cdot}2$ nominal mix, allowing $0{\cdot}4 \times u_w$ where temperature effects are included, there is available a margin of 700 lb/sq. in. to accommodate this.

The effect of temperature on the steel is to produce a tension stress of

$$f_{st} = 0{\cdot}0000055 \times 103 \times 0{\cdot}52 \times 30{,}000{,}000$$
$$= 8{,}800 \text{ lb/sq. in.}$$

The only effect of this is to balance the compression stress already in the steel from self-weight and wind.

(b) Tension side of chimney.

Owing to the chimney-section already being cracked on this side, the restraint causing the temperature stresses will be less than if the shell were uncracked. Accordingly the total compressive

stress in the concrete will be something less than 545 lb/sq. in., and
the total tensile stress in the steel will be less than

$$\underset{\text{(from self + wind)}}{6500} + \underset{\text{(from temperature)}}{8700} = 15,300 \text{ lb/sq. in.}$$

(iii) *Stresses at thickenings*

Special consideration of temperature effects is required at local
thickenings to the shell – as for example the breasts, any head-
feature, and the corbel-rings supporting the brick lining. Corbel-
rings should be slotted vertically, partly to assist ventilation of the
cavity, but also to reduce the effect of temperature stresses set up
by the sudden change in shell thickness.

ART. 13.8. FOUNDATIONS FOR TALL CHIMNEYS

SPREAD FOUNDATION FOR CHIMNEY 250 FT. HIGH

Figure 13.8 shows a spread-foundation, as constructed for a

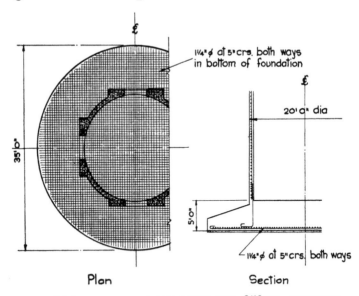

Plan   Section

FIG. 13.8. SPREAD FOUNDATION FOR 250 FT. CHIMNEY

250 ft. chimney. The foundation diameter, D, is 35 ft., and its
thickness is 5 ft.

Total weight of chimney + base, $W = 895$ tons.

Overturning moment at level of underside of base, $M = 5670$ ft. tons.

Thus eccentricity of load, $e = \dfrac{5670}{895} = 6 \cdot 33$ ft.

For the whole of the underside of the base to be in compression, the eccentricity e must not exceed

$$\frac{D}{8} = \frac{35 \text{ ft.}}{8} = 4 \cdot 38 \text{ ft.}$$

In the present case e considerably exceeds this. And since tensile stresses cannot develop under the base, the pressure distribution

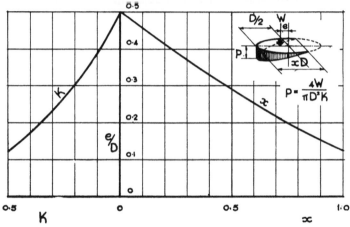

FIG. 13.9. PRESSURE DISTRIBUTION UNDER ECCENTRICALLY LOADED CIRCULAR BASE

must be such as to present a resultant force at an eccentricity of $6 \cdot 33$ ft. Clearly the labour to determine such a distribution under an unknown extent of circular base could be considerable.

Accordingly, the authors have prepared the design curves given in Fig. 13.9. For any value of $\dfrac{e}{D}$, the constant K can be read off directly, and the maximum edge pressure p determined by substitution in the formula

$$p = \frac{4W}{\pi D^2 K}.$$

Thus in our case $\qquad \dfrac{e}{D} = \dfrac{6 \cdot 33}{35 \cdot 0} = 0 \cdot 181,$

so that, from Fig. 13.9, $K = 0\cdot39$

and $$p = \frac{4 \times 895}{\pi \times 35^2 \times 0\cdot39} = 2\cdot4 \text{ tons/sq. ft.}$$

Figure 13.9 also gives x, the proportion of base diameter under compression. Thus, in our case, $x = 0\cdot81$, so that the length of base under compression is $0\cdot81 \times 35$ ft. $= 28\cdot3$ ft. From this information the shear forces and bending moments can be calculated without difficulty.

PILED FOUNDATION FOR CHIMNEY 350 FT. HIGH

Figure 13.10 shows a suitable arrangement of piled foundation for a tall chimney: in this case the chimney was 350 ft. high. The piles are arranged in four concentric circles, the pitch-circles and number of piles per circle being as follows:

Pitch-circle diameter	No. of piles
ft.	
18	12
27	18
36	24
45	30

Total No. of piles $= 84$

The piles in the two outer circles were set to a rake of 1 in 5. Thus under eccentric loading from wind, the more heavily loaded piles on the leeward side derived support over a greater area of the better ground lower down. And the loads in the piles, having a horizontal component, were available to resist the wind shear.

Total weight of chimney + pile-cap, $W = 2100$ tons
Overturning moment at level of underside of
 pile-cap, $M = 14{,}700$ ft. tons
Horizontal shear (due to wind), $H = 90$ tons

Therefore load per pile from direct load $= \dfrac{2100}{84} = 25\cdot0$ tons.

If the piles in each pitch-circle are treated as an annular ring of unit thickness, we have the second moment of area about the chimney axis

$$I = \Sigma\pi(\text{rad})^3 t$$
$$= \pi(22\tfrac{1}{2}^3 + 18^3 + 13\tfrac{1}{2}^3 + 9^3) \times 1$$
$$= 64{,}300 \text{ ft}^4.$$

Thus maximum load per foot of outer ring due to the overturning moment is

$$\pm \frac{M}{z} = \pm 14,700 \times \frac{22\frac{1}{2}}{64,300} = \pm 5\cdot14 \text{ tons/ft.}$$

As the piles in the outer ring are at 4·75 ft. centres, the load per pile due to overturning $= \pm 5\cdot14 \times 4\cdot75 = \pm 24\cdot4$ tons.

The loads in the extreme piles therefore vary from maximum load on leeward side $= 25\cdot0 + 24\cdot4 = 49\cdot4$ tons, to minimum load

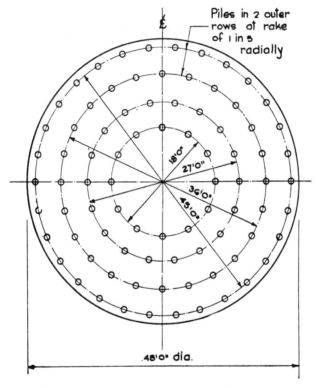

FIG. 13.10. PILED FOUNDATION FOR 350 FT. CHIMNEY

on windward side $= 25\cdot0 - 24\cdot4 = 0\cdot6$ tons. Note that none of the piles are in tension: a desirable condition where oscillating structures are supported.

The process of checking the horizontal reaction provided by the raking piles is simple, though laborious. It will not be given here *in extenso*. Simply, the load in each pile depends on its distance from the chimney axis, and each can be determined by direct proportion. For piles set at a rake of 1 in 5, the horizontal component is one-fifth, and this has then to be resolved to find the useful component directed against the wind force. The piles on the leeward side carry the greatest load, and accordingly contribute favourably against the wind: the piles on the windward side act in the wrong direction, but the force they give is relatively small. In the case in question, the net lateral reaction available is more than 120 tons, whereas the force to be resisted is only 90 tons.

ART. 13.9. CONSTRUCTION DETAILS

Owing to their extreme exposure, special precautions are necessary in the construction of tall chimneys. The concrete mix should be

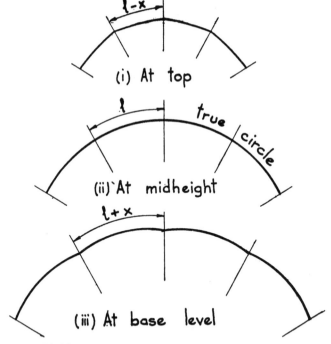

FIG. 13.11. METHOD OF FORMING TAPERED CHIMNEY

given close attention, and the water content carefully controlled (see Article 2.2). A smooth unpitted face will prevent the start of atmospheric corrosion, and for this reason formwork properly lined is better than timber formwork. Reinforcements should be given 2 in. cover: more than this is undesirable, as it reduces the effectiveness of the reinforcement in resisting temperature cracking.

No vertical construction joints should be allowed. Horizontal construction joints will be arranged at intervals to make for an exact number of shutter lifts between the corbel rings supporting the brick lining. Laps in reinforcements will in turn be made to coincide with the shutter lifts. All reinforcement laps should be staggered so that only one bar in every three lap at any level. For economy and comfort, scaffolding is provided as an internal cage only, generally incorporating a hoist: because of this, and the danger of wind snatching at shutter panels while they are being erected or struck, shutter lifts should be limited to about 3 ft. or 3 ft. 6 in.

Formwork for tapered circular chimneys is achieved as follows. Each shutter panel is made to a slight taper. Only at its mid-height is the chimney truly circular: below and above this, the chimney is made of a series of cusps as shown in Fig. 13.11. At the bottom, the chimney cusps provide an excess of curvature making for slightly re-entrant intersections on the external face: whereas at the top of the chimney the intersections produce slight arrises. These effects are only perceptible to those who know. The shutters are of course made first to suit the lowest lift of the chimney, and for each subsequent lift a narrow strip is removed from one side of each shutter panel. Needless to say, there is considerable skill required in this technique, and some contractors are better at it than others.

Every tall chimney should be provided with effective lightning protection: also a full-height metal ladder (with removable bottom length to prevent unauthorised and irresponsible use).

REFERENCES

1. A.C.I. STANDARD (505–54). *Specification for the Design and Construction of Reinforced Concrete Chimneys.* American Concrete Institute, (1954).
2. *Report on the Emission of Soot, Ash, Grit and Gritty Particles.* Electricity Commissioners. (1932).
3. STANTON, T. E. Experiments on Wind Pressure. *Proc. I.C.E.* Vol. CLXXI. (1907–1908).
4. DRYDEN and HILL. *Wind Pressures on Circular Cylinders and Chimneys.* U.S. Department of Commerce. No. 221. (1930).
5. TAYLOR, GLENDAY and FABER. The Design of Ferroconcrete Chimneys. *Engineering.* (March, 1908).
6. TAYLOR and TURNER. *Reinforced Concrete Chimneys.* Concrete Publications Limited. (1948).

CHAPTER 14

ROADS AND PAVINGS

ART. 14.1. INTRODUCTION

THE main requirements of good road construction are three-fold. The road has to present the user with a permanent, hard, even surface, laid to falls and gradients so as to be self-draining, and of a texture as far as possible free from slip. It has also to serve the function of spreading the wheel-loads so that the subgrade soil is not overstressed. It has further to protect the subgrade from effects of weather, such as rain which would cause softening, and other seasonal influences which in certain strata cause swelling or settlement or frost-heave.

Although riding quality is of first interest to the road-user, the success of road design depends fundamentally on efficient protection of the subgrade soil from traffic overstress and the effects of weather. Concrete has advantage over other road-making materials in that it is considerably more rigid than the subgrade, and has some tensile strength. Thus a thin concrete slab has sufficient structural rigidity to spread wheel-loads over a considerable area of subgrade, whereas other road-making materials require a greater depth of construction to achieve the same effect. Concrete also forms a satisfactory umbrella to protect the subgrade from detrimental effects of weather. Unfortunately, concrete is also subject to movements resulting from temperature and moisture-changes and initial shrinkage. Due to friction on the underside of the concrete slab, these movements can all produce tensile stresses even when the slab is not loaded by traffic, and if the slabs are laid continuously in uninterrupted lengths, cracking will occur. In certain conditions of extreme temperature, concrete roadways may fail by compression and upward buckling, unless expansion joints are provided at suitable intervals. Temperature and moisture gradients through the slab also induce warping, and if the resulting strains cannot relieve themselves through the provision of suitable joints at frequent intervals, cracking will ensue.

Thus in the design of concrete roads attention must be given to the provision of suitable joints for expansion, contraction and the effects of warping. And since on contraction the slab can be in considerable tension, the reinforcement (if any) and the distances

432

between joints must be interrelated. This aspect of concrete road design is as important as any other. The joints themselves – whether for expansion, contraction or warping – require careful treatment. While being a most necessary feature in road design, they are also the most undesirable. They tend to spoil the riding-qualities of the road, interfere with the process of laying the concrete, frequently contain the poorest concrete of any part of the road, and are without question the greatest source of weakness. To be successful, the joints require sealing in order to prevent the ingress of water and grit, while at the same time allowing the joints to act in a flexible manner. The questions of the spacing and construction of the various forms of joint are dealt with later.

The other major feature in road design is the structural strength of the slab itself. This depends on the quality of the concrete, its thickness, and the weight and arrangement of reinforcement. The quality and thickness of the concrete will influence the amounts of linear and warping movements, as well as the maximum advisable distances between the joints. The strength required of the slab will depend on the traffic intensity and the strength of the sub-grade soil. Traffic intensity is in no way a precise factor: it can be measured in terms of the number of vehicles passing in unit time, or in terms of the heaviest vehicle ever likely to pass and whether this is likely to be when the slab is in its most unfavourable condition as a result of temperature, moisture and shrinkage conditions. The strength of the subgrade soil is likely to vary over a considerable range in short distances along any road, partly as a result of the condition of the soil as presented in nature and partly as a result of the degree of disturbance caused by the excavation work carried out in setting the levels of construction.

It will be appreciated then that road design is exceedingly complex, being concerned with so many indeterminate variables. In practice it is generally recognised that for economy concrete roads cannot be designed to a standard aimed at perfection. A certain amount of cracking is generally considered acceptable provided the amount of cracking is suitably restricted. There is no need for a high factor of safety in the sense that is provided in designing other structures: failure of a concrete road is unlikely to have the dire physical consequences that one would associate with the failure of an important superstructure. The effect of road failure will normally be restricted to the cost and inconvenience of repair, and design is concerned with striking a suitable compromise between first construction cost and subsequent maintenance and repair costs.

Various attempts have been made from time to time to produce

29

calculation methods of design for concrete roads, notably by Westergaard[1,2]. Westergaard's theory applies to slabs of finite extent, and assumes that the concrete slab is homogeneous with uniform elastic properties, and that the reaction of the subgrade is proportional to its deflection. This latter assumption is probably the weakest point in the theory, since it is known that warping strains resulting from shrinkage and temperature and moisture gradients can interfere considerably with the uniformity of contact between the slab and the subgrade. Westergaard derived

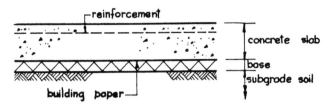

FIG. 14.1. FUNDAMENTAL MAKE-UP OF CONCRETE ROAD

expressions for the stresses in the slab due to loading at three vital and typical positions, these being at the interior, on one edge and at one corner. The results of his findings are certainly of qualitative interest, and show for example that the tensile stress in the slab due to a corner load is roughly equal to the stress due to an edge load, whereas the stress due to interior loads are not more than about half such values. Corner-load stresses, of course, become more severe with acute-angled slabs which should be avoided in all cases.

But most engineers feel that the variation in quality and thickness of the concrete slab is likely to be greater than any painstaking attempts at calculation would justify; though the theory is of interest as a basis for the comparison and analysis of practical experience. It is therefore general in Great Britain at the present time to design concrete road construction almost entirely by experience, and it is the aim of the present chapter to give certain indications in this regard. In what follows reference is made to the *concrete slab*, *base*, and *subgrade soil*. The definition of these is indicated in Fig. 14.1.

ART. 14.2. SUBGRADE SOIL

The specific strength of the subgrade soil has less influence on its suitability for supporting a concrete road than does the *uniformity* of support it provides. The stresses in a concrete slab of uniform thickness increase roughly in proportion as the uniformity of support from the subgrade decreases. Lack of uniformity arises mainly from warping of the slab due to shrinkage and temperature and moisture gradients, and also from variation in the condition of the soil. Certain soils, notably clays, vary in volume according to their moisture content, so that varying conditions of moisture (particularly in the period immediately following the construction of the road) may cause uneven support. Other soils, notably sands and gravels, compact in time under repeated loading, and intense traffic loads in limited areas or along limited tracks are liable to cause settlements and uneven support. The actual strength of the subgrade soil is far less important because the resultant increased deflections normally produce more gradual bending in the road slab over larger areas and consequently very little increase in slab stress.

The most important consideration is the achievement of thorough and uniform compaction to avoid variable settlements. With clay soils it is advisable to skim off the last 3 in. immediately prior to putting down the base: this will reduce disturbing the natural moisture content of the soil.

ART. 14.3. BASE

The function of a base is to give a clean working surface, easily trimmed to levels and gradients, on which the concrete slab can be laid without damage to the subgrade. The structural benefit from a base of any reasonable thickness is negligible in comparison with the structural merits of the concrete slab: and it is normally uneconomic to thicken the base with a view to reducing the thickness of the slab, though exceptional occasions may occur where large quantities of stone can be delivered easily and cheaply. Normally 3 in. is considered an adequate thickness for the base, —except in certain soils, particularly silts, chalk and some clays, where additional thickness may be advisable in order to keep the underside of the base 14 in. or 16 in. below ground level as a precaution against frost-heave. The conditions most likely to produce frost-heave are severe and prolonged frost in locations where ground-water level is close to the surface.

The base should consist of broken stone or other hard and durable material, suitably graded, and mechanically compacted, to form a solid compacted mass. The upper surface of the base should be well blinded to give a smooth even surface which should in turn be covered with waterproof paper generously lapped at all joints. The combined effect of the smooth surface to the base and the waterproof paper considerably reduces the friction force between the base and the underside of the concrete slab, thus allowing the slab to shrink relatively freely on initial hardening of the concrete before it has attained its full tensile strength. The waterproof paper also prevents loss of grout and mortar into the base, which would weaken the lower part of the concrete slab. The use of paper on a rough unblinded base is of little value, since the weight of the wet concrete causes the stones in the base and the aggregate in the concrete to puncture the paper, allowing the mortar to run through into the base, thus forming a mechanical key and a weak underside to the slab.

ART. 14.4. REINFORCEMENT

In Great Britain, it is general now to reinforce all concrete roads. (However, the question of unreinforced road slabs is referred to later in this chapter.) The amount of reinforcement is determined in the main by experience and not by calculation; the amount required depends as much on the spacing of joints as on any other consideration. Its purpose is not to prevent cracks, but to control the formation of cracks so that they shall be many in number but each negligible in width. The reinforcement then further serves to hold the concrete on both sides of such hair-cracks closely together so that relative movement of the slab on either side of the crack is prevented and in this way wear of the crack faces and spalling of the concrete is avoided. There is no objection to such hair-cracks so long as these do not impair the shear strength of the slab nor allow water to seep through to the subgrade. Recommended weights of reinforcement are given in Table 14.2.

It is not the prime purpose of the reinforcement to enhance the structural strength of the slab in flexure; accordingly there is generally no advantage in using high tensile steel as against ordinary mild steel, since it is the elastic modulus rather than the ultimate strength of the steel which determines the quantity required. Based on the above considerations, it would appear that it can make little difference whether the reinforcement is placed near the top of the slab or near the bottom, and this has to some extent been borne out in practice. However, where the weight of reinforcement required is considerable, it may prove convenient to reinforce the slab in both the top and bottom faces: indeed in severe conditions where a considerable weight of reinforcement is required such as on embankments or subgrades of very plastic clays or organic matter, there may be some slight structural advantage in a double system of reinforcement.

However, where only one layer of reinforcement is provided, this is frequently arranged near the top of the slab, and for the following reasons. Westergaard's analyses show that the most severe tensile stresses occur in the slab due to corner- and edge-loading where cantilever action results. And the severe warping which can result on the initial hardening of the slabs will again cause cantilever action. (These two effects are indicated in Fig. 14.2.) Furthermore the reinforcement near the top of the slab is less likely in the

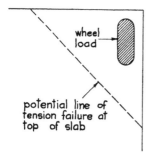 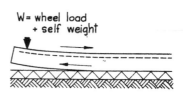

(a) Plan of Load at corner of slab.

(b) Warping due to resisted and unequal shrinkage.

FIG. 14.2. ARGUMENTS FOR REINFORCING TOP FACE OF SLAB

disturbance of placing and vibrating the concrete to be worked down on to the base.

Reinforcement for road work is normally provided in made-up fabric which should always be obtained in flat mats and never in rolls which are liable to spring in the plastic concrete before setting and so move considerably out of the required position. Fabric is normally made up into sizes to suit the widths of construction and positions of joints with a view to avoiding waste and unnecessary lapping. The proportion of reinforcement to be arranged in any direction will depend on the design and spacing of joints. These are discussed later. In rectangular panels, the greater weight of the reinforcement should be arranged parallel to the longer side.

ART. 14.5. THE CONCRETE SLAB

Good concrete roads are only achieved with properly designed concrete mixes (see Art. 1.8), and the concrete well compacted to give a minimum compressive strength of not less than 4,000 lb/sq. in. at 28 days. Whatever attention is given to other aspects of design and construction work, the road will have little chance of success if the concrete slab itself is not of uniform thickness and quality throughout. Formwork has to be substantial and accurately set to line and level, compaction of the concrete has to be thorough both in degree and extent, and the finishing and smoothing processes, for reasons of drainage and ridability, require careful control within fine limits. A skid-resistant finish is then normally given by brushing. Curing is commonly effected by covering the concrete with an impervious membrane of resins and gums applied in solution by spray equipment: this membrane retains the moisture in the concrete for about fourteen days without in any way disturbing the finish or weakening the surface mortar; and later the membrane removes itself entirely by evaporation.

In America great use has been made of air-entrained concrete (*see* Art. 1.9) for road construction work, but this technique has so far been little employed in Great Britain. Initially air-entrained concrete was used in America because of its resistance to attack by frost: but later it was found to have also the properties of easier workability, and less segregation and bleeding in transport; and the sand content in the mix could be somewhat reduced. These latter properties may yet be found sufficient to encourage the use of air-entraining agents in concrete work in Great Britain.

Slab thicknesses are based principally on experience, and following the collection and examination of data over a very wide range of roadwork and over a considerable period of time by the Roads Research Laboratory, the question of slab thickness need no longer be settled in any haphazard manner. In the Road Note, No. 19, *The Design Thickness of Concrete Roads*, published by H.M.S.O., 1955[3], recommended slab thickness are given for different soil conditions and various traffic intensities. Suitable base thicknesses and reinforcements are also given. Tables 14.1 and 14.2 are based on tables given in Road Note No. 19. Table 14.1 classifies traffic intensities into six main groups; and recommended thicknesses for slabs and bases, and amounts of reinforcements for these different traffic intensities are given in Table 14.2. These recommendations are based on transverse and longitudinal cracking being properly controlled by correctly-spaced joints with suitable load-transfer devices where necessary.

ART. 14.6. JOINTS IN SLAB

All joints in concrete road slabs are undesirable to the road user because, unless they are perfectly formed, they spoil the riding-quality of the road. Furthermore, they are the weakest part in the whole of the construction, and if the slabs on both sides of the road are not at the same level, or if there is irregularity of the concrete surface at this point, considerable impact-forces will occur on the very part of the slab which is least able to cope with such forces.

TABLE 14.1. *Classification of traffic*

Type of traffic	Total daily average flow of commercial vehicles* (both directions)	Approx. total weight of all traffic (tons/day)	Column of slab thicknesses and weights of reinforcement to be used in Table 14.2
Very heavy	> 4,500	> 36,000	For traffic flows of about 4,500 commercial vehicles/day, the thicknesses given in column (a) should be used. For traffic flows much in excess of this amount it may be necessary to strengthen the concrete slabs either by thickening or by using heavier reinforcement or both.
Heavy	3,000–4,500	24,000–36,000	Column (a)
	1,500–3,000	12,000–24,000	Column (b)
Medium-heavy	450–1,500	4,000–12,000	Column (c)
Medium	150–450	1,500–4,000	Column (d)
Light	45–150	< 1,500	Column (e)
Very light	15–45	—	Column (f): these two categories are combined because, owing to the need to allow for construction traffic, it would be inadvisable to use thinner slabs for the "exceptionally light" category.
Exceptionally light	0–15	—	

* The description "commercial vehicles" used in this Table is intended to include all public service vehicles, and all commercial vehicles having an unladen weight of more than 30 cwt.

All joints require to be sealed against ingress of water and grit with sealers as described later. All rebates to receive sealers should normally be about ¾ in. wide by 1 in. deep, the top arrises of the rebate being rounded off to a pencil radius. In rounding the arrises,

TABLE 14.2. *Design thicknesses of slabs and bases, and weights of reinforcement for different soil conditions and traffic intensities*

Description of subgrade	Thickness of base	Thickness of slab and amount of reinforcement Expected intensity of traffic:					
		(a) Very heavy and heavy	(b) Heavy	(c) Medium-heavy	(d) Medium	(e) Light	(f) Very light and exceptionally light
		Reinforced slabs					
(1) *Normal subgrades* [i.e. all those not included in categories (2)–(6) below]	3 in. except on gravel, sand or gravel-sand-clay subgrades which can be thoroughly compacted and where no base is needed	10 in. with not less than 8½lb/ sq.yd. of reinforcement	9 in. with not less than 7½lb/ sq.yd. of reinforcement	8 in. with not less than 7½lb/ sq.yd. of reinforcement	7 in. with not less than 4½lb/ sq.yd. of reinforcement	6 in. with not less than 4½lb/ sq.yd. of reinforcement	5 in. with not less than 3½lb/ sq.yd. of reinforcement
(2) *Very stable subgrades* Well-compacted and undisturbed foundations of old roads, solid rock, well-graded gravels compacted to an air-void content of 5 per cent or less and having a C.B.R. of not less than 100 per cent at the highest moisture content likely to occur in the road	No base needed except where a levelling course is required	Slab thickness may be 1 in. less than that needed for normal subgrades. Reinforcement to be the same as for normal subgrades					
(3) *Subgrades very susceptible to non-uniform movement* Organic or highly plastic clays having a C.B.R. of 2 per cent or less, or subgrades containing pockets of peat within a depth of 15 ft. below the surface	Up to 6 in.	Slab thickness to be 1 in. greater than that needed for normal subgrades. Reinforcement to be the same as for normal subgrades					
(4) *Embankments more than 4 ft. high*	No base needed if embankment consists of a gravel, sand or gravel-sand-clay which can be thoroughly compacted. On other materials a 3-in. base is needed	Slab thickness to be 1 in. greater than that needed for normal subgrades. Reinforcement to be the same as for normal subgrades					

(5) Subgrades where the water-table may rise to within 2 ft. of the formation	No base needed on a gravel, sand or gravel-sand-clay which can be thoroughly compacted. Base up to 6 in. thick needed on subgrades which are very susceptible to non-uniform movement [as in (3) above]. Base of 3-in. thickness on other subgrades	Slab thickness to be 1 in. greater than that needed for normal subgrades. Reinforcement to be the same as for normal subgrades	
(6) Chalks and other soils susceptible to frost heave	Where no unreasonable inconvenience or danger to traffic can be permitted [as on roads carrying the types of traffic listed under headings (a), (b), (c) and (d)], a granular base must be used to make up the total thickness of slab plus base to 14 in.	Slab thickness and reinforcement to be as for (1) or (3) above, whichever is appropriate	
	Where some occasional inconvenience to traffic is acceptable [as on roads in the interest of economy [as on roads under headings (e) and (f)]: (i) on chalk no base is needed		(i) Slab thickness and reinforcement to be as for (2) above
	(ii) on other soils the base should be as in (1) or (3) above, whichever is appropriate		(ii) Slab thickness and reinforcement to be as for (1) or (3) above, whichever is appropriate
	Unreinforced slabs		
All categories of subgrade	Base to be as needed for reinforced slabs	Unreinforced slabs are not recommended for traffic categories (a), (b), (c) and (d)	For traffic categories (e) and (f) the thickness of unreinforced slabs should be 2 in. greater than for reinforced slabs

a check should be made by straight edge across all joints to see that the road surface is truly flush and has in no way been interfered with by the formation of the joint. The sealer has, of course, to be watertight to keep water from getting down to the subgrade soil. It also has to adhere to the edge of the concrete slab, under conditions of tension, and be able to stretch without cracking. It further has to be hard enough to keep out grit and stones which would become crushed in the joint, when under compression, causing spalling of the concrete or, in the case of expansion joints, reducing their efficiency. The sealer has further to be able to melt on heating so that it can be readily applied, but on the other hand must not soften sufficiently in hot weather for it to squeeze up and out of the joint when it would be spread by traffic crossing the surface causing disfigurement of the road and being no longer available to act when the joint opens up in cold weather.

The most satisfactory durable material found for the purpose of sealing joints is a rubber-bitumen compound. The concrete surface to which it is to be applied should be dry and coated with a special primer to assist adhesion. It is found that seals 1 in. deep generally give adequate protection, with good adhesion and suitable stretching characteristics, without producing excessive upward-squeeze. Different grades and hardnesses of rubber bitumen compounds are available for different situations: for example a harder bitumen is desirable in tropical climates, or on roads where grit or stones are likely to be brought in by industrial traffic or otherwise, whereas on susceptible clay soils the certainty of successful sealing against water ingress may outweigh other considerations and a softer more pliable sealer may be appropriate.

Various forms of joint have been tried, but the types shown in Fig. 14.3 known as *expansion joints, complete contraction joints* and *partial contraction joints* fulfil all requirements and are generally most satisfactory.

EXPANSION JOINTS

Expansion joints are required at intervals depending on the net expansion of the concrete likely due to extreme conditions of temperature, and on the thickness and compressibility of the flexible filler. The purpose of the expansion joints is to prevent compression failures which may appear in the form of upward buckling of two adjacent slabs. In addition to the temperature effect, expansion can also be caused by moisture content of the concrete, but it is not normal to consider these two effects as being additional, since moisture and extreme conditions of temperature do not normally go hand-in-hand. In considering the net expansion of the concrete, some allowance is made for the initial shrinkage of the concrete on hardening which generally exceeds the amount of expansion. But the cracks which form as a result of shrinkage may in time become filled with grit so that after a period of perhaps

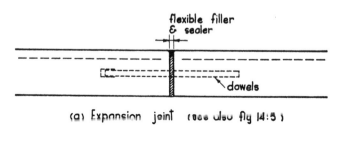

(a) Expansion joint (see also fig 14:5)

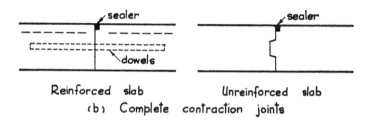

Reinforced slab   Unreinforced slab
(b)   Complete   contraction   joints

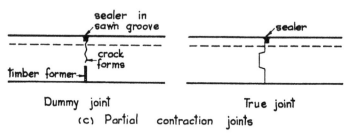

Dummy joint   True joint
(c) Partial   contraction   joints

FIG. 14.3. TYPES OF JOINTS FOR SLABS

10 or 20 years they are no longer free to accommodate the movements arising from temperature differences.

It is of interest that the temperature gradient in a slab in the hottest weather will also be such as to cause the slab to warp in a hogging fashion as shown in Fig. 14.4 so that the eccentric end-loading, due to the linear compression, will contribute favourably to the effect of the weight of the slab in acting contrary to upward buckling.

Expansion joints spaced widely apart in reinforced concrete roads cause considerable tensile stresses in the slab on cooling. However, in the case of unreinforced roads this is not so, since it is normal practice to provide complete contraction joints at about

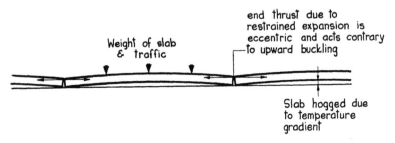

FIG. 14.4. SLAB EXPANDED DUE TO TEMPERATURE RISE

15 ft. centres in order to prevent the development of shrinkage stresses or warping stresses which would cause the unreinforced slab to crack. It is, therefore, normally permissible to space expansion joints in unreinforced slabs further apart than in reinforced slabs. Indeed the shrinkage at 15 ft. centres between plain concrete slabs may be sufficient to avoid the need for expansion joints altogether, provided the joints are properly sealed against ingress of grit, and except when concreting is carried out in extreme cold conditions.

In Great Britain expansion joints are frequently $\frac{3}{4}$ in. thick and spaced apart some 100 ft. or more for normal road work, but for thin slabs, which generally have less reinforcement available for drawing the ends in on cooling, the spacing of joints should be reduced to about 50 ft. intervals. Since the edges of slabs are their weakest points, it is normally necessary to provide load-transfer devices as shown typically in Fig. 14.5. The dowel bars are concreted into the slab at one side of the joint with direct bond between the dowel and the concrete, whereas on the other side of the joint there is complete freedom between the dowel and the concrete to allow for taking up the expansion movement. This freedom is obtained by dipping the dowels first into a special bitumen solution and then leaving them to dry in an upright position before incorporating in the work. The bitumen solution is a mixture as follows:

200 penetration bitumen – 66 per cent by weight;
Light creosote oil   – 14 per cent by weight;
Solvent naphtha   – 20 per cent by weight.

To allow freedom at the free end of the dowels, a metal or card-board sheath 3 in. long is provided with the end space filled with ¾ in. of flexible material to accommodate the necessary movement.
The dowel bars are normally 1 in. diameter, 24 in. long, arranged at 12 in. centres except for the very heaviest conditions of loading where it may be desirable, depending on the subgrade, to close the dowels to about 9 in. centres. For very light traffic conditions where thin slabs are used it is not normal to provide dowels, and indeed they might do more harm than good. Great care is required in setting the dowels accurately in position and at right angles to

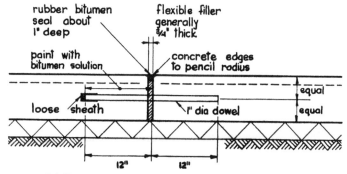

FIG. 14.5. EXPANSION JOINT: SHOWING DOWELS USED AS LOAD-TRANSFER DEVICE

the line of the expansion joint, otherwise they bind on the concrete when the joint moves, reducing the efficiency of the joint and in time spalling the concrete. Extra care is required in compacting the concrete between the dowels since the obstruction caused by the dowels is liable to interfere with successful compaction, yet this is the very position where perfect compaction is essential.

Other methods of strengthening joints have been tried such as the provision of a separate continuous strip of slab under the joint, or the thickening-up of the road slab itself, but these methods all interfere with the proper and uniform compaction of the base, so giving weakness to the whole construction at the very point where enhanced strength is required.

The flexible filler provided at expansion joints clearly must be readily compressible at small unit pressure in order that the joint will readily close. On the other hand, it must be sufficiently resilient to recover a reasonable amount when the joint re-opens, otherwise the sealer will not be given proper support and may be worked down into the joint, allowing grit and stones to become jammed in the upper part of the joint; furthermore, considerable movements may well have occurred before the seal is first applied and in its molten condition it must be prevented from running down the side of the flexible filler. The filler, which has been found most suitable for the purpose, being efficient and yet inexpensive, is composed of long springlike cellular cane fibres, and is protected

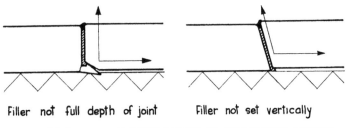

Filler not full depth of joint   Filler not set vertically

FIG. 14.6. POOR WORKMANSHIP AT EXPANSION JOINTS

against rot by impregnating with a bitumen composition, the quantity of which is balanced to preserve the cellular structure and retain the air spaces. Care is required to ensure that the flexible filler is arranged continuously and for the full depth of the joint (except at the top where the sealer is poured), otherwise concrete will run into the intervening spaces, preventing the joint from closing up on expansion, and setting up very high stresses which may result in spalling and damage to the road structure. Care is also required to ensure that the flexible filler is arranged truly vertically since otherwise, on expansion, the slab on one side of the joint may ride up over the slab on the other side. These two defects are illustrated in Fig. 14.6.

CONTRACTION JOINTS

Potential discontinuity of the concrete is required at all contraction joints. This is achieved in various ways. On small-scale work, where the concreting is split up into small bays, a bitumen solution is sometimes painted on the hardened face of the panel concreted first, or alternatively a strip of building paper is interposed; though it is doubtful whether any such care is justified, since provided the hardened face is not hacked it seems clear that the interruption in the concreting procedure must create a plane of weakness at this point, so that the slab will certainly form a clean crack here without any further encouragement. On more extensive road work a dummy contraction joint is formed by sawing along the surface of the concrete a continuous groove of appropriate depth to receive the sealing compound. This groove, in combination with a timber former previously cast in the under-side of the slab, causes a plane of weakness along the slab which subsequently breaks through giving the desired freedom for con-traction (*see* Fig. 14.3(*c*)). It is important that the timber former is set truly vertical.

Where contraction joints are formed by a definite break in the concreting operation, it is normal to form the rebate for the sealer entirely on one side of the joint as shown in Figs. 14.3(*b*) and (*c*). If an attempt is made to construct the rebate centrally about the joint the dimensions of the formwork for the two halves of the rebate become so small that the likelihood of success of uniformity and soundness of concrete is reduced although the cost is increased.

(i) *Complete contraction joints*

By reference to Fig. 14.3 it will be noted that in the reinforced slabs the reinforcement at complete-contraction joints is inter-rupted, whereas at partial-contraction joints it is continued through. This has the effect that whereas on expansion in hot weather the whole of the movement is accommodated at the expansion joints, on contraction in cool weather the movement is shared equally by the opening-up of the expansion and complete-contraction joints. Very little opening-up occurs at the inter-mediate partial-contraction joints because the reinforcement holds these together. This arrangement is unsatisfactory, because the complete-contraction joints contain no flexible filler to support the sealer, and in time the joint becomes inefficient and a source of trouble. In view of the greater movement at complete-contraction joints, as against the movement at partial-contraction joints, no simple rebated arrangement will act efficiently in load-transfer without causing spalling with eventual failure, so that the full dowel bar treatment as described for expansion joints becomes necessary though without the flexible filler (*see* Fig. 14.3(*b*)). It will be seen, therefore, that in reinforced roads, complete-contrac-tion joints are very nearly as expensive to form as expansion joints, but are unsatisfactory in operation. It is, therefore, better to avoid the use of complete-contraction joints, and rely entirely upon expansion joints.

On the other hand, in unreinforced roads, complete-contraction joints are required at regular intervals to prevent the development of uncontrolled shrinkage cracks. Shrinkage cracks would occur at

intervals of about 30 ft. if an unreinforced slab were concreted for
an indefinite length, but with the incidence of traffic this interval
is likely to be reduced to about 15 ft. Accordingly it is normal in
unreinforced slabs to provide complete-contraction joints at 15 ft.
intervals. These joints also serve to relieve the slab of stresses
which would otherwise arise as a result of restrained warping
effects. Since the complete-contraction joints in unreinforced slabs
are so closely spaced, the objection which arises with complete-
contraction joints in reinforced slabs is absent, and it is satisfactory to provide simple tongue and groove rebating (see Fig. 14.3(b)).
With continuous-laying machines, the joints in unreinforced slabs
can be of the dummy type.

(ii) *Partial contraction joints*

In reinforced slabs it is necessary to provide partial-contraction
joints every 15 ft. for the reason as given immediately before for
the complete-contraction joints in unreinforced slabs. Here again
with small-scale work it is convenient to provide each joint with a
tongue-and-groove rebated face, whereas with continuous-laying
equipment, dummy joints are formed (see Fig. 14.3(c)).

LONGITUDINAL JOINTS

For reasons of convenience, longitudinal joints are normally
provided wherever the road width exceeds about 16 ft. And where
the road exceeds about 30 ft. width, the longitudinal joints are
spaced about 10 ft. to 12 ft. apart to coincide with the lanes of traffic.
Longitudinal joints in road-work do not require provision for
expansion unless the road is hemmed in on both sides between
retaining walls or substantial buildings : indeed in all other circumstances longitudinal expansion joints are undesirable. This is
because, due to lack of restraint at the edges of the road, there is a
fear that the moving to and fro of the slabs in alternate conditions
of hot and cold may result in the slabs drifting apart so that the
longitudinal joints become improperly wide, and the seals and load
transfer devices cannot function effectively. The drifting apart of
slabs may be aggravated by the settlement of the subgrade at the
edges of the road. For these reasons the only satisfactory form of
joint is as illustrated in Fig. 14.3(c) for the partial contraction joint
with the reinforcement running through to tie the slabs together.
Even with otherwise unreinforced slabs, it may be desirable to
provide a fabric tie between longitudinal joints.

ARRANGEMENT OF JOINTS

At the outset, detailed plans should be prepared of all concrete
road work and pavings, indicating the positions of each type of

joint. All joints should be arranged in-line, with no staggering. This may involve the introduction of radial or splayed joints at curves and intersections. Where in the past staggered joints have been permitted, sympathetic cracking has occurred in the unjointed slab as a continuation of the joint in the abutting slab.

As far as possible, joints should be arranged at right angles to one another, and where at intersections this cannot be achieved, the variation from a right angle should be an increase in angle rather than a decrease. Acute angles at the corners of slabs make for higher stresses with corner wheel-loads, and may result in the corner of the slab breaking off with a wide crack, allowing water through to the subgrade with consequent spalling of the concrete due to movements. Small panels of slab are objectionable since clearly in the limit they would be unable to spread wheel loads on to the subgrade: generally a minimum size of panel of 70 sq. ft. should be provided.

Having prepared a detailed plan showing all joints in accordance with the above, the reinforcing mats can be ordered to the sizes required, eliminating waste and unnecessary laps.

ART. 14.7. CURBS

Curbs are normally precast, and laid in 3 in. plain concrete so as to come against the edge of the road slab. A better construction is to cast the curbs in situ on the slabs, since this gives a monolithic joint between the curb and the slab at the point where rainwater collects and is most likely to seep down to jeopardise that part of the road which is inherently already the weakest; however, for convenience this method of construction is seldom adopted. Where precast curbs are laid at the edge of the road slab, it is important that the joint between the road slab and the curb is properly sealed.

The joints between the ends of adjacent precast curbs should be left clear by about one-sixteenth of an inch, with no buttering or pointing, and no special provision is required in the line of curbing for expansion. It is however necessary to see that no curbs stand astride joints in the road slab, since otherwise sympathetic cracking may appear in the curbs. Where in situ curbs are provided monolithic with the slab, joints must be provided consistent with the joints in the road slab, with flexible fillers at the expansion joints, and all joints properly sealed.

REFERENCES

1. WESTERGAARD, H. M. *Stresses in concrete pavements computed by theoretical analysis.* Public Roads, Washington. Vol. 7. No. 2. (1926).

30

2. TELLER, L. W. and SUTHERLAND, E. C. *Structural design of concrete pavements. An experimental study of the Westergaard analysis of stress conditions in concrete pavement slabs of uniform thickness.* Public Roads, Washington. Vol. 23. No. 8. (1943).
3. *The Design Thickness of Concrete Roads.* Road Note No. 19. Road Research Laboratory, H.M. Stationery Office. (1955).
4. SPARKES, F. N. and SMITH, A. F. *Concrete roads.* The University Press, Glasgow. (1952).
5. COLLINS, A. R. The design of concrete roads. *The Struct. Engr.* Vol. 34. No. 2. (February, 1956).
6. COLLINS, A. R. and SHARP, D. R. The design and construction of concrete roads overseas. *Proc. Inst. C.E.* Vol. 9. (January, 1958).

CHAPTER 15

PRESTRESSED CONCRETE

ART. 15.1. FUNDAMENTALS OF PRESTRESSED CONCRETE

In Chapter 3 we considered the behaviour of ordinary reinforced-concrete members (not prestressed) as they suffer the effects of bending, bending combined with direct compression, and shear and torsion. The chapters following Chapter 3 demonstrate the use of reinforced concrete members in various practical applications, including frames, bunkers, silos, reservoirs, tall chimneys, foundations, and so on. In all the foregoing, the fundamental principle has been, that, owing to its intrinsic weakness in tension, concrete is incapable (or at any rate unreliable) in resisting tensile stresses, and as a result, structural members in plain concrete would crack. Reinforcement is therefore built into the members to provide the tensile strength they otherwise would lack.

If the working tensile strength of a concrete is (say) 175 lb/sq. in., the stress in mild steel reinforcements would be $175 \times 15 = 3060$ lb/sq. in. if the concrete were not to crack. Clearly this would make poor use of the steel; and in all reinforced concrete members the steel is stressed far beyond this, so that in the tension zones, the concrete cracks, and the reinforcements resist the whole of the tension. In certain applications this cracking may be undesirable; for example, in reservoirs, silos and the like. Furthermore, with the materials behaving in this manner, very little benefit results from the use of high-strength concretes or steels in members subjected to bending: (see Effects of Varying Design Stresses in Article 5.2). Once the concrete has cracked, the analysis of stress in the member becomes complex; whereas prior to cracking the analysis is more simple, as was shown, for example, in the cases where tension does not develop for Bending and Direct Compression Combined in Article 3.2, or for Tall Chimneys in Article 13.5.

The underlying principle behind all *prestressed concrete* work is to stress the members artificially to a state of compression, so that under conditions of superimposed loading, the normal tendency for tensile stress to develop only goes towards nullifying the artificial *prestress*. Consider for example a rectangular beam, prestressed by jacks operating concentrically on both ends of the beam so that a compressive stress of + 1000 lb/sq. in. is developed uniformly over the total cross-section (*see* Fig. 15.1): if now the beam is subjected to bending stresses of ± 1000 lb/sq. in., the extreme top-fibre stress will be 1000 + 1000 = 2000 lb/sq. in., and the extreme bottom-fibre stress will be 1000 − 1000 = zero. Thus the beam is nowhere stressed in tension; though clearly the compressive stress in the top of the beam is higher than we have discussed previously for ordinary reinforced-concrete, and something

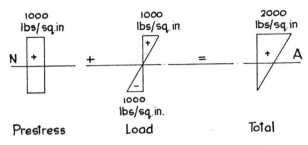

FIG. 15.1. SIMPLE EXAMPLE OF BEAM WITH CONCENTRIC PRESTRESS

better than the nominal mix strengths given in Table 1.3 will be required.

Suppose now the self-weight of our beam is such that it would produce a bending stress of ± 500 lb/sq. in. when the soffite-formwork is released, even before any useful load is applied. To carry the same superimposed load as before, we can of course increase the amount of applied prestress from 1000 lb/sq. in. to 1500 lb/sq. in.: then under full-load conditions, we have a top-fibre stress of

$$\underset{\text{(prestress)}}{1500} + \underset{\text{(dead)}}{500} + \underset{\text{(super)}}{1000} = 3000 \text{ lb/sq. in.}$$

and a bottom-fibre stress of

$$\underset{\text{(prestress)}}{1500} - \underset{\text{(dead)}}{500} - \underset{\text{(super)}}{1000} = \text{zero}$$

(*see* Fig. 15.2*a*). Thus the concrete strength needs to be even greater.

Alternatively, we can apply a smaller prestressing force at some suitable eccentricity below the horizontal axis of the beam so as to produce a hogging prestress giving − 500 lb/sq. in. at the top-fibre, and + 1500 lb/sq. in. at the bottom-fibre. Then, on striking the formwork, we have a top-fibre stress of

$$- 500 \quad + \quad 500 \quad = \text{zero}$$
$$\text{(prestress)} \quad \text{(dead)}$$

and a bottom-fibre stress of

$$+ 1500 \quad - \quad 500 \quad = + 1000 \text{ lb/sq. in.}$$
$$\text{(prestress)} \quad \text{(dead)}$$

and, under full-load conditions, a top-fibre stress of

$$- 500 \quad + \quad 500 \quad + \quad 1000 \quad = + 1000 \text{ lb/sq. in.}$$
$$\text{(prestress)} \quad \text{(dead)} \quad \text{(super)}$$

and a bottom-fibre stress of

$$+ 1500 \quad - \quad 500 \quad - \quad 1000 \quad = \text{zero (see Fig. 15.2}b\text{).}$$
$$\text{(prestress)} \quad \text{(dead)} \quad \text{(super)}$$

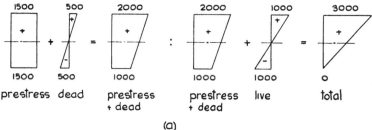

prestress   dead   prestress   prestress   live   total
                + dead   + dead

(a)

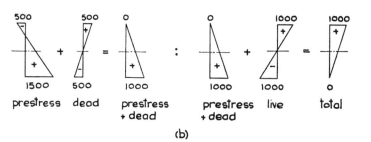

prestress   dead   prestress   prestress   live   total
                + dead   + dead

(b)

FIG. 15.2. SIMPLE EXAMPLE TO SHOW ADVANTAGE
OF ECCENTRIC PRESTRESS

Thus, in our rather over-simplified example, we have still achieved the condition of no tension : and the maximum concrete stress has been kept down to 1000 lb/sq. in. Other considerations arise, as will be discussed later.

In a practical application, where the applied bending-moment varies along the length of the member, clearly the matter is not as simple as suggested above. Either the amount of prestress has to show a surplus at intermediate positions along the member, or the eccentricity of the prestressing force has to be varied. Before considering the possibilities of the latter, the general mechanism of achieving the prestress needs to be considered. High-tensile steel wires or cables of very high ultimate strength (up to about 150 tons/sq. in.) are stretched from one end of the member to the other, and anchored at both ends. The force in stretching the wires is then reacted by the concrete member. There are two ways in which this can be done: *post-tensioning* and *pre-tensioning*.

POST-TENSIONED MEMBERS

Except for units precast in factories for repetitive work, as for example railway sleepers, floor-slab units, and precast piles, the greater engineering interest in prestressed work undoubtedly lies in post-tensioned work. This chapter deals mainly with such.

In some cases the wires or cables are assembled and lie freely in tubes or ducts carefully located in the formwork before the concrete is cast. In other cases, the ducts are left empty in the members as the concrete is cast, and the wires threaded through later: this way the ducts are sometimes formed by embedding in the concrete rubber cores which are drawn out in the early stages of the concrete hardening. Either way, the object is the same – to leave the wires free in the member, and unfixed by bond. When the concrete has acquired sufficient strength (generally after about 14–28 days), the cables are tensioned by hydraulic jacks acting against the ends of the member itself: (exceptions are referred to later). The ends of the cables are then anchored with the special devices described in Article 15.5, and the ducts filled completely with cement grout injected under pressure. The wires are thus protected against rusting, and are now doubly secured, primarily by the mechanical anchorage devices, and secondly by bond by the grouting. Indeed, it has been claimed that the anchorages can then safely be released and re-used, but invariably they are retained, and concreted over for protection against corrosion.

An elaboration of the simple principle of post-tensioning technique described above has been used with great success, by threading cables through a number of precast units and then prestressing the units together. This arrangement comes closest to the well-worn analogy of carrying a dozen books from one book-case to another: if you press the ends in (prestress the twelve books) they act like a single unit, resisting the bending and shears arising from their own weight. Thus a long member like a bridge of 300-ft. span, can be factory-made, in segments of perhaps 15 ft. length, each segment being cast with ducts accurately registering for later receiving cables passing through all segments from one end of the bridge to the other. The segments are assembled on staging, the wire drawn through and tensioned and anchored, and the ducts grouted as previously described. The compressive stresses set up by the prestressing force produce sufficient friction between the units to enable the composite structure to stand without risk of any individual units becoming displaced. The joints between the ends of the units require to be formed with particular care, either with cast-in-situ concrete, or with semi-dry mortar rammed well in. The advantage of precasting the work in segments in this manner, is that the casting and curing can be done under factory conditions where great strength can be obtained, and where the trials of cold, rain, and drying winds are reliably avoided. The use of high strength concrete in this way leads to a reduction in the quantity of concrete required, with reduction also in the steel quantity, so reducing the dead weight of the structure – a most important consideration in long-span constructions.

In post-tensioned works, the cable-ducts can be cast in the members to any desired curved paths; and by sweeping the cables up at the ends of beams, the eccentricity can be reduced as required to suit the tailing off of the bending moments; and the upward component of the cable force acts also to assist in carrying the end shear forces.

PRE-TENSIONED MEMBERS

Pre-tensioning is applied to relatively simple units precast in factories laid out specially for the purpose. By this method the prestress is applied in one direction only, all wires running parallel to the longitudinal axis of the members. It lends itself therefore particularly to items such as precast piles, floor beams, slab-units and the like.

The steel wires are either stretched the length of unit moulds, with the moulds providing the reactive forces; or alternatively the wires are stretched the full length of the casting shop (say 400 or 500 ft.) between anchorages rigidly held in the ground. In the latter case the wires pass freely through a whole series of moulds, each mould being mounted on rollers and spaced with small end-gaps between adjacent moulds to allow the wires to be cut later and the individual units freed. The concrete is now placed in the moulds and cured in ideal factory conditions. When the concrete has reached the required strength, the wires are released from the end anchorages: and by virtue of high bond stresses between the wires and concrete at the ends of each unit, a compressive pre‑ stress is applied equal approximately to the initial tensile force in the wires.

THE AFTER-EFFECT OF PRESTRESSING

Whichever method of prestressing is adopted, the object in all cases is the same; the very high tension stress from the wires (up to about 100 tons/sq. in.) induces compression into the concrete, so that *applied* tension effects do no more than relieve the com‑ pressive *prestress*. In normal designs it is arranged that no net tension stresses occur under working-load conditions. The merits of prestressed concrete therefore include complete absence of cracking, shallower concrete sections, and reduced quantities of steel. On the other hand, the concrete and steel both require to be of considerably higher quality than is used for ordinary reinforced concrete work; the steel is very much more expensive per ton, and the techniques involve special equipment and more highly skilled and responsible operatives, as well as closer supervision. Neverthe‑ less the system is applied with advantage in suitable circumstances, notably for large-span construction, liquid-retaining structures and silos, and also for numerous precast units.

It has to be remembered that over a period of time concrete shrinks (Article 2.8), and under conditions of stress it creeps (Article 2.7). Steel also creeps, particularly at the stresses required for prestressed concrete work. It is therefore necessary for the *stretch* of the steel wires to be considerable in relation to the shrinkage and creep strains, otherwise too great a proportion of the prestress would be lost in the first few months' life of the struc‑ ture. This explains the need for the use of *high strength* steels for all prestressed concrete work; and the need for the stress in the steel at the time of prestress to be higher than the final stress relied upon in working-strength calculations. This need was evidently not appreciated in pioneering prestressed concrete work. Losses also arise from slip of the wires at the ends where they are anchored, whether the anchorage is achieved by bond development in pre‑ tensioned work, or by take-up of mechanical anchorages in post‑ tensioned work.

In post-tensioned work, other complications arise from friction of the wires in the ducts, at the anchorages and in the jack itself at the time of prestressing. In pre-tensioned work, a major loss is the elastic deformation of the concrete when the prestress is first applied: this does not arise in post-tensioned work, except where a number of cables are stressed progressively one after the other, because the prestress is only achieved by the very act of imparting elastic compression into the concrete. Pre-tensioned work also suffers more severely the effects of shrinkage, about half of which takes place in the first three weeks from date of casting. In post-tensioned work the overall losses in prestress may amount to about 20 per cent; but in pre-tensioned work the losses may be as great as 25 per cent. Losses are considered in greater detail in Article 15.4.

ART. 15.2. MATERIALS

The need for working to high stresses in prestressed concrete work has already been given in Article 15.1. Such stresses are only safe where the materials are of high quality, prepared under closely controlled conditions.

CONCRETE

Concrete for prestressed work is normally made with ordinary or rapid-hardening Portland cement. High strength is obtained using good aggregates, mechanical vibration and a low water/cement ratio. Good workability is of course important, though to a lesser extent than with ordinary reinforced concrete work, owing to the relative freedom from congestion of reinforcements. Thus the water/cement ratio can more easily be kept down, making for higher strength, with reduced creep and shrinkage characteristics (see Arts. 2.7 and 2.8). Creep and shrinkage jointly make up the greater part of the prestress loss. To assist in keeping the water/cement ratio low, workability agents are sometimes introduced into the mix. The use of calcium chloride as an accelerator requires care, owing to its tendency to cause corrosion to the tendons.

C.P. 115 (1959) *The Structural Use of Prestressed Concrete*[1] only recognises concretes with 28-day works cube strengths of at least 4500 lb/sq. in. for post-tensioned work. More normal works strength values are 6000 to 8000 lb/sq. in. Clearly then we are in the class of designed concrete mixes. The mixes have to be designed to achieve specified cube-strength requirements; and the need for strictest control of all stages of concrete manufacture as described in Article 1.8 becomes paramount. C.P. 115 (1959) defines two classes of control, known as the *normal method of quality control*, and the *special method of quality control*: with special control, the *mean* works cube strength is not required to be as high as with normal quality control.

STEEL

Two types of steel are currently in use for prestressing: (a) hard-drawn steel wire, normally used plain for post-tensioned work, but often used indented or crimped for pre-tensioned work; and (b) high-tensile steel bars.

(i) *Hard-drawn steel wire*

The high tensile strength of wire is achieved by the process of cold-drawing, which changes the metallurgical properties of the

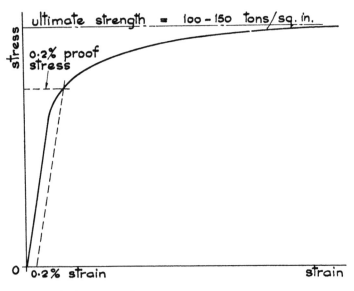

FIG. 15.3. STRESS/STRAIN CURVE FOR HIGH-TENSILE
STEEL WIRE

skin. Thus the smaller wires, which have a higher proportion of skin, have higher ultimate strengths. The wire is either used "as-drawn", or after carefully controlled heat-treatment at low temperatures. Unlike ordinary mild steel, high-tensile wires have no well-defined yield-points, and it is necessary to define the proof-stresses – i.e. the stresses which cause specified permanent strains (*see* Fig. 15.3). "Heat-treated" wires of the same *ultimate strengths* as "as-drawn" wires, will have higher *proof stresses*. Heat-treated wires also suffer less stress-relaxation (steel creep). The following table shows typical strength characteristics of as-drawn and heat-treated wires for the two diameters most normally used in post-tensioned work:

Diameter	Type	Ultimate strength	0·1% proof stress	0·2% proof stress
in.		tons/sq. in.	tons/sq. in.	tons/sq. in.
0·200	As-drawn	110–120	72	84
0·200	Heat-treated	110–120	97	99
0·276	As-drawn	100–110	70	80
0·276	Heat-treated	100–110	90	92

The elastic modulus for such wires is taken as the secant modulus at working stress: this is normally 28×10^6 lb/sq. in. with sufficient accuracy for most purposes.

Wires in small coils become highly stressed in the process of being wound, and consequently suffer permanent strain. Accordingly it is normal to have the deliveries in large coils, which pay out the wire straight.

Wire-cables are now achieving some popularity. These cables are made up of wires wound helically on a core. They give higher tendon forces per cross-sectional area of duct, and should therefore make for smaller and lighter members in cases where the cover to the ducts forms a controlling feature.

(ii) *High-tensile steel bars*

C.P. 115 (1959) recognises only steel bars of 60 tons/sq. in. minimum strength, with a 0·2 per cent proof stress between 75 and 92 per cent of the actual ultimate strength. High tensile alloy-steel bars are normally used in 1 or $1\frac{1}{8}$ in. dia. and have typically an ultimate strength of 68 tons/sq. in. with a 0·2 per cent proof stress of 58 tons/sq. in. The elastic modulus for these bars is frequently taken as 25×10^6 lb/sq. in.

GROUT

After the steel tendons have been prestressed and anchored, the ducts which contain the tendons are filled completely with grout. This protects the steel against corrosion and prevents any free water in the ducts freezing with consequent expansion and cracking. The grout also enables proper bond to be developed, and this reduces deformation under conditions of overloading.

Sand-cement grouts are used with a water/cement ratio of 0·6, sometimes with air-entraining agents (Art. 1.9) incorporated in the mix. When the cable detail makes for congestion within the duct, neat-cement grouts are used, with a water/cement ratio of about 0·45. The grout is preferably of special colloidal type made by very high-speed mixing.

ART. 15.3. DESIGN STRESSES

BENDING AND DIRECT STRESSES IN CONCRETE

Often it is undesirable to allow any part of the concrete to be stressed in tension. Nevertheless, in some circumstances, low tensile stresses may be admissible. In any event, for all cases of loading (due to prestress, dead load, applied load) a close watch needs to be kept for tensile stresses.

(i) *Working stresses*

Under working load conditions, permissible stresses for post-tensioned work, properly grouted to achieve bond, are given in Table 15.1, related to specified works cube strengths u_w.

TABLE 15.1. *Permissible concrete stresses at working loads in post-tensioned work (based on C.P. 115 (1959))*

Description	Permissible stress
	lb/sq. in.
COMPRESSIVE:	
Bending	$0·33\, u_w$
Direct	$0·25\, u_w$
TENSION:	
Bending	
(a) under load often occurring and/or of long duration	$\dfrac{u_w}{60} + 100$
(b) under load rarely occurring and of short duration	$\dfrac{u_w}{60} + 200$
Direct	
(a) under load often occurring and/or of long duration	$\dfrac{u_w}{120} + 50$
(b) under load rarely occurring and of short duration	$\dfrac{u_w}{120} + 100$

It is important to note that under no circumstances should tension be permitted across mortar joints; and the authors would not consider it desirable to have tension across any construction joints in cast-in-situ work. C.P. 115 (1959) allows compressive stresses to be increased by 25 per cent above the values given in Table 15.1, where the increase is due solely to wind forces.

(ii) *Stresses at transfer*

It will be shown later, in Article 15.4, how the stresses occurring at the time of prestressing subsequently fall-off due to certain losses. It is reasonable therefore at the initial stage, to allow stresses which are higher in relation to the cube strength of the concrete at the age considered, than under working conditions. Further, the conditions at transfer then serve to some degree to pre-test the structure; and if the structure proves unsatisfactory at this stage, the consequences of failure are likely to be less severe than when the working load is being carried. Permissible stresses at the time of transfer are given in Table 15.2, related to the works cube strengths at the age of transfer, u_{wt}.

TABLE 15.2. *Permissible concrete stresses at time of transfer in post-tensioned work (based on C.P. 115 (1959))*

Description	Permissible stress
	lb/sq. in.
COMPRESSION: For triangular or roughly triangular distribution of prestress }	$0.5\ u_{\mathrm{wt}}$*
For uniform or approximately uniform distribution of prestress }	$0.4\ u_{\mathrm{wt}}$*
TENSION	$\dfrac{u_{\mathrm{wt}}}{60} + 100$

* But not greater than 3000 lb/sq. in.

C.P. 115 (1959) permits tensile stresses at transfer to be increased for a short period not exceeding 48 hours, but not more than twice the values given in Table 15.2.

(iii) *Handling stresses*

Stress increases are permitted during handling and during construction, provided no permanent damage or greater loss of prestress results.

SHEAR STRESSES IN CONCRETE

Shear stress calculations are based on the principal tensile stresses resulting from transverse and axial forces (from loading and prestress) in much the same way as described in Article 3.4. Maximum permissible principal stresses at working loads and at ultimate loads where the section remains uncracked, are given in Table 15.3 related to the specified works cube strength u_w.

TABLE 15.3. *Permissible principal tensile stresses in concrete (based on C.P. 115 (1959))*

Description	Principal tensile stress
	lb/sq. in.
At working loads	$\dfrac{u_w}{60} + 50$
At ultimate loads in uncracked sections	$\dfrac{u_w}{30} + 150$

C.P. 115 (1959) recommends that where the principal stresses given in Table 15.3 are exceeded, shear reinforcement should be provided. At working loads, the proportion of shear to be resisted by the reinforcement should be assumed to vary linearly with the principal stress from 0 for the stress given in Table 15.3 to 1·0 for a stress 1·5 times that given: beyond this the total shear should be carried by the reinforcement. At ultimate loads in uncracked sections, the whole of the shear in excess of that resisted by inclined tendons should be resisted by shear reinforcement acting at a stress not exceeding 80 per cent of the yield stress.

STEEL STRESSES

Steel stresses are limited to provide a margin of safety on the steel, and also to avoid excessive stress-relaxation due to creep of the steel. In this way, the risk of permanent deformation from overload is also reduced.

For hard-drawn steel wire, C.P. 115 (1959) recommends the initial tensile stress should not exceed 70 per cent of the ultimate strength, nor the 0·2 per cent proof stress, whichever is less. To reduce loss of prestress due to creep of as-drawn wires, a 10 per cent overstress is sometimes held for two minutes.

For alloy-steel bars the initial tensile stress should not exceed 70 per cent of the ultimate strength, nor 85 per cent of the 0·2 per cent proof stress, whichever is less.

ART. 15.4. PRESTRESS LOSSES

Losses in achieving the final prestress force arise from a number of causes. The following detailed descriptions apply to post-tensioned work, where the losses fall conveniently into three groups:

(a) losses occurring during the process of tensioning;
(b) the loss occurring at the stage of anchoring; and
(c) losses arising subsequently.

LOSSES IN TENSIONING

When the tendons are being extended (to produce the prestress) friction losses arise on the walls of the duct; where the wires fan-out at the anchorages; and within the jacking device. The losses in the jack and at the anchorages will clearly vary with different prestressing systems, and appropriate loss-coefficients can be obtained from the proprietors of the systems.

Cables passing through steel or plastic spacers in ducts have the smallest friction losses.

The friction within the duct causes the prestress force in the steel to diminish at points further and further from the jacking point, and with long members it may for this reason be necessary to jack from the two opposite ends. The rate at which the friction reduction occurs depends on many factors. Ducts lined with steel clearly produce less friction than bare concrete surfaces: duct-wobble (the inherent non-straightness of the former) makes for additional friction – hence the merit of rigid, stiffened duct-profiles: and short curves make for less friction than long sweeping ones.

The friction in curved ducts depends on the amount of the curvature, and on μ, the coefficient of friction between the tendons and the duct surface. Then the tension in the cable at any point is given by

$$P_x = P_0 e^{-\frac{\mu x}{R}} \tag{15.1}$$

where P_x is the tension at any distance x along the curve from the tangent point,
P_0 is the tension at the tangent point,
R is the radius of curvature of the duct.

Similarly, from tests it has been shown that the effect of duct wobble (in both straight and curved ducts) can be taken as

$$P_x = P_0 e^{-Kx} \tag{15.2}$$

where P_x is the tension at any distance x from the jack,
P_0 is the tension at the jacking point,
K is a duct-factor, depending on the nature of duct surface, and how truly the duct former is held straight during concreting.

Values of K, per foot length, are normally not less than 10×10^{-4}; though it is possible with well supported rigid formers to get factors as low as 5×10^{-4}.

In practice it is normal to combine the effects of equations (15.1) and (15.2) – instead of treating the two effects separately – whence

$$P_x = P_0 e^{-\left(Kx + \frac{\mu x}{R}\right)},$$

and since

$$e^{-\left(Kx + \frac{\mu x}{R}\right)} = 1 - \left(Kx + \frac{\mu x}{R}\right) + \left(Kx + \frac{\mu x}{R}\right)^2 - \text{etc.},$$

we have, for small values of Kx and $\dfrac{\mu x}{R}$,

$$P_x \approx P_0 \left[1 - Kx - \frac{\mu x}{R}\right]. \tag{15.3}$$

The term in brackets is known as the tension ratio.

ANCHORAGE LOSSES

Anchorage losses arise from slip or take-up of the wires when the force is transferred from the jack to the anchorage device. In post-tensioned work, this slip normally amounts to $\frac{1}{8}$ in. with most systems, but more exact estimates can be obtained from the proprietors of the system.

LOSSES OCCURRING AFTER STRESSING PROCEDURE HAS BEEN COMPLETED

The subsequent losses are due to:

(i) shrinkage of concrete;
(ii) creep of concrete;
(iii) elastic strain of concrete; and
(iv) stress relaxation (creep) of steel.

With (i), (ii) and (iii), the losses arise from the shortening of the concrete reducing the initial stretch of the steel. The effect (iii) is greatest in pre-tensioned work: with post-tensioned work it only occurs where a number of cables are stressed progressively one after the other.

(i) *Shrinkage of concrete*

Shrinkage has been discussed in Article 2.8. With the low water/cement concretes used in prestressed work it is usual to assume that one-third of the total shrinkage occurs in the first 2–3 weeks, half has occurred by the end of 4 weeks, and three-quarters has occurred after 6 months. Thus for normal post-tensioned work, the shrinkage per unit length likely to occur after tensioning is about 200×10^{-6}. The loss from this amounts to about 4 per cent of the initial prestress.

(ii) *Creep of concrete*

Creep is described in Article 2.7, and depends on the intensity of applied stress and the age of the concrete when first loaded. For prestressed work it is normal to assume that creep is directly proportional to the initial prestress; and in post-tensioned work it is taken per unit length as

$$0.25 \times 10^{-6} \times \frac{6000}{u_{wt}} \times \text{initial prestress in concrete (lb/sq. in.)},$$

but $\dfrac{6000}{u_{wt}}$ is not to be taken as less than unity.

The initial prestress in the concrete is taken at the level of the tendons. For this purpose it is sufficiently accurate to assume that the wires act at the point of their centroid. With post-tensioned work the creep loss is frequently of the order of 8 per cent of the initial prestress.

For pre-tensioned work (where the concrete receives the prestress earlier), the creep per unit length is taken as

$$0.33 \times 10^{-6} \times \frac{6000}{u_{wt}} \times \text{initial prestress in concrete (lb/sq. in.)}.$$

(iii) *Elastic strain of concrete*

C.P. 115 (1959) recommends that for calculating the elastic strain of the concrete, values for the modulus of elasticity should be taken as in Table 15.4.

For pre-tensioned work (where this effect is greatest) the loss in prestress at transfer is equal to $\dfrac{E_s}{E_c}$ times the stress in the adjacent concrete. For pre-tensioned work, the prestress loss due to elastic compression of the concrete may be about 4 per cent. For post-tensioned work, with progressive tensioning, it is normal to consider the loss equally shared from all tendons, and equal to $\dfrac{E_s}{2E_c}$ times the stress in the adjacent concrete.

(iv) *Steel relaxation loss*

The amount the steel creeps, over a period of time, depends on the quality of the steel and how highly it is stressed. As-drawn wire suffers greater creep-loss than heat-treated wire. Stress loss due to steel creep with as-drawn wire varies up to about 15,000 lb/sq. in.

TABLE 15.4. *Values of modulus of elasticity of concrete (based on C.P. 115 (1959))*

Cube strength of concrete at stage considered, u_w	Modulus of elasticity, E_c
lb/sq. in. $\times 10^{-3}$	lb/sq. in. $\times 10^{-6}$
4	4
5	4·5
6	5
8	6
10	6·5

when stressed to 70 per cent of the ultimate strength: but a 10 per cent overstress held for 2 minutes may reduce this to 10,000 lb/sq. in. These are the figures recommended by C.P. 115 (1959). 10,000 lb/sq. in. is also taken for the loss in heat-treated wires. The manufacturers of steel-alloy bars claim that the relaxation-loss is relatively small; though it is normal to allow 10,000 lb/sq. in. An average value for relaxation-loss due to steel creep is about 6 per cent.

ART. 15.5. POST-TENSIONING SYSTEMS

In all post-tensioning systems, the basic principle is the same: to induce compression into the concrete member cast previously, by pulling on steel wires which in general are accommodated in pre-formed ducts running within the member. In pulling the wires, they stretch; and the prestress force is obtained by achieving the desired extension, using a hydraulic jack acting against the end of the concrete member. Before the pull in the jack is released, the wires are anchored in their extended position against the end of the concrete, generally by some wedging device. Variations on the wire arrangements and jacking and anchorage devices constitute the differences between the various systems. An exception to the foregoing is the Lee–McCall system, where high-tensile alloy-steel bars are used (ultimate strength 64–72 tons/sq. in.) instead of the higher strength wires used by the other systems (ultimate strengths often about 120 tons/sq. in.) : and instead of a wedge-action anchorage, the Lee–McCall bars are anchored with nuts operating on a special thread at the ends of the bars.

To ensure the correct prestress being achieved, it is generally necessary to measure both the extension of the steel, and the load applied by the jack. Measurement of the extension will give a useful indication as to how much of the steel is being properly tensioned: for example, friction between the steel and the sides of the ducts may be hampering the extension of the steel, and in this way preventing the steel at the end remote from the jack receiving its proper tension. This is most likely to occur with curved cables. On the other hand, extension may have occurred by overstressing only a part of the length of the wire; and for this reason it is necessary that the prestress load should also be measured.

The anchorage devices bear on the concrete at the ends of the members, producing high compressive stresses, and additional reinforcement is required locally to prevent the concrete splitting or failing by shear. End bearing stresses of 3500–4000 lb./sq. in. are commonly adopted.

There are now available a number of stressing systems, each varying in detail, and each protected by patent rights. The types here described are widely used in Great Britain; but there are other systems, used in this country and abroad, and the proprietors of these would claim their systems to be equally as satisfactory when used in appropriate circumstances. The wire-systems are all suitable, with adaptation, to the post-tensioning of circular structures such as tanks and silos, but the Preload System (described later) has been developed specially for this single purpose.

FREYSSINET SYSTEM

The Freyssinet prestressing system (Fig. 15.4) was the first method of post-tensioning to be developed. The Freyssinet cable consists of a number of high-tensile steel wires or strands grouped around a central helix core and taped together to form a cable. All the wires are stressed simultaneously, and for this reason no form of spacer or separator is required, and the final cable is both compact and flexible, and can be used with slender sections. The wires are anchored by being held between two concrete cones which fit one inside the other. The female part of the anchorage is a conical steel-wound lining heavily reinforced with high-tensile steel spirals to resist bursting forces. The mesh-reinforced concrete male cone is fluted to space evenly the requisite number of wires. A central tube passing axially through the male cone permits grout to be injected through it, and no other provision for cable grouting is normally necessary. The male and female cones constitute a pure friction anchorage which is therefore able to work to the full ultimate tensile strength of the wire. Anchorages are available for the follow-sizes of cables: 12/0·200 in., 12/0·276 in., and 12/0·500 in. strands. The last size is the most recent development and provides a cable with a working force of about 115 tons. The stressing of the cables

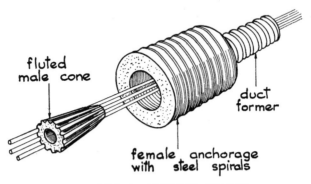

FIG. 15.4. FREYSSINET SYSTEM

is carried out by a special tensioning jack, which first tensions the wires, after which a subsidiary ram forces the male cone home to achieve the anchorage.

MAGNEL BLATON SYSTEM

In the Magnel Blaton system (Fig. 15.5) the cable is rectangular in section. It is composed of layers of wires, 0·200 in. or 0·276 in. diameter, arranged four wires per layer and up to sixteen layers deep for cables of 64 wires. Each wire is separated from neighbour-ing wires in the same layer and from those in adjacent layers by a space of $\frac{3}{16}$ in. The wire pattern is maintained throughout the length of the cable by means of grilles spaced at regular intervals along it. These grilles offer little frictional resistance to the wires which are free to move relative to each other during tensioning. Anchorage is by means of wedging the wires two at a time into sandwich plates. These are steel plates about an inch thick, with two wedge-shape grooves in their upper and lower faces. The wires are arranged two in each groove and, after tensioning, a steel wedge is driven between the wires to anchor them against the plate. Each complete anchorage is composed of from one to eight sand-wich plates depending on the number of wires in the cable, each plate anchoring 8 wires. The sandwich plates are placed in the form of a bank, one above the other, against a distribution plate, and held thus by means of a temporary bolted clamp during tensioning operations. Distribution plates are either cast into the member at the extremity of the duct or placed on thin mortar beds after the concrete has hardened, prior to assembling the anchorage. Tensioning is carried out by jacking two wires at a time by means of a temporary wire grip. For circular tank work, special sandwich

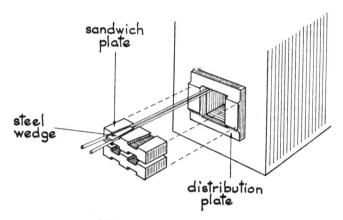

FIG. 15.5. MAGNEL BLATON SYSTEM

plates are available of opposed type, similar to the ordinary sandwich plate except that one groove (and wedge) on each side of the plate is reversed in direction: this enables the anchorage to secure pairs of wires pulling in opposite directions.

P.S.C. MONO-WIRE SYSTEM

The P.S.C. Mono-wire prestressing system (Fig. 15.6) provides a means of applying prestressing forces by tensioning the wires individually. Single-wire systems may have advantages where the larger systems may lack the necessary range of prestressing forces. Mono-wire anchorages are available for 1-, 2-, 4-, 8-, and 12-wire cables of sizes up to and including 0·276 in. diameter. For 8- and 12-wire cables, a high-strength plastic spacer is used to make up the cable and separate the wires for stressing; for 2- and 4-wire cables, the use of a special sheath makes the use of spacers unnecessary. In all cases cable grouting takes place by direct access through the anchorage and no special provision need normally be made. The P.S.C. Mono-wire anchorage is based on the use of a single piece "collet" sleeve wedging in a conical hole: the sleeves are serrated internally to give positive gripping. In all cases the prinicipal component of the various anchorage assemblies is a steel trunked guide, which leads each wire from its cable position to its point of anchorage through a gentle curvature. This guide provides reinforcement to the heavily stressed anchorage zone. In addition to

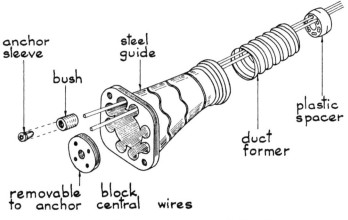

FIG. 15.6. P.S.C. MONO-WIRE SYSTEM

the guide, individual or group anchorages, seating co-axial with the wire, complete the assembly.

C.C.L. STANDARD SYSTEM

The C.C.L. Standard system (Fig. 15.7) is another single-wire stressing system. Any number of wires from 1 to 12 can be used in a circular cable. The size of duct required is $1\frac{1}{2}$ in. diameter for 8-wire cables and 2 in. diameter for 12-wire cables. Spacers are used to index the wires and are placed up to 4 ft. intervals, depending on the curvature of the ducts. The purpose of the spacers is to separate the wires and to lift the outer wires from the sheathing, thus facilitating grouting and reducing friction to a minimum. Each wire is anchored individually by a C.C.L. anchor grip, which consists of hard steel wedges fitting over the wire and within a steel barrel with a central tapered hole. The teeth of the wedges are machined and are in contact with the wire, the back of the wedges fitting against the tapered part of the barrel. As the load is transferred from the wire, the wedges are drawn forward and the taper forces the teeth to indent the wire, thus providing a positive anchorage by combining a wedging action with a mechanical grip.

The barrels are $\frac{13}{16}$ in. diameter and 1 in. long, and the wedges are either 2 separate pieces or 4 segments held together by a steel circlip to form one unit. The clips are seated on a drilled anchor plate through which the wires pass, and this bears on a thrust ring which is cast into the concrete to form an even bearing. The anchor

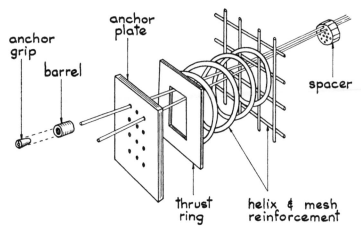

anchor grip

barrel

anchor plate

spacer

thrust ring

helix & mesh reinforcement

FIG. 15.7. C.C.L. STANDARD SYSTEM

plate is in mild steel, $\frac{3}{4}$ in. thick and has $\frac{11}{32}$ in. diameter holes drilled at 1 in. centres in parallel rows to accommodate the wires. The thrust ring is also in mild steel, $\frac{1}{4}$ in. thick, rectangular, with a central rectangular hole. To deal with local bursting forces a helix is placed behind the thrust plate, and vertical mesh reinforcements are placed behind the helix. The wires are tensioned one at a time.

GIFFORD UDALL SYSTEM

The Gifford Udall system offers three main methods of pre-stressing:

(i) In the original plate anchorage system the wires are stressed and anchored individually by small wedge-type Udall grips seating against a bearing plate much in the same manner as in the C.C.L. Standard system. The bearing plate locates against a thrust ring cast into the concrete and the end of the duct is encircled by a helix. Provision is made for grout entry about 3 in. behind the thrust ring. Anchorages of this type are supplied for cables of 2, 4, 6, 8 and 12 wires, and various types of jack are available to suit different operating conditions.

(ii) The most recent Gifford Udall wire system anchors the wires by wedges which fit directly into tapered recesses in the bearing plate (Fig. 15.8) so eliminating the need for separate grips. The bearing plate also includes a threaded hole for grouting purposes. The bearing plate seats against a tube unit incorporating the thrust ring and helix in a single element which bolts into the formwork and is cast into the concrete. This anchorage is supplied for cables of 8 to 12 wires; it is compact and reduces the congestion of steel in the anchor block.

(iii) Another new Gifford Udall system, now in use, has a tendon of $1\frac{1}{8}$ in. diameter H.T. strand. The strand is slightly stronger than twelve 0·276 in. wires and only requires a $1\frac{1}{2}$ in. diameter duct. It is anchored in a $3\frac{1}{2}$ in. diameter grip with a three-segment interlocking wedge. The barrel of the grip seats directly against a tube unit which is cast into the concrete. The thrust ring of the tube unit is provided with a threaded grout entry hole. Stressing of the strand is carried out using a large hollow ram jack incorporating an electrical load cell. The wedges of the final

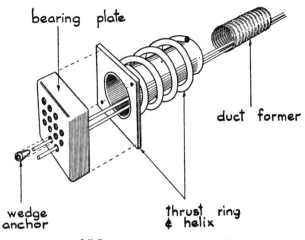

FIG. 15.8. GIFFORD UDALL SYSTEM

grip are pressed home hydraulically when anchoring off. This system of prestressing permits a high concentration of prestress and is suitable for large members where considerable prestressing forces are required.

LEE–MCCALL SYSTEM

Whereas in the methods described previously high-tensile steel wires are used, in the Lee–McCall system high-tensile alloy steel bars are used as the prestressing tendons. Macalloy steel is produced to B.S. 970: 1955 and is of EN 45 quality (United States equivalent A151 9260). It is an open-hearth Silico-Manganese steel, hot-rolled into bars, and subsequently processed to give the required physical properties. These may be summarised as follows:

Ultimate tensile strength: 64–72 tons/sq. in.;
0·2% proof stress: 58 ± 4% tons/sq. in.;
Elastic modulus: (secant modulus at 42 tons/sq. in.): approx. 25 × 10⁶ lb/sq. in.;
Elongation on 8 in. gauge length: 6 per cent minimum.

The recommended initial jacking stress is 45 tons/sq. in., and losses due to creep and shrinkage are usually taken as 15 per cent

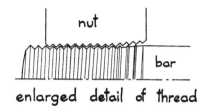

enlarged detail of thread

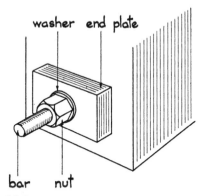

FIG. 15.9. LEE–MCCALL SYSTEM

in average designs. Bars are supplied in ⅞ in., 1 in., 1⅛ in., and 1¼ in. diameter and in lengths up to 60 ft. Normally the bars are anchored by screwing on special threaded nuts which bear against the concrete through washers and a steel distribution plate (*see* Fig. 15.9). The bar thread has a slow run-off or taper up on to the plain bar. The nut is bell-mouthed, and if it is run up to the hand-tight position and then tightened with a spanner so that the threads in the bell-mouth of the nut are forced a further $\frac{1}{16}$ in. or $\frac{1}{8}$ in. up the taper on the bar to what is called the High Efficiency or spanner-tight position, a strength of nearly 100 per cent of that of the plain bar will be obtained for the assembly. This efficiency arises from the transfer of load from the bar to the nut being as gradual as possible, and a very great deal depends on the accuracy of the position of the nut on the bar. Standard thread lengths are provided to allow a casting tolerance of $\pm$ 2 in. on the length of the unit; nevertheless the system necessitates the bars with their special tapered threads being manufactured to correct lengths to suit the requirements of each particular job, and the concrete work has to be cast to reasonably accurate lengths. To overcome this disadvantage, a wedge grip has now been devised comprising a single-piece conical steel wedge which fits into a conical hole machined in the end plate. The wedge grip is not used for stressing short lengths where complete accuracy of prestress is provided by the threaded end, but is useful for larger lengths where it is difficult to forecast the finished length of unit.

PRELOAD SYSTEM

The Preload system has been developed in America for pre-stressing circular structures such as tanks. In this system a cradle is slung outside the tank wall from a trolley which runs along the top of the wall. As the cradle makes its rounds of the tank, it pays out the prestressing wire through a die slightly smaller than the diameter of wire. The drag of the wire in passing through the die imparts the necessary tension, so that the wire is wrapped on to the tank already tensioned. This has the great merit of eliminating the high friction losses which arise with systems which stress the wires only at end anchorages arranged intermittently round the tank, giving not only a *greater* prestress force for a given wire size, but also a more *uniform* prestress. The paying-out cradle starts its journey at base level, where the wire is first anchored to the tank; and then as the cradle makes each circuit of the tank it raises itself on its supporting slings by an amount equal to the centres required between the prestressing wires. This interval can be adjusted as required, so that by maintaining a constant tension in the wire and increasing the wire centres, the prestress force is reduced to suit the reduction in ring tension. The platform is held off the face of the wall by four wheels, and pulls itself round by teeth from the drive mechanism engaging in a continuous chain wrapped once round the tank. On completion of the prestressing operation, the wires are protected by a layer of pneumatic mortar about one inch thick.

ART. 15.6. THEORY OF BENDING
(Elastic and Ultimate Theories)

C.P. 115 (1959) requires that members in bending should comply with the permissible stresses during normal working, at transfer, and during handling, all as given before in Article 15.3: in addition, the member should be capable of carrying, without collapse, a total load of $1\frac{1}{2}$ times the dead-load plus $2\frac{1}{2}$ times the superimposed load, but the total load need not exceed twice the sum of the dead plus superimposed loads. Thus the design calculations necessarily fall into two parts: firstly, the safe conditions arising during construction and during the normal use of the structure, at which stages the behaviour is elastic, and we work with a *factor of safety*; and secondly, the ultimate conditions with a *load factor*, when the structure is approaching collapse, when the behaviour becomes plastic. The second calculation on the basis of the ultimate load is extremely important, and may control the design, notwithstanding an inadequate structure complying perfectly well with the stress requirements under normal working conditions.

WORKING CONDITIONS: (Elastic Behaviour)

Under working-load conditions, compressive and tensile stresses are kept within the range at which the concrete behaves elastically; and this enables design to be made on the straightforward elastic theory for uncracked sections. Two further assumptions are made as follows:

The horizontal component of the prestress force does not vary along the length of the cable.

The horizontal component of the prestress force does not vary as the loading conditions are changed.

The assumptions are near enough true for most practical cases.
The following symbols are used in the analysis:

A is the area of the section,

d is the overall depth of the section,

e_s is the eccentricity of the prestress force below the neutral axis,

f' is the bottom-fibre concrete stress,

f'' is the top-fibre concrete stress,

f_c is the compressive stress under total load,

f_t is the tensile stress under total load,

I is the second moment of area of the section,

M_D is the moment due to Dead Load,

M_1 is the positive (sagging) moment due to Superimposed Load,

M_2 is the negative (hogging) moment due to Superimposed Load,

M_L is the total variation of moment due to Superimposed Loads, i.e. $M_1 - M_2$,

H is the horizontal component of the prestress force,

P is the prestress force,

V is the vertical component of the prestress force,

y_1 is the bottom-fibre distance,

y_2 is the top-fibre distance.

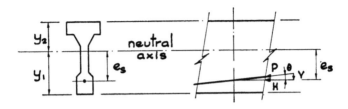

(a) Section   Elevation

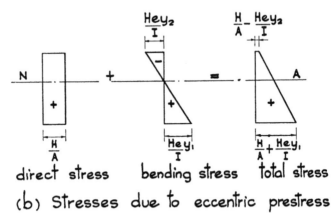

direct stress   bending stress   total stress

(b) Stresses due to eccentric prestress

FIG. 15.10. STRESS DIAGRAMS FOR BEAM SHOWING
BASIC PRESTRESS EFFECT

First consider the general case as in Fig. 15.10a of a member asymmetrical about the neutral axis, and subject to a prestress

force of horizontal component H acting at some eccentricity e below
the neutral axis. Then the extreme fibre-stresses in the concrete are

$$\frac{H}{A} \text{ (direct)} \pm \frac{He \cdot y}{I} \text{ (bending)},$$

as shown in Fig. 15.10b.

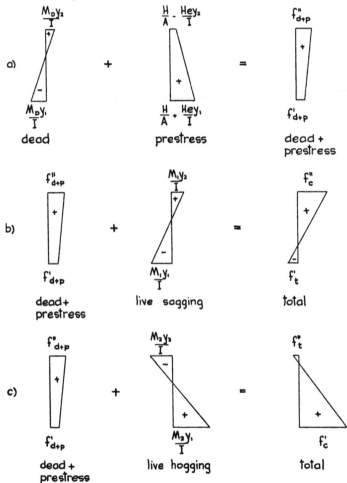

FIG. 15.11. STRESS DIAGRAMS FOR BEAMS SHOWING
THE THREE DESIGN CASES

Now suppose the same member acts as a beam. There will be
bending stresses set up from the Dead Load and any Superimposed

Loads, and by selecting a suitable prestress force and eccentricity, we can set $\dfrac{H}{A}$ and $\dfrac{He \cdot y}{I}$ against these stresses, within certain limits, to suit our own design purpose, in the manner earlier depicted in Figs. 15.1 and 15.2.

Normally there are two design cases to cater for:

(a) Stresses due to $H.e$ and M_D (Fig. 15.11a); and

(b) Stresses due to $H.e$, M_D and M_1 (Fig. 15.11b).

If it is possible for a negative moment to occur at the same section, a third case can arise:

(c) Stresses due to $H.e$, M_D and M_2 (Fig. 15.11c).

Normally one would aim in design at a minimum concrete stress of zero: but in what follows, the general case is given for a small permissible tensile stress p_t developing in the concrete. Later, we can simplify the general expressions by putting p_t and M_2 equal to zero, giving the case where no tensile stress develops, and where there are no negative moments from superimposed load.

(i) *Determination of section modulus*

From Fig. 15.11 it is evident that the greatest compressive stress in the top fibres occurs under condition (b). This stress is

$$f_c'' = \frac{M_D y_2}{I} + \frac{H}{A} - \frac{He \cdot y_2}{I} + \frac{M_1 y_2}{I}. \tag{15.4}$$

And the greatest tensile stress in the top fibres occurs under condition (c). Then

$$f_t'' = \frac{M_D y_2}{I} + \frac{H}{A} - \frac{He \cdot y_2}{I} + \frac{M_2 y_2}{I}. \tag{15.5}$$

By subtracting equation 15.5 from equation 15.4; and, substituting M_L for $M_1 - M_2$, and z_2 for $\dfrac{I}{y_2}$, and rearranging, we have

$$f_c'' = \frac{M_L}{z_2} + f_t''. \tag{15.6}$$

Similarly, for the bottom fibres we have

$$f_c' = -\frac{M_D y_1}{I} + \frac{H}{A} + \frac{He \cdot y_1}{I} - \frac{M_2 y_1}{I} \tag{15.7}$$

and $$f_t' = -\frac{M_D y_1}{I} + \frac{H}{A} + \frac{He \cdot y_1}{I} - \frac{M_1 y_1}{I}; \tag{15.8}$$

and putting z_1 for $\dfrac{I}{y_1}$, we have

$$f_c' = \frac{M_L}{z_1} + f_t'. \tag{15.9}$$

It is clear from equations 15.6 and 15.9 that if f_t' and f_t'' are limited to equal values (often zero), the compressive fibre stress will be a maximum at the top or bottom of the section depending on which of z_1 or z_2 is the smaller. And if p_c and p_t are maximum permissible compressive and tensile stresses respectively,

$$p_c = \frac{M_L}{z_{min}} + p_t. \tag{15.10}$$

From equation 15.10 we can determine z_{min} and thus select a suitable section.

It is of interest to note that the section required is independent of the dead-load moment. This gives rise to the saying that "a prestressed member carries its own weight": which means only that a suitable prestressing force can always be applied at some eccentricity so as to counteract the dead-load moment. But as soon as superimposed loads are applied, which can vary in amount, the prestress has to cater for a *range* of conditions, and this can only be achieved within permissible stresses by adopting a minimum section as given by equation 15.10.

Note that if there can be no negative moment from superimposed loads, M_L becomes M_1: and if p_t is reduced to zero (for no tension), equation 15.10 reduces to

$$p_c = \frac{M_1}{z_{min}}.$$

(ii) *Determination of prestress force*

In practical problems, the applied bending moments vary from point to point along the length of the member. Indeed in any one position the applied moment will vary depending on the magnitude and direction of loading. Thus, for a given prestress force, in order to limit p_t to permissible values, the cable eccentricity must be properly determined. But if we can once find an eccentricity to suit the *worst* applied moment, it should be easier to find an eccentricity to cope with any lesser moments – whether these are due to tailing off of the bending-moment diagram, or due to less taxing conditions of loading. Indeed we shall find that a little latitude becomes available.

Taking again our general case, the extreme conditions are:

(b) Stresses due to $H.e$, M_D and M_1; and
(c) Stresses due to $H.e$, M_D and M_2.

Let us now define e_a and e_b as the cable eccentricities to limit the tensile fibre stresses, under conditions (b) and (c) respectively, to the permissible value p_t. Then from equation 15.8, we have

$$p_t = -\frac{M_D y_1}{I} + \frac{H}{A} + \frac{He_a \cdot y_1}{I} - \frac{M_1 y_1}{I}$$

which, by rearrangement, gives

$$e_a = \left(\frac{M_D + M_1}{H}\right) + \frac{I}{y_1}\left(\frac{p_t}{H} - \frac{1}{A}\right). \tag{15.11}$$

Similarly, from equation 15.5 we have

$$p_t = \frac{M_D y_2}{I} + \frac{H}{A} - \frac{He_b \cdot y_2}{I} + \frac{M_2 y_2}{I},$$

yielding

$$e_b = \left(\frac{M_D + M_2}{H}\right) - \frac{I}{y_2}\left(\frac{p_t}{H} - \frac{1}{A}\right). \tag{15.12}$$

It is seen from equation 15.11 that a decrease in eccentricity below e_a will give a tensile stress in excess of p_t in the *bottom* fibre: and from equation 15.12, an increase in eccentricity above e_b will give

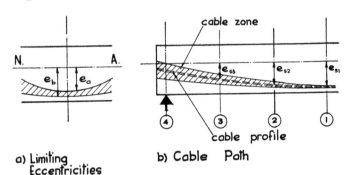

a) Limiting
   Eccentricities

b) Cable Path

FIG. 15.12. THE CABLE PATH

a stress in excess of p_t in the *top* fibre. Thus we have at every point along the beam a zone (known as the *cable zone*) within which the cable must lie if the permissible stress p_t is not to be exceeded (*see* Fig. 15.12a).

Now for maximum economy the cable zone should be limited to

a minimum; and if at the critical section (where the worst applied moments occur), we equate e_a and e_b, we have from equations 15.11 and 15.12

$$\left(\frac{M_D + M_1}{H}\right) + \frac{I}{y_1}\left(\frac{p_t}{H} - \frac{1}{A}\right) = \left(\frac{M_D + M_2}{H}\right) - \frac{I}{y_2}\left(\frac{p_t}{H} - \frac{1}{A}\right).$$

Substituting M_L for $M_1 + M_2$, and d for $y_1 + y_2$, and rearranging, we have

$$H = A\left(\frac{y_1 y_2}{Id} \cdot M_L + p_t\right). \tag{15.13}$$

Having previously determined our section from equation 15.10, we can now calculate H.

Note that if there is no applied negative moment, and if p_t is limited to zero, equation 15.13 reduces to

$$H = A \cdot \frac{y_1 y_2}{Id} \cdot M_1.$$

(iii) *Determination of eccentricity*

Equation 15.13 can be rearranged as

$$\frac{I}{y_2} = \frac{M_L y_1}{d\left(\dfrac{H}{A} - p_t\right)},$$

and since, at the critical section,

$$e_b = e_a = e_s,$$

by substituting for $\dfrac{I}{y_2}$ in equation 15.12, we have for the critical section

$$e_s = \frac{1}{H}\left[M_D + \frac{M_1 y_1}{d} + \frac{M_2 y_2}{d}\right]. \tag{15.14}$$

Where there is no applied negative moment, the last term within the bracket vanishes.

If the position of the cable is determined to suit the critical section, it will not in general be possible to achieve zero cable-zone elsewhere. Thus the designer can fix the path of the cable at his discretion (*see* Fig. 15.12b), bearing in mind friction losses and the end-shear requirements. The latter are described in Article 15.7.

ULTIMATE CONDITIONS: (Plastic Behaviour)

The following studies apply to post-tensioned work where *good bond* is achieved between the tendons and concrete by efficient grouting: then as the safe working load is exceeded, and tension develops until the modulus of rupture is passed, the cracks that appear will be many in number and well distributed along the member. Where only poor bond is achieved, failure generally follows from the development of only one crack, and at about 75 per cent of the ultimate strength of an identical member where the cables are properly bonded. As the load applied to the well-grouted member is further increased, the cracks penetrate deeper in from the tension flange, and the depth to the neutral axis diminishes and plastic behaviour sets in. The section now acts as an ordinary reinforced-concrete member, the tendons functioning as the reinforcement. The plastic behaviour of reinforced sections has already been described in Article 3.3, and the basic theory will not be repeated here. However, in prestressed work the following slight differences arise. Firstly, with the high-strength concrete used for prestressed work, the stress-strain curve does not bend over as smartly as shown in Fig. 3.19; and the concrete may continue carrying a relatively high stress while yet the strain increases – indeed the stress may actually *increase* as the ultimate strain is reached. Secondly, at ultimate conditions, the steel – having no well-defined yield point – will function at slightly *increased* stresses as the steel strain increases considerably before failure occurs.

(i) *General case*

Let the maximum steel stress be f_m. Since the stress/strain relationship at this range is not linear, we must first determine the steel strain at collapse of the member, before we can find f_m. The strain diagram for the condition of collapse is given on the right in Fig. 15.13.

$\varepsilon_{c(ult)}$ is the ultimate concrete strain,

ε_{s1} is the steel strain when f'' is zero (generally taken as the strain at the time of prestress – allowing for losses),

ε_{s2} is the increase in steel strain as f'' increases from zero to its ultimate value,

ε_{st} is the total steel strain.

Thus
$$\varepsilon_{st} = \varepsilon_{s2} + \varepsilon_{s1}$$
$$= \varepsilon_{c(ult)}\left(\frac{1}{n_1} - 1\right) + \varepsilon_{s1}.$$

Or
$$\varepsilon_{st} = \varepsilon_{c(ult)}\left(\frac{1}{n_1} - 1\right) F' + \varepsilon_{s1} \tag{15.15}$$

where F' is a *bond factor* with values ranging between 0·9 and 1·0 for well-grouted post-tensioned work. Normally $\varepsilon_{c(ult)}$ is taken as 0·003.

Before we can use equation 15.15 to determine ε_{st} – and hence f_m –, we must first know n_1.

Now from Fig. 15.13 we have

$$F_C = k_1 k_3 u_w n_1 d_1 b \qquad (15.16)$$

which must equal

$$F_T = A_{st} f_m. \qquad (15.17)$$

Thus we have three equations in which ε_{st}, f_m and n_1 are dependent upon one another. The simplest method of solution is to try a value

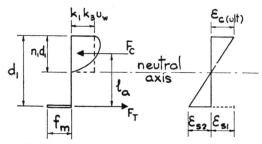

FIG. 15.13. STRESS AND STRAIN DIAGRAMS FOR BEAM AT COLLAPSE

for n_1, and so determine ε_{st} and thence f_m: the values of F_C and F_T can then be calculated, and these should of course be numerically equal. The matter is thus a process of trial and error. When the appropriate values of n_1 and f_m have thus been found, the ultimate moment the member can resist is calculated from the equation

$$M_{ult} = A_{st} f_m (1 - k_2 n_1) d_1. \qquad (15.18)$$

(ii) *Special case with low percentage of steel*

Where the percentage of steel in prestressed beams is low, the member will behave in much the same fashion as an ordinary "under-reinforced" section. At failure of the section, the steel stress is then so nearly approaching the ultimate tensile strength, that this value may be used in determining the ultimate resistance-moment of the section (*see* Fig. 15.14). If then, we put f_u for the ultimate steel stress, equation 3.46 becomes

$$n_1 = \frac{0·01 r_{p1} f_u}{k_1 k_3 u_w}. \qquad (15.19)$$

Now $M_r = A_{st}f_{st}l_a$, and since the depth of the centre of compression is $k_2n_1d_1$, we have

$$M_{(ult)} = A_{st}f_u \left(1 - \frac{k_2}{k_1k_3} \cdot \frac{0.01r_{p1}f_u}{u_w}\right) d_1. \qquad (15.20)$$

Values of k_1k_3 and k_2 can again be read from Fig. 3.21.

NOTE: If the value for M_{ult} given by either of equations 15.18 or 15.20 is less than will give the necessary *load factor* against

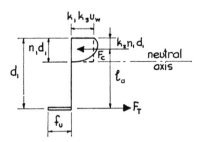

FIG. 15.14. STRESS DIAGRAM FOR "UNDER-REINFORCED"
BEAM AT COLLAPSE

collapse, then the remedy will be either to increase the section, or to increase the amount of steel in the tension zone. Such additional steel may most conveniently be mild steel or ordinary high-tensile steel bars. In pretensioned work, un-tensioned wires are sometimes used to meet the need.

(iii) *Case given by C.P.* 115 (1959)

An alternative approach, catering for any percentage of steel, is described in C.P. 115 (1959). This method is direct and simple to apply, but is unfortunately limited to the special cases of beams which at failure are rectangular above the neutral axis. The method is based on the results of tests to destruction of beams, which were stressed initially to 0·6 to 0·7 times f_u. The following assumptions are made:

    (i) at failure, $f_{c(av)} = 0.4u_w$ (i.e. $k_1k_3 = 0.4$);
    (ii) the centre of compression is $0.4d_n$ below the top fibre (i.e. $k_2 = 0.4$);
    (iii) $f_t = $ zero;
    (iv) in calculating the stress in steel in the compression zone, the compressive strain in the concrete is proportional to its distance from the neutral axis, and has a maximum value of 0·002.

For post-tensioned work with good grouting, the stress in the tendons, and the depth of the neutral axis are taken from Table 15.5 and then substituted in the equation

$$M_u = f_m A_{st}(d_1 - 0.4d_n). \qquad (15.21)$$

TABLE 15.5. *Conditions at failure for beams with post-tensioned steel having effective bond (based on C.P. 115 (1959))*

$\dfrac{f_u}{u_w} \cdot \dfrac{A_{st}}{bd_1}$	$\dfrac{f_m}{f_u}$	$\dfrac{d_n}{d_1}$
0.025	1.0	0.06
0.05	1.0	0.125
0.10	1.0	0.25
0.15	1.0	0.375
0.20	0.95	0.475
0.25	0.90	0.56
0.30	0.85	0.64
0.40	0.75	0.75

Where ordinary steel reinforcement is provided in the tensile zone, C.P. 115 (1959) recommends that the effect of this be taken into account by assuming the effectiveness of its area as

$$A_{su}\frac{f_y}{f_u}$$

where A_{su} is the area of the steel reinforcement in the tensile zone,
and f_y is the yield stress of mild steel reinforcement, or the 0.2% proof stress of high-tensile steel.

ART. 15.7. RESISTANCE TO SHEAR AND TORSION

SHEAR

The calculation of shear resistance for prestressed members at safe working conditions is more straightforward than for ordinary reinforced-concrete members, because the section (by virtue of the prestress) remains uncracked. The stress distribution is therefore parabolic, interrupted only by straight-line changes where the breadth of the section varies (*see* Fig. 15.15). The above does not apply to cases where the section in shear may crack at ultimate load conditions. Then the analysis follows the theory given at Article 3.4 for normal reinforced-concrete work, and it becomes necessary to provide stirrups.

SHEAR AT SAFE-WORKING CONDITIONS

Design analysis for the case of the uncracked section is taken in three stages. First the balancing effect of up-swept cables is allowed for: next, we determine the amounts of the unbalanced transverse shear stress and the direct stresses due to bending: and

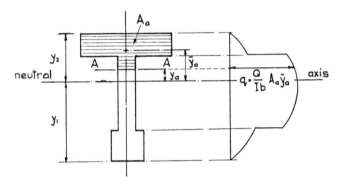

FIG. 15.15. SHEAR STRESS DISTRIBUTION FOR
PRESTRESSED BEAM

finally we combine the latter to give the principal tensile stresses. The principal tensile stresses generally benefit from the effect of the bending stresses – these being compressive from the prestress.

(i) *Determination of cable slope*

Before the balancing effect of the up-swept cable can be determined, we must first decide on the cable slope; and from this the vertical component of the prestress force can be calculated. The available latitude of cable eccentricity (the *cable-zone*) was demonstrated in equations 15.11 and 15.12: clearly for a simply supported beam of rectangular section, where the end moment is zero, the cable eccentricity can be as much as $\pm \dfrac{d}{6}$ without anywhere producing tension stresses. Now there must be an optimum path for the cable, so as to give the best compromise of shear balance under the extreme conditions of dead and superimposed loads.

If Q_D is the shear due to dead load,
   Q_1 is the shear due to superimposed load causing positive moments,

Q_2 is the shear due to superimposed loads causing negative
    moments (if any),

V is the vertical component of the prestress force,
then maximum shear values will be either $Q_D + Q_1 - V$, or
$Q_D + Q_2 - V$. If these are to be equal and opposite, we have

$$Q_D + Q_1 - V = - (Q_D + Q_2 - V)$$

whence $$V = Q_D + \frac{Q_1}{2} + \frac{Q_2}{2}. \qquad (15.22)$$

Now since $V = H \tan \theta$ (see Fig. 15.16), we have the slope of the
cable given by

$$\tan \theta = \frac{1}{H} \left(Q_D + \frac{Q_1}{2} + \frac{Q_2}{2} \right). \qquad (15.23)$$

Note in this that Q_2 may be zero.

Suppose the worst shear conditions occur with the same loading
as will produce the worst moment conditions (as frequently they

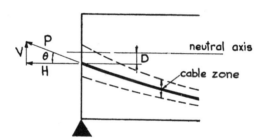

FIG. 15.16. CABLE SLOPE TO SUIT CONSIDERATIONS
OF SHEAR

do, though not always), and if e_s is the cable eccentricity, and x
is the distance along the member, then

$$\tan \theta = \frac{de_s}{dx},$$

so that $$V = H \frac{de_s}{dx}. \qquad (15.24)$$

And since, from simple beam theory, we have $Q = \frac{dM}{dx}$, we can
rewrite equation 15.22 as

$$V = \frac{d}{dx} \left(M_D + \frac{M_1}{2} + \frac{M_2}{2} \right). \qquad (15.25)$$

Equating equations 15.24 and 15.25, and integrating, we have

$$e_{\mathrm{s}} = \frac{1}{H}\left(M_{\mathrm{D}} + \frac{M_1}{2} + \frac{M_2}{2} \right) + D \tag{15.26}$$

which is the equation for the "shear" cable. This is seen to be similar to equation 15.14 for the cable eccentricity considering bending, with the constant D added. In fact for the case of a simple rectangular beam, where $\dfrac{y}{d} = \dfrac{1}{2}$, the equations are identical except for the constant D, which in that case must be zero. For an asymmetrical section, the constant D will have some value (*see* Fig. 15.16): and this is found by substituting in the shear-cable equation the values for e_{s}, H and M as calculated for bending at the critical section.

(ii) *Determination of shear stresses*

Having now determined the amount of the unbalanced shear force, Q, the shear stress at any section on a plane AA (distant y from the centroid) is given by the well-known formula

$$q = \frac{Q}{Ib} \int_y^{y_2} yb \cdot dy$$

$$= \frac{Q}{Ib} \cdot A_{\mathrm{a}} \bar{y}_{\mathrm{a}} \tag{15.27}$$

where A_{a} is the area of section above the plane AA, and $\bar{y}_{\mathrm{a}}$ is the distance of the centroid for A_{a} above the centroid for the total section (*see* Fig. 15.15).

(iii) *Determination of principal tensile stresses*

Failure due to shear will in fact take the form of tensile failure on lines of principal stress. Thus the stresses due to shear have to be combined with the direct stresses due to bending, and the appropriate formula is

$$f_r = \tfrac{1}{2} f_{\mathrm{cb}} \pm \tfrac{1}{2} \sqrt{f_{\mathrm{cb}}{}^2 + 4q^2} \tag{15.28}$$

where f_r is the principal tensile stress,
f_{cb} is the bending stress at the point considered,
q is the shear stress at the point considered.

SHEAR AT FULL LOAD-FACTOR CONDITIONS

As ultimate conditions are approached, beams become very sensitive to the effects of shear. Shear cracks can seriously reduce the resistance of the member to bending.

If bending is causing tension in the beam at the lower fibres only, the effect of shear may give rise to actual cracking, with a redistribution of the bending stresses: the neutral axis will rise, and the concrete strain will increase. As the load is increased, the rate of increase of principal tensile stress then becomes more and more rapid; and with shock-loads a momentary rise of the neutral axis position may be sufficient to precipitate complete failure of the member. On the other hand, where cracks due to bending extend up to above the level of the cables, the members may be analysed for shear quite simply as for normal reinforced-concrete work, the tendons being considered as the reinforcement. Then, as before,

$$q = \frac{Q}{bl_a}.$$

In view of the uncertainties of the matter and the cumulative nature of the effects of cracking under these conditions, the authors would exercise special care in regard to shear wherever tension can develop due to bending at the design *load-factor conditions*. This does not necessarily mean at *ultimate conditions*, because the ultimate capacity of the member may be greater than necessary to carry the design loads with the necessary overload allowance: and the difference between *load-factor condition* and *ultimate condition* may be considerable in terms of this critical issue of shear effect. Where tensions arise from bending, there are strong arguments for carrying the whole of the shear forces by stirrup reinforcements working at the higher stresses allowed for these conditions: and, of course, stirrups are required elsewhere if the allowable principal tensile stresses in the concrete are exceeded. As stirrups are normally provided anyhow for securing the cable ducts, they are already available for use as shear reinforcements, and act in combination with the longitudinal fixing-reinforcements in the top and bottom of the beam. The fixing-stirrups need to be closed up, centre to centre, as the shear forces increase from point to point along the beam.

TORSION

Since failure by torsion is abrupt, design should be on a conservative basis. The principal stresses may be calculated in accordance with the elastic theory, and these should be kept well within permissible values. In cracked sections at *load-factor conditions*, the reduced area of concrete above the neutral axis should be used when calculating the principal stresses.

Tests have shown that the torsional resistance of concrete is increased by the effect of prestressing, and this is not greatly

affected by moderate bending stresses within the member. Clearly
longitudinal and stirrup reinforcements must contribute to resistance to torsional failure, much as already described in Article 3.5,
and even though no precise assessment of their effect is yet
available, they will certainly go some way towards remedying the
abruptness associated with torsional failure of otherwise unreinforced sections.

ART. 15.8. STRESSES AT ANCHORAGE

Where the anchorage devices of post-tensioned systems bear
against the ends of the concrete, very high local bearing stresses
result; and as these spread out into the concrete they set up complicated stress patterns. A variety of analyses of this problem have

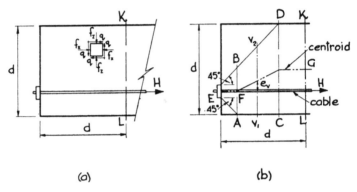

(a)   (b)

FIG. 15.17. ANCHORAGE STRESSES

been developed. The following is due to Professor Magnel[2]. The
stresses local to the anchorage bearing are assumed to have
dissipated over a length of beam equal to the beam depth as
shown in Fig. 16.17a. Within this end zone, any elemental unit is
acted on by three forces f_x, f_z and q being respectively horizontal,
vertical and shear stresses. Each of these will now be considered in
turn.

(i) Horizontal stresses

The compression from the anchorage device is assumed to
spread out at 45° cutting the top and bottom surfaces of the
member at A and D as shown in Fig. 15.17b. The line for the centroid of the stressed zone is thus known and can be drawn as EFG.
Therefore on any vertical plane, say $v_1 v_2$, of height equal to v, the

cable eccentricity e_v is known, and the stress distribution due to the cable force can be determined. At any point, distant y from the centroid, the magnitude of the horizontal stress is then given by

$$f_x = \frac{H}{bv} \pm \frac{He_v \cdot y}{I_v} \qquad (15.29)$$

where I_v is the second moment of area of the zone under stress.

(ii) *Shear stresses*

Magnel's theory shows that the distribution of shear on any horizontal plane, say, WW in Fig. 15.18a, is given by the equation

$$q = K_q \cdot \frac{S}{bd} \qquad (15.30)$$

where K_q is a shear stress factor, with values as given in Fig. 15.19, and S is that part of the prestress force at KL, which acts

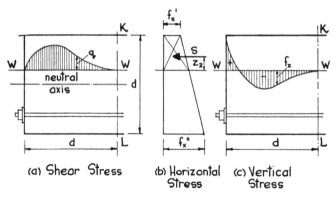

(a) Shear Stress   (b) Horizontal Stress   (c) Vertical Stress

FIG. 15.18. VERTICAL AND HORIZONTAL SHEAR
STRESSES AT ANCHORAGES

above WW. S is derived from the stress diagram (Fig. 15.18b) by summing all stresses f_x above WW. In determining S, the amount of the opposing cable force must be deducted, when this occurs above the plane being considered.

For that part of the end-block which occurs over the end-seating for the member, the shear stresses produced by the bearing pressure have also to be taken into account. Normally it is considered sufficiently accurate to assume that the bearing pressure is uniform over the length of the seating.

(iii) *Vertical stresses*

Magnel's theory shows that the distribution of vertical stress on any horizontal plane, say WW in Fig. 15.18c, is given by the equation

$$f_z = K_z \cdot \frac{M}{bd^2} \qquad (15.31)$$

where K_z is a vertical stress factor, with values as given in Fig. 15.19, and M is the total moment due to $S \times z_2$, where z_2 is the

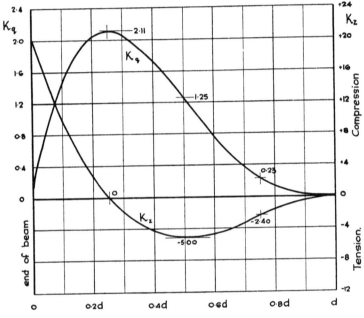

FIG. 15.19. MAGNEL'S VERTICAL AND HORIZONTAL
SHEAR STRESS FACTORS

distance of centroid S above WW. When the plane WW is below the level of the prestressing cable, M has to be reduced by the amount of the cable force times its distance above WW.

(iv) *Determination of principal stresses*

Having now determined f_x, f_z and q, we apply the well-known formula

Principal stress $= \frac{1}{2}(f_x + f_z) \pm \frac{1}{2}\sqrt{(f_x - f_z)^2 + 4q^2}.$

The angle of inclination of this stress is given by

$$\tan 2\theta = \frac{2q}{f_z - f_x}.$$

For simplicity, the whole of the above has been related to a member with a single cable. The method applies equally to members with any number of end-arrangement of cables. An example of the use of this theory is given in Article 15.9.

ART. 15.9. APPLICATIONS OF PRESTRESSED CONCRETE

BEAMS

There is no simple answer to the problem of determining the best profile for any particular prestressed beam: yet the economics of the matter are closely associated with the skill of this choice. Clearly, for safe-working conditions, in all but the smallest beams, an I-type section with wide flanges will have the merit of providing the greater part of the concrete where it is most required; and for any given variation of superimposed load will give the smallest concrete area, and thus the least dead weight. However, where the ratio of moments from dead to superimposed loads is greater than about 1 to 3, it may not be economical to accommodate the cables in an I-section at the desired eccentricity; and advantage is gained from using a T-section, or at any rate a top-heavy I. Conversely where the proportion of dead load is particularly low, there may be merits in an inverted T-section. At full load-factor conditions, the normal T-section is generally best – and for the same reasons as for ordinary reinforced concrete.

As an example, let us consider the design of roof beams of 42 ft. span required to carry superimposed loads – from roof construction and live loads – of 620 lb/ft. run. The beams are to be precast and post-tensioned on site before being lifted into position. Great care will be required throughout the lifting operations; and the beams will be steadied laterally by purlins before the roof construction and any live loads are applied.

$$M_{\mathrm{L}} = M_1 = 620 \times 42^2 \times \frac{12}{8} = 1,640,000 \text{ in. lb.}$$

For concrete with $u_{\mathrm{w}} = 6000$ lb/sq. in.,

$$p_{\mathrm{c}} = 0.33 \times 6000 = 2000 \text{ lb/sq. in.}$$

Allowing no tension, i.e. p_t = zero, we have from equation 15.10

$$z_{min} \text{ required} = \frac{1,640,000}{2000} = 820 \text{ in.}^3$$

This is given suitably by the section shown in Fig. 15.20: the bottom flange is small, but necessary to provide stiffness against buckling at the initial conditions; and the large top flange is required for the full load-factor condition. Ignoring the small cable holes, we have:

$$A = 161 \text{ sq. in.,} \qquad I = 12,200 \text{ in.}^4,$$
$$y_1 = 14.6 \text{ in.,} \qquad y_2 = 11.4 \text{ in.,}$$
$$z_1 = z_{min} = 840 \text{ in.}^3,$$
$$M_D = 161 \times 42^2 \times \frac{12}{8} = 425,000 \text{ in. lb.}$$

Now from equation 15.13 we have the tendon force

$$H = A \frac{y_1 y_2}{Id} \cdot M_L$$
$$= 161 \times \frac{11.4 \times 14.6}{12,200 \times 26} \times 1,640,000$$
$$= 137,000 \text{ lb.}$$

The area of 3 cables, each of 12/0.200 in. wires, is

$$36 \times 0.0314 = 1.13 \text{ sq. in.,}$$

requiring a steel stress, after losses of

$$\frac{137,000}{1.13 \times 2240} = 54.3 \text{ tons/sq. in.}$$

[In passing, note this is about one-quarter the steel and about one-third the concrete of an ordinary reinforced concrete beam to perform the same duty. But the quality and unit costs of the materials for the prestressed beam are both higher.]

To determine the cable eccentricities at mid-span, we have from equation 15.14

$$e_s = \frac{1}{H} \left(M_D + \frac{M_L y_1}{d} \right)$$
$$= \frac{1}{137,000} \left(425,000 + \frac{1,640,000 \times 14.6}{26} \right)$$
$$= 9.85 \text{ in.,}$$

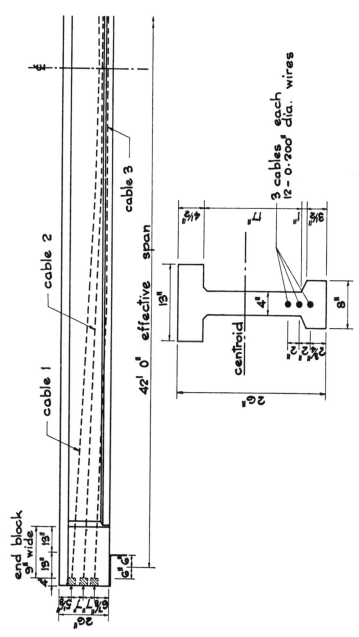

FIG. 15.20. EXAMPLE OF PRECAST POST-TENSIONED ROOF BEAM

495

giving the depth from top of beam to centroid of cables

$$= 11\cdot4 + 9\cdot85 = 21\cdot25 \text{ in.}$$

The next step is to determine the cable path. The extreme cable eccentricities allowable at different points along the beam are given by equations 15.11 and 15.12, and may be set out in tabular manner as below:

	Position along beam from support			
	$\frac{1}{8}L$	$\frac{1}{4}L$	$\frac{3}{8}L$	$\frac{1}{2}L$
	in.	in.	in.	in.
e_a	1·40	6·09	8·94	9·85
e_b	8·09	9·04	9·64	9·85

At the ends of the beam, where the section is rectangular (26 in. × 13 in.) the cable centroid has to lie within the middle-third so that, for no tension

$$e_a = e_b = \frac{26}{6} = 4\cdot33 \text{ in.}$$

The cable zone is shown diagrammatically in Fig. 15.21.
Now from the point of view of shear, the best position for the

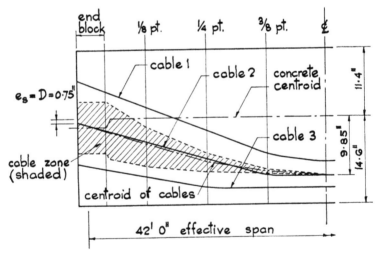

FIG. 15.21. BEAM EXAMPLE: CABLE ZONE

cable centroid at the end of the beam, as given by equation 15.26, is

$$e_8 = \frac{1}{H}\left(M_D + \frac{M_L}{2}\right) + D.$$

We already have $e_8 = 9 \cdot 85$ in., $M_D = 425,000$ in. lb., and $M_L = 1,640,000$ in. lb. Thus we can determine D as

$$D = 9 \cdot 85 - \left(\frac{425,000 + 820,000}{137,000}\right)$$

$$= 0 \cdot 75 \text{ in.},$$

so that at the end of the beam, where M_D and M_L are both equal to zero, we have

$$e_8 = D = 0 \cdot 75 \text{ in.}$$

The cables are therefore arranged as shown in Fig. 15.21 with their centroid falling within the cable zone and giving an end-eccentricity of $0 \cdot 75$ in. In tabular form, the eccentricities are:

	Position along beam from end				
	0	$\frac{1}{8}L$	$\frac{1}{4}L$	$\frac{3}{8}L$	$\frac{1}{2}L$
	in.	in.	in.	in.	in.
Cable 1	$- 6 \cdot 25$	$- 2 \cdot 00$	$2 \cdot 25$	$6 \cdot 00$	$7 \cdot 85$
Cable 2	$0 \cdot 75$	$3 \cdot 78$	$6 \cdot 82$	$9 \cdot 85$	$9 \cdot 85$
Cable 3	$7 \cdot 75$	$9 \cdot 80$	$11 \cdot 85$	$11 \cdot 85$	$11 \cdot 85$
Centroid	$0 \cdot 75$	$3 \cdot 86$	$6 \cdot 97$	$9 \cdot 23$	$9 \cdot 85$

We can now go back and check the actual fibre stresses, using equations 15.4 and 15.7. The results are given below:

	Position along beam from support			
	$\frac{1}{8}L$	$\frac{1}{4}L$	$\frac{3}{8}L$	$\frac{1}{2}L$
	lb/sq. in.	lb/sq. in.	lb/sq. in.	lb/sq. in.
Due to dead load and pre-stress:				
Top-fibre stress	$+ 535$	$+ 260$	$+ 40$	0
Bottom-fibre stress	$+ 1260$	$+ 1610$	$+ 1885$	$+ 1955$
Due to dead load, prestress and super load:				
Top-fibre stress	$+ 1205$	$+ 1410$	$+ 1480$	$+ 1535$
Bottom-fibre stress	$+ 400$	$+ 140$	$+ 45$	0

All the foregoing relates to conditions at safe working conditions, *after* the various losses have occurred. It is of course necessary to know the *initial tension stress* required in the cable, and approximately (allowing 20 per cent loss) this will be

$$54 \cdot 3 \times \frac{120}{100} = 65 \cdot 5 \text{ tons/sq. in.}$$

For those who like an attempt at greater accuracy, we can work as follows.

Shrinkage loss: If E_s is 28×10^6 lb/sq. in., and shrinkage per unit length is 200×10^{-6}, then loss in tendon due to shrinkage is

$$E_s \times \text{shrinkage strain}$$
$$= 28 \times 10^6 \times 200 \times 10^{-6}$$
$$= 5600 \text{ lb/sq. in.}$$

Creep loss: Assuming a total loss of 20 per cent (we can check back on this later), the initial tendon-stress would be $65 \cdot 5$ tons/sq. in., giving an average concrete stress at tendon level of 1600 lb/sq. in. Taking creep per unit length as $0 \cdot 25 \times 10^{-6}$ times the stress, the loss in tendon stress due to creep is

$$E_s \times \text{creep strain}$$
$$= 28 \times 10^6 \times 0 \cdot 25 \times 10^{-6} \times 1600$$
$$= 11,200 \text{ lb/sq. in.}$$

Relaxation loss: This is taken as 10,000 lb/sq. in. for heat treated wires.

Thus total loss of stress is:

$$
\begin{aligned}
& 5,600 \\
& 11,200 \\
& 10,000 \\
\hline
& 26,800 \text{ lb/sq. in.} = 12 \cdot 0 \text{ ton/sq. in.}
\end{aligned}
$$

Therefore initial steel stress required is

$$54 \cdot 3 + 12 \cdot 0 = 66 \cdot 3 \text{ ton/sq. in.}$$

This compares satisfactorily with the allowable value of

$$0 \cdot 7 \times f_u = 0 \cdot 7 \times 100 = 70 \text{ tons/sq. in.}$$

The above allows nothing for friction loss, nor for the effect of elastic loss with stressing in sequence: but there is a margin here to cover for these.

With an initial tendon-stress of $66 \cdot 3$ ton/sq. in., we have

$$H_1 = 66 \cdot 3 \times 1 \cdot 13 \times 2240 = 168,000 \text{ lb.}$$

Then top-fibre stress is

$$\frac{M_D y_2}{I} + \frac{H}{A} - \frac{H e \cdot y_2}{I}$$

$$= \frac{425{,}000 \times 11 \cdot 4}{12{,}200} + \frac{168{,}000}{161} - \frac{168{,}000 \times 9 \cdot 85 \times 11 \cdot 4}{12{,}200}$$

$$= 100 \text{ lb/sq. in. tension,}$$

and bottom-fibre stress is

$$\frac{- M_D y_1}{I} + \frac{H}{A} + \frac{H e \cdot y_1}{I}$$

$$= - \frac{425{,}000 \times 14 \cdot 6}{12{,}200} + \frac{168{,}000}{161} + \frac{168{,}000 \times 9 \cdot 85 \times 14 \cdot 6}{12{,}200}$$

$$= 2500 \text{ lb/sq. in. compression.}$$

For a triangular distribution of stress at transfer, the permissible compressive stress is $0 \cdot 5 u_{wt}$ (see Table 15.2). Thus we require $u_{wt} = \dfrac{2500}{0 \cdot 5} = 5000$ lb/sq. in. The permissible tensile stress is then

$$\frac{u_{wt}}{60} + 100$$

$$= \frac{5000}{60} + 100 = 183 \text{ lb/sq. in.}$$

which is greater than the actual stress value.

Before leaving the matter of bending, a check has to be made that the ultimate moment is not less than

$$2(M_L + M_D) = 2(1{,}640{,}000 + 425{,}000)$$
$$= 4{,}130{,}000 \text{ in. lb.}$$

This is best done here by using equations 15.15, 15.16, and 15.17. We will take $k_1 k_3$ as $0 \cdot 5$, and $\varepsilon_{c(ult)}$ as $0 \cdot 003$. Also we will assume the stress/strain relationship for the tendons, as given by the wire manufacturer, to be as shown in Fig. 15.22, so that at working conditions (when $f_{st} = 54 \cdot 3$ tons/sq. in.), $\varepsilon_{s1} = 0 \cdot 0038$. Trying $0 \cdot 4 \times 21 \cdot 25 = 8 \cdot 5$ in. for the neutral axis depth, and taking a bond factor of unity for F', we have from equation 15.15

$$\varepsilon_{st} = \varepsilon_{c(ult)} \left(\frac{1}{n_1} - 1 \right) F' + \varepsilon_{s1}$$

$$= 0 \cdot 003 \left(\frac{1}{0 \cdot 4} - 1 \right) + 0 \cdot 0038$$

$$= 0 \cdot 0083.$$

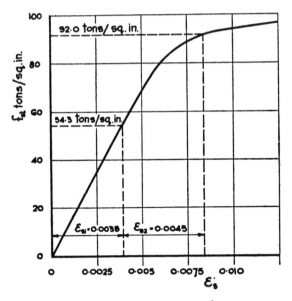

FIG. 15.22. BEAM EXAMPLE: STRESS/STRAIN CURVE
FOR STEEL WIRES

From Fig. 15.22 again, we see this strain corresponds to a stress of
92 tons/sq. in., giving

$$F_T = 92 \times 1{\cdot}13 \times 2240 = 233{,}000 \text{ lb. tension:}$$

and with the neutral axis depth of 8·5 in., we have the stressed
concrete   area $= (13 \times 4\frac{1}{2}) + (4 \times 4) = 74{\cdot}5$ sq. in. (see   Fig.
15.23), so that

$$F_C = 0{\cdot}5 \times 6000 \times 74{\cdot}5 = 224{,}000 \text{ lb. compression.}$$

This shows reasonable agreement of F_T and F_C, and as a compromise we will take $F_T = F_C =$ say 228,000 lb. The centroid of the
stressed area works out here to be 3·1 in. below the top of the
section, so that

$$\begin{aligned} M_{\text{ult}} &= 228{,}000\ (21{\cdot}25 - 3{\cdot}1) \\ &= 4{,}150{,}000 \text{ in. lb.} \end{aligned}$$

This is not less than 4,130,000 in. lb., and therefore the section
chosen is satisfactory. Checks made at other points along the
beam are also found to be satisfactory.

Let us now consider the matter of shear over the length of beam excluding the end-block. (Stresses at the anchorages will be considered last.) At working conditions, the maximum shear at start of end-block is:

Due to self-weight 18·85 ft. $\times$ 161 lb/ft. =   3,100 lb.
Due to superload 18·85 ft. $\times$ 620 lb/ft. = 11,700 lb.

14,800 lb.

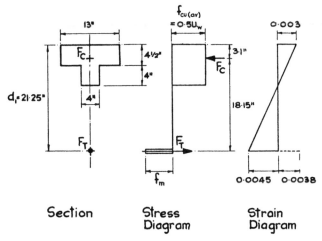

Section   Stress   Strain
           Diagram   Diagram

FIG. 15.23. BEAM EXAMPLE: STRESS AND STRAIN DIAGRAMS AT ULTIMATE CONDITIONS

The opposing shear due to cable inclination is

$$137{,}000 \left(\frac{3 \cdot 86 \text{ in.} - 0 \cdot 75 \text{ in.}}{5 \cdot 25 \times 12 \text{ in.}} \right) = 6750 \text{ lb.}$$

Therefore max upward shear is $6{,}750 - 3{,}100 = 3{,}650$ lb.
and max downward shear is $14{,}800 - 6{,}750 = 8{,}050$ lb.
At the section centroid

$$q = \frac{Q}{Ib} \cdot A_a \bar{y}_a$$
$$= \frac{8050}{12{,}200 \times 4} \times 86 \cdot 2 \times 7 \cdot 64$$
$$= 108 \text{ lb/sq. in.}$$

From Table 15.3, the limiting principal tensile stress is

$$\frac{u_w}{60} + 50$$

$$= \frac{6000}{60} + 50 = 150 \text{ lb/sq. in.,}$$

so that clearly no stirrup reinforcement is required on this account. Checks made at other points along the beam show that the balance of reduced shears from applied load and cable inclination are everywhere within allowable values.

At full load-factor conditions, tension stresses from bending first occur at a point 2 ft. 10 in. from the centre-line of the support. At this point the net shear is 21,650 lb. The provision of $\frac{3}{8}$ in. stirrups at 6 in. centres (at the stress allowed for this condition) gives a shear resistance of

$$Q_R = A_{st} \times 0.8 f_{sy} \times \frac{l_a}{s}$$

$$= (2 \times 0.11) \times (0.8 \times 38,000) \times \frac{24}{6}$$

$$= 26,600 \text{ lb.}$$

At 4 ft. from the support, where the shear is 20,000 lb., the stirrups can be opened out to 8 in. centres; and 8 ft. from the support, where the shear is 13,300 lb., the stirrups can be spaced at 12 in. centres.

There now remains the question of anchorage stresses. Consider the section VV at mid-length of the 9 in. wide end-block (Fig. 15.24a). From equation 15·29

$$f_x = \frac{H}{bv} \pm \frac{He_v \cdot y}{I_v}$$

$$= \frac{137,000}{9 \times 26} \pm 137,000 \times 0.75 \times 13 \times \frac{12}{9 \times 26^3}$$

$$= 585 \pm 100 \text{ lb/sq. in.}$$

The f_x stress diagram is given in Fig. 15.24b.

From equation 15.30, $q = K_q \cdot \frac{S}{bd}$; and taking K_q from Fig. 15.19 as 1·25, we have

$$q = 1.25 \times \frac{S}{9 \times 26} = 0.00534S.$$

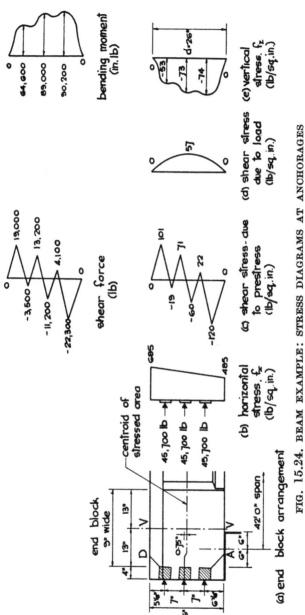

FIG. 15.24. BEAM EXAMPLE: STRESS DIAGRAMS AT ANCHORAGES

The shear stresses due to horizontal effects are shown in Fig. 15.24c. Due to vertical loading, the shear at VV is

$$(620 + 161)\ 20 = 15{,}600\ \text{lb.},$$

less the opposing shear due to the inclined cables $= 6750$ lb.; giving a net shear of 8850 lb., and shear stress of

$$q = \frac{3}{2} \cdot \frac{Q}{bd}$$

$$= \frac{3}{2} \cdot \frac{8850}{9 \times 26}$$

$$= 57\ \text{lb/sq. in. (see Fig. 15.24}d\text{).}$$

From equation 15.31, $f_z = K_z \cdot \dfrac{M}{bd^2}$; and taking K_z from Fig. 15.19 as $-5{\cdot}0$, we have

$$f_z = -5{\cdot}0 \times \frac{M}{9 \times 26^2} = -0{\cdot}000822M.$$

The f_z stress diagram is given in Fig. 15.24e.

All stresses are now known. By inspection of the stress diagrams it is seen that the worst principal stress at section VV occurs about 10 in. from the top of the member, where

$$f_x = 607\ \text{lb/sq. in.,} \quad f_z = -62\ \text{lb/sq. in.,}$$
$$q\ (\text{due to prestress}) = 70\ \text{lb/sq. in.,}$$
$$q\ (\text{due to load}) = 54\ \text{lb/sq. in.,}$$

giving a principal stress of 144 lb/sq. in.

Shear stresses predominate about $\frac{1}{4}d$ from the end of the member, where the maximum principal stress occurs about 3 in. from the top, and equals 164 lb/sq. in. acting on a plane at $14°\ 33'$ to the horizontal. This stress is greater than permissible and stirrup reinforcement is required. The tension per inch length of beam is

$$F_T = f_r b \sec \theta$$
$$= 164 \times 9 \times 1{\cdot}033$$
$$= 1530\ \text{lb.}$$

The area of steel required for this is $\dfrac{1530}{20{,}000} = 0{\cdot}076$ sq. in./in. which is provided by $\frac{3}{8}$ in. dia. stirrups in pairs (4 strands) at 6 in. centres (area $= 0{\cdot}073$ sq. in./in.).

LIQUID-RETAINING STRUCTURES

In addition to the basic principles for prestressed-concrete design set out earlier in this chapter, the British Standard Code

C.P. 2007 (1960)[3] includes special requirements for liquid-retaining structures as follows:

(a) A load-factor of not less than 1·25 is to be provided against cracking on the liquid-retaining face. In calculating the actual cracking resistance, values for concrete strengths are to be taken as in Table 15.6.

TABLE 15.6. *Concrete strengths to be assumed in calculating resistance to cracking. After C.P. 2007 (1960)*

Minimum works cube strength of concrete at 28 days u_w	Direct tensile strength	Bending tensile strength
lb/sq. in.	lb/sq. in.	lb/sq. in.
5,000	225	450
6,000	250	500
8,000	275	550
10,000	300	600

(b) A load factor of not less than 2·0 is to be provided against failure.

(c) Provision is to be made for the elastic distortions which arise as the work is being prestressed.

(d) The following requirements apply to cylindrical prestressed concrete tanks:

(i) *Steel stress:*

The tensile stress in the steel is not to exceed $0.65f_u$.

(ii) *Concrete stresses:*

The principal compressive stress in the concrete is not to exceed $0.33u_w$.

The average shear stress, based on the gross cross-sectional area is not to exceed $0.02u_w$.

When the tank is full, there should be a compression in the concrete at all points of at least 100 lb/sq. in.

(iii) *Friction losses:*

Values to be taken for μ are:
0·45 on smooth concrete,
0·25 on fixed steel bearers
0·10 on steel rollers.

As an example, let us consider the design of a circular tank 80 ft. diameter, 18 ft. deep with walls 6 in. thick (*see* Fig. 15.25). The joint where the wall sits in a rebate on the floor will be left free during construction so that the wall can slide: later, when the wall has been distorted by the prestressing, the rebate will be filled with grout to form a pinned joint at the base.

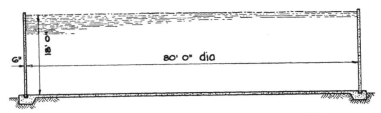

FIG. 15.25. EXAMPLE OF PRESTRESSED TANK 80 FT. DIA. AND 18 FT. DEEP

Ignoring for the present all question of base restraint, we have the maximum ring tension at 18 ft. depth

$$(62 \cdot 4 \times 18 \text{ ft.}) \times 40 \text{ ft. rad.} = 44,900 \text{ lb/ft.}$$

In order that the minimum concrete
stress is never less than 100 lb/sq. in.,
we must additionally prestress by
100 lb/sq. in. $\times$ 12 in. $\times$ 6 in.   = 7,200 lb/ft.

Therefore total prestress required   = 52,100 lb/ft.

This can be provided with twelve 0·200 in. dia. wires per foot, and the wire spacing can of course be increased higher up the tank. If the full prestress is immediately applied over a narrow band at the bottom of the tank, this will cause severe vertical bending in the wall, and excessive tensile stress: this effect can be reduced by taking the prestressing in easy stages – stressing only a proportion of the wires first, and later coming back to stress the remainder. The wires are normally placed outside the wall; and where they are pulled between anchorages, the anchorages can be formed using opposed sandwich plates (Magnel Blaton system), or stiffened steel channels (C.C.L. and Gifford Udall systems), or by pilasters built on the external face of the tank (Freyssinet system). Each wire then normally extends only halfway round the tank (to reduce friction losses) with its anchors diametrically opposed: and the anchors for adjacent groups of wires are staggered at 90 degrees, so that the part of one group of wires which is stressed to a maximum (near to its anchorages) coincides with the part of the adjacent group of wires which, being furthest from their anchorages, are stressed the least because of friction losses. The friction losses are kept to a minimum by running the wires over $\frac{1}{2}$ in. dia. mild steel bars suitably spaced on the outer face of the concrete wall.

Still ignoring the matter of base restraint, we have the compressive stress in the bottom of the wall due to horizontal prestress as indicated in Fig. 15.27a, and made up of:

To resist ring tension $\dfrac{44,900 \text{ lb.}}{12 \text{ in.} \times 6 \text{ in.}}$ = 625 lb/sq. in.

To allow minimum positive compression   100 lb/sq. in.

                                                         725 lb/sq. in.

Now comes the ticklish question of base restraint. C.P. 2007 (1960) very properly says that the ring prestress should be designed on the assumption that the wall-foot is free to slide withont frictional resistance. Here then is full provision against rign failure, as we have already made. C.P. 2007 also says that even though the wall-foot is free to slide, the wall should be designed to take vertical bending as though a restraint is provided equal to one-half of that provided by a pinned foot: and here is provision for the effect of the wall-foot being prevented by friction from sliding in a mathematically free manner. The vertical bending effect of such restraint at a pinned foot has been carefully investigated by many workers, and perhaps the best conclusions are presented most conveniently in one place in a Cement and Concrete Association Technical Report by Reynolds and Morice[4] dated July 1956, from which equations 15.32, 15.33 and 15.34 have been taken.

The radial restraint, $2H$, at a pinned foot is given by

$$2H = \frac{2pR^2\alpha\beta^3}{t} \qquad (15.32)$$

where $\alpha = \dfrac{t^3}{12(1 - \sigma^2)}$ and $\beta = \dfrac{[3(1 - \sigma^2)]^{\frac{1}{4}}}{\sqrt{Rt}}$,

where R is the tank radius,
      t is the wall thickness,
      p is the pressure at base level,
      σ is Poisson's ratio.

Taking σ as $0\cdot15$, we have, for our case, $\alpha = 18\cdot42$ and $\beta = 0\cdot0244$. Also, with the total ring prestress of $52,100$ lb/ft., we have a unit lateral pressure of

$$\frac{52,100 \text{ lb/ft.}}{40 \text{ ft. rad.}} = 1310 \text{ lb/sq. ft.} = 9\cdot05 \text{ lb/sq. in.}$$

Then from equation 15.32 we have

$$2H = \frac{2 \times 9\cdot05 \times (40 \times 12)^2 \times 18\cdot42 \times 0\cdot0244^3}{6}$$

$$= 186 \text{ lb. per inch of wall.}$$

For half this restraint, we have

$$H = 93 \text{ lb. per inch of wall.}$$

Now the ring stress f_{H} at any height x up the wall due to the restraint H at the foot is given by

$$f_{\mathrm{H}} = \frac{H}{2\alpha\beta^3 R} \cdot \Gamma(\beta x). \qquad (15.33)$$

Values of Γ for different values of (βx) are given in Fig. 15.26. In our case we have

$$f_{\mathrm{H}} = \frac{93}{2 \times 18\cdot42 \times 0\cdot0244^3 \times 40 \times 12} \cdot \Gamma(\beta x) = 363\Gamma(\beta x)$$

and the distribution of ring stress due to restraint is shown in Fig. 15.27b. Figure 15.27c shows the summation of Figs. 15.27a and b, and by superposition gives the actual ring stresses for the case of the restrained foot, after prestress.

The vertical bending moment M_{v} at any height x up the wall due to the restraint H at the foot is then given by

$$M_{\mathrm{v}} = \frac{H}{\beta} \cdot \Delta(\beta x). \qquad (15.34)$$

Values of Δ are given in Fig. 15.26. In our case we have

$$M_{\mathrm{v}} = \frac{93}{0.0244} \cdot \Delta(\beta x) = 3810\Delta(\beta x).$$

The bending stresses are then determined by dividing the bending moment by the section modulus z, which here is

$$\frac{1 \times 6^2}{6} = 6 \text{ in.}^3/\text{in. width.}$$

The results are shown in Fig. 15.27d.

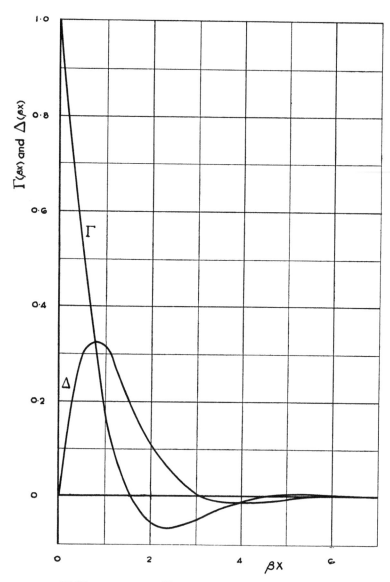

FIG. 15.26. FUNCTIONS Γ AND Δ FOR CYLINDRICAL WALL
RESTRAINED AT FOOT. (AFTER REYNOLDS AND MORICE.)

The above all relates to the prestress pressure acting *inwards*. Calculations of the effect of water pressure acting *outwards* can be done in identical manner, but all stresses are of opposite sense; and the curves are shown in Fig. 15.27 to the same numerical scale, so that the degree of balance achieved can be compared. It is clear from Fig. 15.27c that the restraint against sliding has halved the

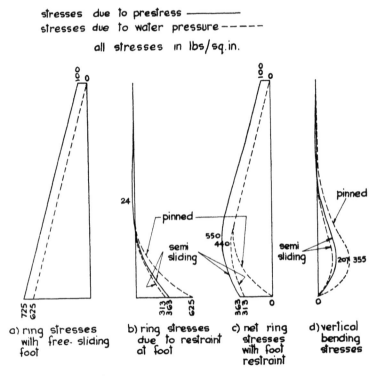

FIG. 15.27. TANK EXAMPLE: STRESS DIAGRAMS

residual prestress compression to only 50 lb/sq. in.: and in Fig. 15.27d there is clearly net vertical tension stressing on the liquid-retaining face when the tank is full. These effects can be taken care of either by increasing the prestress, or better, by pinning the foot-joint by filling the rebate in the floor with grout after the circumferential prestress has been applied. The effect of pinning the foot is also shown dotted in Figs. 15.27c and d. In the latter it should be noted that the net vertical bending is increased, but the liquid-retaining face is now in compression.

The maximum stresses due to vertical bending (lb/sq. in.) are therefore as follows:

	Tank empty	Tank full
Inner face	− 207	+ 148
Outer face	+ 207	− 148

The 148 lb/sq. in. tension at the outer face when the tank is full is not permissible: indeed a 100 lb/sq. in. compression is required. Therefore a vertical prestress of 148 + 100 is necessary. Final stresses (lb/sq. in.) are then:

	Tank empty	Tank full
Inner face	+ 41	+ 396
Outer face	+ 455	+ 100

The vertical prestressing can be achieved with cables looped at the bottom and stressed at both ends at the top of the wall, or by straight cables or bars anchored at the base and stressed at the single free end at the top.

A check on the resistance to cracking is made as follows. For ring tension, the water pressure would produce a stress in the concrete of 625 lb/sq. in.; whereas the prestress is 725 lb/sq. in. compression as against a permissible direct tension (Table 15.6) of 225 lb/sq. in.: thus the load factor is $\dfrac{725 + 225}{625} = 1.52$ which is satisfactory. For vertical bending, the tensile stress due to water pressure would be 355 lb/sq. in.; whereas the vertical stress due to the effect of circumferential prestress is 207 lb/sq. in. compression, the vertical prestress is 248 lb/sq. in. compression, while yet the permissible tensile bending stress (Table 15.6) is 450 lb/sq. in.: thus the load factor is $\dfrac{207 + 248 + 450}{355} = 2.55$ which is satisfactory.

The shear stress due to restraint at the wall-foot is a maximum when the tank is empty, and equals $\dfrac{93}{1 \text{ in.} \times 6 \text{ in.}} = 16$ lb/sq. in. average.

The wires are protected from rusting by the application of pneumatic mortar. This requires to be thick enough to accommodate the $\frac{1}{2}$ in. dia. vertical hold-off bars and the cables, as well as some margin to give at least 1 in. cover protection. The mortar is normally reinforced with a light fabric reinforcement to control shrinkage cracking. The mortar should not be applied when the tank is empty, because tensile stresses are then set up in the mortar when the tank is filled later; and these combined with shrinkage stressing tend to cause cracking. On the other hand if the mortar is applied and allowed to harden when the tank is full, better protection to the wires is ensured; and in certain instances benefit may be gained by the prestress force being partly resisted by the mortar in addition to the share taken by the original concrete.

PILES

Prestressed concrete piles are normally of smaller cross-section than ordinary reinforced-concrete piles to perform the same duty. For friction-piles, where a minimum face area is required, this may offer but little advantage; but in other cases the reduced sections, being easier to handle and pitch, and more efficient in the driving, can make for some economy. If the pile is adequately prestressed, tension cracks can be completely avoided during handling; and due to the precompression the pile is more resilient, and can better stand the shock of driving.

In Chapter 9 it is indicated that a reinforced-concrete pile 40 ft. long required to carry 50 tons might reasonably be 14 in. square and reinforced with four 1 in. dia. rods. In prestressed concrete, a suitable pile would be 12 in. square. If lifted at the fifth points, the maximum bending moment is

$$M = \frac{WL}{40} = \frac{144 \times 40^2 \times 12}{40} = 69,000 \text{ in. lb.}$$

The pile z is 288 in.^{3}, and the bending stress is therefore $\pm$ 240 lb/sq. in. For no tension we require a prestress of $240 \times 144 = 34,600$ lb., and this can be attained with 0·200 in. dia. wires (area 0·0314 sq. in.) stressed to 55 tons/sq. in. after losses, in which case the number of wires required is

$$\frac{34,600}{55 \times 2240 \times 0·0314} = 9.$$

In practice we should provide twelve wires, giving a direct concrete stress of 320 lb/sq. in. When the pile is carrying its working load of 50 tons, the concrete stress is:

Due to prestress   =   320 lb/sq. in.

Due to load $\dfrac{50 \times 2240}{144}$ =   780 lb/sq. in.

                              1100 lb/sq. in.

ROADS

Prestressing has been successfully applied to roads. The advantages claimed include longer stretches between joints, thinner slabs for same bending strength, and greater resilience and flexibility. Generally the slabs have been stressed individually between joints spaced at intervals up to as much as 600 ft. The main cables are normally arranged parallel to the direction of the road, and secondary stressing (or reinforcement) is required transversely: alternatively, the cables have been arranged on a diagonal grid. The resistance (caused by subgrade friction) to the slabs compressing under the prestress load is overcome to some extent by the provision of a sliding layer of building paper or polythene sheet on a thin smooth-finished layer of sand.

Prestress design has been based on overcoming this subgrade restraint, and then applying a further 200 lb/sq. in. prestress as a margin. The slab thickness is mainly governed by the need to accommodate the cables with adequate cover: and often about 6 in. has proved satisfactory.

RAILWAY SLEEPERS

Pretensioned prestressed concrete railway sleepers have been manufactured and used in great quantities with complete success

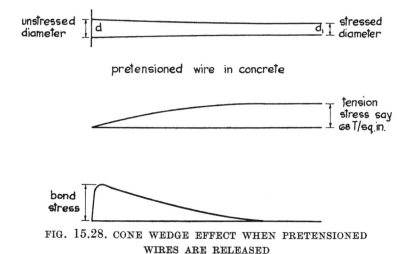

FIG. 15.28. CONE WEDGE EFFECT WHEN PRETENSIONED WIRES ARE RELEASED

for about 20 years. Whereas ordinary reinforced concrete sleepers crack and ultimately fail in crushing, prestressed sleepers are more resilient and recover on removal of the load.

Near the ends of the sleepers, where the maximum bending moments occur, bond stress problems arise: and these are overcome by using small diameter wires with crimped surfaces. Nevertheless average bond stresses of about 425 lb/sq. in. occur, and these can only be justified as follows. In the first place, the cover of concrete is usually about three diameters; and the enhanced bond given by generous covers has already been explained in Article 3.6. Secondly the wires, when stressed to about 68 tons/sq. in., reduce in diameter in accordance with Poisson's ratio some 0·15 per cent; so that when the prestress tension is released and has to be taken up by the concrete through bond, the unstressed wires revert to their unstressed diameter as shown in Fig. 15.28 and so present a conical wedging action. This wedging is quite satisfactory with the three-diameter's cover referred to previously. To improve the bond, the wires may be degreased in soda before casting.

REFERENCES

1. BRITISH STANDARD CODE OF PRACTICE, C.P. 115 (1959), General Series. The Structural Use of Prestressed Concrete in Buildings. British Standards Institution, (1959).
2. MAGNEL, GUSTAVE. *Prestressed Concrete*. Concrete Publications Ltd., (1954).
3. BRITISH STANDARD CODE OF PRACTICE, C.P. 2007 (1960). Design and Construction of Reinforced and Prestressed Concrete Structures for the Storage of Water and other Aqueous Liquids. British Standards Institution.
4. REYNOLDS, G. C. and MORICE, P. B. *Analysis of Cylindrical Prestressed Concrete Tanks*. Cement and Concrete Association, (1956).

INDEX